The Dry Forests and Woodlands of Africa

Managing for Products and Services

Edited by
Emmanuel N. Chidumayo and Davison J. Gumbo

earthscan
from Routledge

First published in 2010 by Earthscan

For a full list of publications please contact:
Earthscan

Published 2023 by Routledge
4 Park Square, Milton Park, Abingdon, Oxon OX14 4RN
605 Third Avenue, New York, NY 10017

Routledge is an imprint of the Taylor & Francis Group, an informa business

Typeset by MapSet Ltd, Gateshead, UK
Cover design by Susanne Harris

A catalogue record for this book is available from the British Library

Library of Congress Cataloging-in-Publication Data

The dry forests and woodlands of Africa: managing for products and services / edited by Emmanuel N. Chidumayo and Davison J. Gumbo.
p. cm.
Includes bibliographical references and index.
ISBN 978-1-84971-131-9 (hardback)
1. Forest management–Africa, Sub-Saharan. 2. Forest products–Africa, Sub-Saharan. 3. Forests and forestry–Africa, Sub-Saharan. I. Chidumayo, E. N. II. Gumbo, D. J.
SD242.A357D79 2010
634.9'2096–dc22

2010003283

ISBN: 978-1-84971-131-9 (hardback)
ISBN: 978-0-41585-104-6 (paperback)

To Dr Mário Alberto and Dr Petros Nyathi
who passed away before the book could be published.
Their contributions to the book are invaluable.

Contents

Foreword

This book is a contribution by 30 African scientists associated with at least 16 institutions across Africa and a few outside the African Continent working towards solving African resource management problems in the dry forests and woodlands. It is one of the outputs of a CIFOR (Center for International Forestry Research) project, with support from Sida-Natur, on Achieving the Millennium Development Goals in African Dry Forests: From Local Action to National Policy Reforms. Dry forest vegetation (including woodland) is dominated by woody plants, mainly trees, with tree canopy cover over 10 per cent of the ground surface within climates with a dry season of 3 months or more (i.e. areas characterized by frequent droughts, occasional floods and general vulnerability to climate variability). It covers about 17.3 million km^2 in 31 countries in West Africa, East Africa and southern Africa. It is the home of 505 million people, most of whom depend on the dry forests and woodlands for their livelihoods, mainly through rain-fed crop agriculture, livestock farming and gathering of timber and non-timber forest products, which also support local industries. Earlier books on the dry forests and woodlands focused on the Miombo woodlands of southern Africa, but this book covers the entire area of the dry forests in sub-Saharan Africa.

The value of the book is the link it provides between the available resources of the dry forests and woodlands, and the use and management of their diverse products and services to contribute to poverty reduction and wealth creation. Chapters 2 to 4 cover the detailed description of the dry forest vegetation types in sub-Saharan Africa in terms of their characteristics, floristic diversity and their potential value to provide products and services. Chapters 5 to 10 address the management practices of the African dry forests and woodlands for specific products and services, and their links with policy, tenure, governance, gender and commercialization. These chapters specifically address non-wood forest products (Chapter 5), timber and other wood products (Chapter 6), firewood and charcoal (Chapter 7), livestock, wildlife and rangelands (Chapter 8), forest plantations and woodlots as a means to create alternative resources (Chapter 9), and, finally, environmental and ecosystem services such as carbon sequestration, soil and water conservation and socio-cultural–spiritual values (Chapter 10). Each individual chapter identifies a number of useful key issues and challenges, and Chapter 11 synthesizes the key issues in the management of the dry forests and woodlands, and suggests the way forward in terms of policy and legislative reform, stakeholder participation, resource governance, managing for multiple products and services, and climate and land-use change scenarios.

This book provides a current baseline of knowledge on the current resources of the dry forests and woodlands, their use and value, and possible means to manage them towards sustainable development into the future. It poses a challenge to policy-makers and resource managers to make integrated and sustainable resource management a reality to resources users who depend on good governance and the dry forest resources for their livelihoods and prosperity under threats of environmental change.

Dr Coert J Geldenhuys
Extraordinary Professor in Forest Science
University of Stellenbosch, South Africa

Preface

Since time immemorial dry forests and woodlands of sub-Saharan Africa have provided diverse ecosystems goods and services to large populations of humans and livestock that depend on them. Dry forests and woodlands are profoundly important for local livelihoods, and yet, this role is hardly recognized by the respective sub-Saharan African governments – more so against the backdrop and importance accorded to tropical forests. Further, policy inadequacies are noted and dry forests are a low priority in sub-Saharan Africa. Lately, sub-Saharan Africa's dry forests and woodland have been rapidly declining due to harvesting of wood for commercial and domestic purposes and this has had major implications for the local people. Sub-Saharan Africa is a developing region where deforestation and desertification have remained as major issues of concern. With these changes, biodiversity, which is not only important for ecotourism but also a significant source of non-wood forest products is severely under threat. All the key attributes of the dry forests and woodlands are intricately linked and a change in one will affect the other. Thus, the threat posed by climate change on the forests will invariably affect livelihoods and therefore there is an urgent need to increase natural and human capacity to deal with the problems triggered by this development.

This unique book brings together scientific knowledge on and about dry forests and woodlands from eastern, western and southern Africa, and describes the relationships between forests, woodlands, people and their livelihoods. Dry forest is defined as vegetation dominated by woody plants, primarily trees, the canopy of which covers more than 10 per cent of the ground surface, occurring in climates with a dry season of three months or more. This broad definition – wider than those used by many authors – incorporates vegetation types commonly termed woodland, shrubland, thicket, savanna and wooded grassland, as well as dry forest in its strict sense.

The book provides a comparative analysis of management experiences from the different geographic regions of sub-Saharan Africa, emphasizing the need to balance the utilization of dry forests and woodlands between current and future human needs. Further, the book explores the techniques and strategies that can be deployed to improve the management of African dry forests and woodlands for the benefit of all, and especially the communities that live off this vegetation. This book aims to stoke local, regional, national and international discussion on these forests and woodlands that provide livelihoods to almost 60 per cent of sub-Saharan Africa's population. In this way the book is not only calling for a better understanding of the policy issues surrounding these forests but also the biophysical aspects of the same. Thus, the book lays a foundation for improving the management of dry forests and woodlands for the wide range of products and services they provide.

Emmanuel N. Chidumayo
Davison J. Gumbo
Lusaka, Zambia
July 2010

Contributors

Mário M. M. Aberto (deceased), formerly Lecturer/Researcher, Faculty of Agronomy and Forestry, Eduardo Mondlane University, Maputo, Mozambique

Henri-Noël Bouda, formerly Research Fellow, Center for International Forestry Research, West Africa Regional Office, 06 BP 9478 Ouagadougou, Burkina Faso

Shabani A. O. Chamshama, Professor, Plantation Silviculture, Agroforestry, Tree Improvement. Faculty of Forestry & Nature Conservation Department of Forest Biology, Chuo Kikuu Morogoro, Tanzania

Emmanuel N. Chidumayo, Ecologist/Manager, Makeni Savanna Research Project, P.O. Box 50323, Ridgeway, Lusaka, Zambia

Mwale Chishaleshale, Intern, CIFOR, Zambia Office, Lusaka, Zambia

Michelle Cocks, Researcher at Institute of Social and Economic Research, Rhodes University, Grahamstown 6140, South Africa

Tony Dold, Curator of the Selmar Schonland Herbarium, Grahamstown (GRA) Botany Department, Rhodes University, Grahamstown 6140, South Africa

Hervé Fritz, CNRS/CIRAD HERD Program, Hwange Main Camp Research, Hwange National Park, Hwange, Zimbabwe

James Gambiza, Lecturer, Department of Environmental Science, Rhodes University, Grahamstown 6140, South Africa

Davison Gumbo, Scientist, Centre for International Forestry Research-Zambia Office, Forestry Nursery Site, Elm Road, P. O. Box 50977, Ridgeway, Lusaka, Zambia

Cynthia Kambole, Intern, CIFOR, Zambia Office, Lusaka, Zambia

Sarah Kaschula, Earth Institute Millennium Villages Project: Monitoring and Evaluation at Columbia University, New York, US

Stephano Kingazi, Lecturer, Sokoine University of Agriculture, Faculty of Forestry and Nature Conservation, Morogoro, Tanzania

Emmanuel Luoga, Professor, forestry/agroforestry, ecological economics, ethnobotany, community-based conservation, Sokoine University of

Agriculture, Faculty of Forestry and Nature Conservation, Morogoro, Tanzania

Rogers Malimbwi, Professor, Forest Resource Assessment, growth and yield modelling, Sokoine University of Agriculture, Faculty of Forestry and Nature Conservation, Morogoro, Tanzania

Crispen Marunda, Researcher, forest hydrology section, Division of Forest Research and Development, Forestry Tasmania, 79 Melville Street, Hobart, Australia

Keith Mbata, Professor, Department of Biological Sciences, University of Zambia, Lusaka, Zambia

Guni Mickels-Kokwe, Natural Resources Scientist, Lusaka, Zambia

Salome Misana, Professor, Dar es Salaam University College of Education (DUCE), Dar es Salaam, Tanzania

Jean Nduwamungu, Director of the Centre for Geographic Information and Remote Sensing (CGIS) GIS, National University of Rwanda, Huye, Rwanda

Petros Nyathi (deceased), formerly Lecturer, Department of Forest Resources and Wildlife Management, National University of Science and Technology, Bulawayo, Zimbabwe

Herbert H. T. Prins, Professor, Resource Ecology Group, Wageningen University, Droevendaalsesteeg 3A, 6708 P.B., Wageningen, Netherlands

Patrice Savadogo, Department of Production Forests, National Science & Technology Research Centre, B.P. 7047, Ouagadougou, Burkina Faso

Louis Sawadogo, Sylvopastoralist Researcher, National Agricultural Research Institute (INERA), Ouagadougou, Burkina Faso

Sheona Shackleton, Lecturer, Department of Environmental Science, Rhodes University, Grahamstown 6140, South Africa

Enos Shumba, Miombo Eco-region Leader in the WWF Eastern and Southern Africa Region, 10 Lanark Road, Belgravia, Harare, Zimbabwe

Almeida Sitoe, Tropical Forester working at the Faculty of Agronomy and Forestry of the Eduardo Mondlane University in Maputo, Mozambique

Jonathan Timberlake, Editor of *Flora Zambesiaca* (plant conservation, plant utilization; vegetation survey; applied ecology; *Acacia*), Kew Gardens, London, UK

Graham von Maltitz, Council for Scientific and Industrial Research (CSIR) Pretoria, Natural Resources & the Environment (NRE), Pretoria, South Africa

Eliakim Zahabu, Lecturer, Sokoine University of Agriculture, Faculty of Forestry and Nature Conservation, Morogoro, Tanzania

Acknowledgements

We would like to thank all the authors and contributors for their painstaking efforts in putting this book together. Also the reviewers, Tony Cunningham, Coert Geldenhuys, Jonathan Timberlake, Bruce Campbell, Louis Sawadogo and Mike Kock for taking time to provide valuable comments and constructive criticisms, which led to the improvement of the individual chapters and the book as a whole.

We are grateful to the Swedish International Development Agency (SIDA), which provided support for the Center for International Forestry Research (CIFOR)'s research project Achieving the Millennium Development Goals in African Dry Forests: From Local Action to National Policy Reforms out of which the book emerged. The project was carried out in Burkina Faso, Ethiopia and Zambia and we extend our thanks to the project offices concerned.

We are grateful to Daniel Tiveau, Zida Mathurin (Burkina Faso); Habtemariam Kassa (Ethiopia); Madeleen Husselman, Fiona Paumgarten and Mercy Mwape (Zambia); and for the administrative support of Mireille Karambiri (Burkina Faso) and Jacqueline Chembe-Nkosha (Zambia) as well as the many research assistants that contributed towards making this project a success. The book would not have been possible without the early support of Bruce Campbell, Godwin Kowero, Daniel Tiveau and Cris Marunda, then with CIFOR, whose combined efforts made this book a success.

We are particularly indebted to Bruce Campbell and Cris Marunda, then Director of CIFOR's Forests and Livelihoods Programme and Project Coordinator (Zambia Office), respectively, whose early work and vision made this book part of the Dry Forest Project. The editors also note the important contributions made by administrative staff in CIFOR HQ – Nani Djoko, Ratih Septivita, Henny Linawati and Hiasinta Lestari: they too through their hard work and support made this book a success. Last but not least we would like to thank Tim Hardwick of Earthscan for giving direction as well as encouragement. Finally, we would like, together with the publisher, to thank all those who provided materials for the book including illustrations. We are also grateful to all publishers and organizations for providing copyright permissions to reproduce material, which are acknowledged and cited in relevant captions and reference lists.

We hope that the insights paid in this book contribute towards a better management of sub-Saharan Africa's dry forests and woodlands.

List of Abbreviations

AAC	allowable annual cut
CAMPFIRE	Communal Areas Management Program for Indigenous Resources
CBD	Convention on Biological Diversity
CBNRM	community-based natural resource management
CDM	Clean Development Mechanism
CIFOR	Center for International Forestry Research
CITES	Convention in International Trade in Endangered Species
CO_2	carbon dioxide
dbh	diameter at breast height (1.3m high)
DRC	Democratic Republic of Congo
ES	ecosystem services
FAO	Food and Agriculture Organization of the UN
GDP	gross domestic product
ITTO	International Timber Trade Organization
IUCN	International Union for Conservation of Nature
JFM	joint forest management
NDVI	Normalized Difference Vegetation Index
NGO	non-governmental organization
NTFP	non-timber forest product
NWFP	non-wood forest product
PES	payments for environmental services
R&D	research and development
REDD	Reduced Emissions from Deforestation and forest Degradation
SFI	Sustainable Forestry Initiative (of AF&PA)
SOC	soil organic carbon
SFM	sustainable forest management
TFCA	trans-frontier conservation areas
UNFCCC	UN Framework Convention on Climate Change
WHO	World Health Organization
WWF	World Wide Fund for Nature

1

Dry Forests and Woodlands in Sub-Saharan Africa: Context and Challenges

Emmanuel Chidumayo and Crispen Marunda

Defining Dry Forests and Woodlands of Sub-Saharan Africa

Dry forest and woodland are vegetation types dominated by woody plants, primarily trees, the canopy of which covers more than 10 per cent of the ground surface, occurring in climates with a dry season of three months or more. Dry forests in Africa occupy an area between rainforests in the Congo basin and open woodlands of western and southern Africa. Woodlands in Africa are diverse vegetation formations that include woodland proper, bushland, thicket and, in some cases, wooded grassland. The Center for International Forestry Research (CIFOR) in sub-Saharan Africa has implemented a dry forest programme since 1996 but with a narrower focus on southern Africa. The programme, although called dry forest, includes both dry forest proper and woodlands as defined above. Currently CIFOR's dry forest programme is coordinated by regional offices in Burkina Faso for the West African Region and Zambia for the Southern African Region, and includes all countries in sub-Saharan Africa that have dry forests and woodlands. These vegetation types in sub-Saharan Africa are found in 31 countries in western, eastern and southern Africa and are the dominant vegetation in 63 per cent of these countries. They cover approximately 17.3 million km^2 and are inhabited by nearly 505 million people (2003 estimate).

Table 1.1 *Some indices of human well-being in sub-Saharan African countries with a significant cover of dry forests*

Index of human well-being	*Average*	*Range*
Population living below poverty line (%)	53	35–80
Life expectancy from birth (years)	57	31–58
Budget revenue (US$ per capita)	126	21–1462
Health expenditure (US$ per capita)	22	4–127
Population with access to essential drugs (%)	57	25–87
Agricultural labour (% of total labour force)	69	32–92
Traditional energy use (% of total energy use)	79	25–98

Source: Based on NationMaster.com (2004)

Socio-Economic Context of Dry Forests and Woodlands in Sub-Saharan Africa

Africa entered the 21st century as the poorest, most indebted, marginalized and technologically backward continent in the world. Life expectancies, per capita revenues and expenditures on health are the lowest in the world, while the proportion of the population dependent on traditional energy sources (firewood, charcoal and organic wastes) is among the highest. Various other indices of human well-being (Table 1.1) testify to this predicament that sub-Saharan Africa faces in the new century.

In many countries of sub-Saharan Africa the majority of the people live in rural areas where the main livelihood source is subsistence crop and/or livestock production. The major zone of crop agriculture in sub-Saharan Africa is in the dry forests and woodlands; much of which is rain-fed and is therefore vulnerable to climate variability. The climate of the dry forest and woodland regions is characterized by frequent droughts and occasional floods that frequently cause crop failure. During such times the coping strategies of local people invariably involve gathering of wild foods in the forest. But the reliance on dry forests and woodlands is not only a safety net, important as this may be; these vegetation formations also play a significant role in supporting local industries. Most important, is the diverse range of forest products, including fruits, fish and bush meat, edible insects, beeswax and honey, and traditional medicines, that are indispensible to the lives of communities living in dry forest and woodland zones. Most of these non-wood forest products are produced, traded and consumed outside the formal cash economy and therefore are not adequately captured in national economic statistics.

Importance of Dry Forests and Woodlands to Ecosystem Services, Livelihoods and National Economies

Dry forests and woodlands in sub-Saharan Africa are rich in biodiversity that is important for the supply of ecosystem services, such as regulation of water flows, water quality, climate and protection of land from soil erosion. For example, woodlands in sub-Saharan Africa are of crucial importance to water resources management because all the major river basins in sub-Saharan Africa are either located or have most of their headwaters in the woodlands. Therefore they play a crucial role in sustaining river flows and water supplies. The conservation of these woodlands in watersheds is therefore crucial for maintaining the supply of water for irrigation, sanitation, industry, hydropower and human consumption. People also derive numerous products and services from ecosystems and the biodiversity they contain and this book focuses on a limited number of these products and services.

Non-wood forest products

Poverty in Africa is rife; almost 60 per cent of rural Africans live on less than US$1 a day (Kaimowitz, 2003). In many of the continent's rural areas, poverty appears entrenched and intractable with few opportunities for relief, especially in the context of the huge and devastating impacts of HIV/AIDS (Bryceson and Fonseca, 2006; Shackleton, 2006; Wiegers et al, 2006). The importance of non-wood forest products (NWFPs) for livelihood security, in particular for food security and alleviating dietary deficiencies, and for assisting households to cope with, if not escape, poverty is widely acknowledged in sub-Saharan Africa. In particular, these products have been shown to be important for women and children, both extremely vulnerable groups. The use of NWFPs by urban, in addition to rural households, has also been pointed out, and is likely to grow with the increasing urbanization of Africa's population (UNEP, 2002). In many ways, urban demand helps to create sustainable markets for NWFPs, contributing to their potential as a means for rural people and traders to earn a cash income. Furthermore, NWFPs play a significant role in mitigating some of the devastating impacts of HIV/AIDS. Both plant and insect wild foods are highly nutritious and could assist in meeting some of the nutritional requirements of people living with HIV/AIDS. The demand for traditional medicines has also risen as a result of the AIDS pandemic, with potentially negative outcomes for forest and medicinal plant stocks. Many NWFPs have significant links to culture and identity and contribute to building social capital. Despite this, and the fact that many millions of poor people benefit daily from NWFPs, their crucial importance for livelihood security and significant economic contribution, primarily in the informal sector, is generally poorly recognized and appreciated and sometimes even ignored in terms of national policy and forest management

(Bird and Dickson, 2005; Petheram et al, 2006). Such neglect may undermine the potential of these products to deliver benefits in the future, erode vital safety nets and exacerbate the already persistent poverty endemic to Africa.

Woodfuel

Firewood and charcoal use, especially in urban areas, has socio-economic benefits. The charcoal business employs a large portion of the population along the chain from the producer in rural areas to the consumer in urban areas. Charcoal production contributes significantly, in some cases 60–80 per cent, to rural household income and is therefore important in poverty reduction. Sustainable dry forest and woodland management is thus key to the maintenance of forest-based income generation in rural areas. In some cases, income from charcoal sales is used to buy agricultural inputs and in this way, dry forests and woodlands subsidize agricultural production and therefore contribute to household food security. Income from woodfuel sales cushion rural households against loss of agricultural incomes when producer prices of agricultural crops decline due to economic and other structural adjustment policies. Similarly when people lose jobs, such as in mining and other industries, they find charcoal production an attractive means of income generation.

Timber and wood products

Timber and wood products, such as poles, from indigenous trees in African dry forests and woodlands are used locally to meet basic needs and to generate income. Timber products constitute the base for small-scale industries in many communities, including those centred on woodcrafts, canoe making and the manufacturing of a variety of household tools and utensils. These industries enable communities to generate social and economic benefits. Apart from the direct involvement of communities in the timber business, sustainable forest management creates possibilities for the timber industry to share benefits from the logging operations with local communities. Many countries in dry forest and woodland regions of Africa have adopted measures for benefit sharing between local communities and those conducting logging operations, be they the private sector or the state. The establishment of industries to produce wood products in the rural areas is often accompanied with infrastructure development (hospitals, schools, roads, etc.) and the roads increase accessibility to remote rural areas.

Livestock and wildlife

Rangelands, including those in the woodlands, occupy about 90 per cent of the agricultural land in Africa and sustain the livelihoods of 25 million people.

Livestock production systems in the subsistence economy are usually geared to the production of multiple products, including meat, milk, blood, hides and skins, dung for fuel, transportation, flexible household capital reserves and risk management, while commercial ranching systems are generally geared more narrowly towards meat production (Mearns, 1996). The livelihoods of millions of people in Africa are therefore dependent on livestock (Shackleton et al, 2001, 2005; Dovie et al, 2006). Animal draught power and nutrient cycling through manure compensate for lack of access to modern inputs, such as tractors and fertilizers.

Livestock production makes a significant contribution to gross domestic product (GDP) although its importance in the economy varies among African woodland regions. Livestock are expected to play an important role in fulfilling the Millennium Development Goal of reducing poverty by 2015.

In southern and eastern Africa wildlife management is an important complement to livestock keeping on rangelands. Income from game viewing and/or trophy hunting on private ranches and state game management areas can exceed the income from livestock, and a combination of both provides higher income than livestock or wildlife alone (Kiss, 1990). Although large game animals are now rare in western Africa, smaller game animals, such as duikers, grasscutters and giant rats contribute substantially to local meat supply (Caspary, 1999). Lindsey et al (2007) estimated that sub-Saharan Africa receives about US$201 million per year from trophy hunting, making trophy hunting an important driver of conservation. In spite of the conflicting reports about benefits from community-based wildlife management schemes, this natural resources management approach has the potential to contribute to poverty alleviation and local economic development, especially in woodland areas with a high diversity of wildlife species.

Plantations and woodlots

Plantations and woodlots generate revenue and foreign exchange for national governments. At the local level, they provide jobs offering economic opportunities for rural residents (Whiteman and Lebedys, 2006). In addition, there may be opportunities for local residents to use the residues and by-products left behind after trees have been harvested for woodfuel or timber. Indirect benefits may include government reinvestment of the revenue generated from plantations into education, medicine and infrastructure development in local communities (Morrison and Bass, 1992).

Managing Dry Forests and Woodlands in Sub-Saharan Africa: Challenges and Concerns

With the ever-growing pressure on dry forest and woodland resources to meet human development needs and livelihood demands, these vegetation forms are increasingly being utilized unsustainably in Africa. Often, the high levels of poverty and growing political pressure force governments in Africa to encourage overexploitation and conversion of dry forests and woodlands to other seemingly more profitable land uses at the expense of the environmental and ecological services that they provide.

The issues addressed in this book concern the role and potential of dry forests and woodlands as sources of products and services, and how these can sustainably be harnessed to alleviate poverty and support livelihoods and socio-economic development. Much of the global focus in African forests has been on tropical rainforests; as a consequence of this bias, dry forests and woodlands have received little attention in terms of research, investment and policy development. This book is an attempt to get dry forests and woodlands back on the agenda of national, regional and global debates. However, while recognizing the value and potential of African dry forests and woodlands to livelihoods and national economic development, the book raises a number of concerns about sustainable utilization of these areas. In particular, it discusses the threats posed by resource tenure, governance, international conservation and trade policies. The book highlights the knowledge gap that exists about the inherent ability and capacity of dry forests and woodlands to regenerate, especially in the face of climate change and land-use extension and intensification. The key question is whether there are management practices and models that can be strategically applied to maintain and improve the productivity of African dry forests and woodlands to meet the ever-growing demand for forest products and services in a dynamic global environment.

The main purpose of the book is to lay a firm foundation for improving the management of dry forests and woodlands for a range of products and services, thereby contributing to sustainable livelihoods and poverty reduction. This purpose is achieved by highlighting the socio-economic and environmental importance of African dry forests and woodlands and providing possible approaches to meeting the livelihood needs of dry forest/woodland communities and nations. At the same time, the book points to approaches that can maintain the vitality of these forests and woodlands and secure their capacity to provide goods and services for future generations. Maintaining the balance between utilization to meet current human needs and conservation for meeting future needs will not be easy and demands sacrifices at all levels: individual, community, national, regional and international.

Layout of the Book

The book has ten substantive chapters. Chapter 1 introduces African dry forests and woodlands and their setting in the socio-economic context of sub-Saharan Africa and is immediately followed by a chapter that gives a detailed description of these vegetation types in sub-Saharan Africa, their characteristics and potential to provide products and services. Chapter 3 describes the floristic diversity and value of African dry forests and woodlands and presents possible approaches for managing the rich plant biodiversity. Chapter 4 links the biological resources in African dry forests and woodlands to livelihoods and critically analyses the potential of forest products to contribute to poverty reduction and wealth creation.

Chapters 5 to 10 address the management of African dry forests and woodlands for specific products. They describe opportunities for scaling-up production and expanding trade. They identify possible management practices and link these to institutional and policy issues, such as tenure, governance, gender and commercialization. Non-wood forest products, such as honey, wild foods (plant and insect), oils, medicines and grasses have rarely been comprehensively analysed at such a large geographical scale as sub-Saharan Africa and Chapter 5 is an attempt to fill this gap. The chapter is a synthesis of what is currently known about non-wood forest products at an African scale and considers each of the major categories of non-wood forest products. For each category it presents a description and management practices used to enhance the production and trade of these products. Chapter 6 focuses on timber and other wood products, while Chapter 7 is solely devoted to issues of woodfuel (firewood and charcoal), a resource of paramount importance to both rural and urban communities in sub-Saharan Africa. Chapter 8 focuses on livestock and wildlife, important products of African woodlands. The chapter presents an analysis of livestock and wildlife as sources of livelihoods and economic development and proposes ways of improving the management of rangelands. Finally Chapter 9 focuses on forest plantations. Natural forests may not provide all the products that people in sub-Saharan Africa need and some forms of forest utilization are likely to be unsustainable. Thus, plantations and woodlots might play a critical role in bridging the gap between demand for forest products and depletion of natural supply sources. The chapter also highlights the high potential of plantations and woodlots to provide products and services in the form of wood, browse, fodder and other products.

Chapter 10 is devoted to environmental and ecosystem services that are the focus of emerging global debates. The chapter describes the value of dry forests and woodlands in carbon sequestration and trade in environmental services and briefly explores the role of forests in soil and water conservation for agriculture and hydropower. In addition, the chapter highlights the value of dry forests as spiritual and cultural assets and proposes ways of enhancing the provision of environmental services and the equitable distribution of benefits realized from managing dry forests for these services. The book presents a prognostic synthesis of the management of dry forests and woodlands and proposes the way forward in Chapter 11.

REFERENCES

Bird, N. and Dickson, C. (2005) *Poverty Reduction Strategy Papers: Making the Case for Forestry*, Overseas Development Institute (ODI), London

Bryceson, D. and Fonseca, J. (2006) 'Risking death for survival: Peasant responses to hunger and HIV/AIDS in Malawi', *World Development*, vol 34, pp1654–1666

Caspary, H. U. (1999) 'Utilisation de la faune sauvage en Côte d'Ivoire et Afrique de l'Ouest : potentiel et contraintes pour la cooperation au développement', GTZ, Eschborn, Germany

Dovie, D. B. K., Shackleton, C. M. and Witkowski, E. T. F. (2006) 'Valuation of communal area livestock benefits, rural livelihoods and related policy issues', *Land Use Policy*, vol 23, pp260–271

Kaimowitz, D. (2003) 'Not by bread alone…forests and rural livelihoods in sub-Saharan Africa', in T. Oksanen, B. Pajari and T. Tuomasjukka (eds) *Forestry in Poverty Reduction Strategies: Capturing the Potential*, European Forest Institute, Joensuu

Kiss, A. (ed) (1990) Living with wildlife: Wildlife resource management with local participation in Africa, *World Bank Technical Paper no 130*

Lindsey, P. A., Roulet, P. A. and Ramanach, S. S. (2007) 'Economic and conservation significance of the trophy hunting industry in sub-Saharan Africa', *Biological Conservation*, vol 134, pp455–469

Mearns, R. (1996) 'When livestock are good for the environment: Benefit-sharing of environmental goods and services'. Invited paper for the World Bank/FAO Workshop, Balancing Livestock and the Environment, Washington, DC, an associated event to the Fourth World Bank Conference on Environmentally Sustainable Development

Morrison, E. and Bass, S. M. J. (1992) 'What about the people', in C. Sargent and S. M. J. Bass (eds) *Plantation Politics: Forest Plantations in Development*, Earthscan Publications, London

NationMaster.com (2004) 'Map & Graph: Africa', www.nationmaster.com

Petheram, L., Campbell, B., Marunda, C., Tiveau, D. and Shackleton, S. (2006) 'The wealth of the dry forests. Can sound forest management contribute to the Millennium Development Goals in sub-Saharan Africa?', *Forest Livelihood Briefs*, no 4, October, CIFOR, Bogor

Shackleton, S. E. (2006) 'Forests as safety nets for mitigating the impacts of HIV/AIDS in southern Africa', CIFOR, Bogor

Shackleton, C. M., Shackleton, S. E. and Cousins, B. (2001) 'The role of land based strategies in rural livelihoods: The contribution of arable production, animal husbandry and natural resource harvesting in communal areas in South Africa', *Development Southern Africa*, vol 18, pp581–604

Shackleton, C. M., Shackleton, S. E., Netshiluvhi, T. R. and Mathabela, F. R. (2005) 'The contribution and direct-use value of livestock to rural livelihoods in the Sand River catchment, South Africa', *African Journal of Range and Forage Science*, vol 22, pp127–140

UNEP (2002) *Africa Environment Outlook: Past, Present and Future Perspectives*, UNEP, Nairobi

Whiteman, A. and Lebedys, A. (2006) 'The contribution of the forestry sector to African economies', in A. J. Pottinger, D. Okali and P. Sall (eds) Special Issue: Africa – its forests and future, *The International Forest Review*, vol 8, no 1

Wiegers, E., Curry, J., Garbero, A., and Hourihan, J. (2006) 'Patterns of vulnerability to AIDS impacts in Zambian households', *Development and Change*, vol 37, pp1073–1092

2

Distribution and Characteristics of African Dry Forests and Woodlands

Jonathan Timberlake, Emmanuel Chidumayo and Louis Sawadogo

Introduction

It is important to define and characterize dry forests and woodlands as these forms of vegetation mean different things to different people depending on their discipline and background. This chapter provides the reader with a description of the dry forest and woodland types of sub-Saharan Africa and illustrates some of their basic biological features. Here we define dry forest and woodland as vegetation dominated by woody plants, primarily trees, the canopy of which covers more than 10 per cent of the ground surface, occurring in climates with a dry season of three months or more. Such a broad definition – wider than those used by many authors – incorporates vegetation types commonly termed woodland, shrubland, thicket, savanna and wooded grassland, as well as dry forest in its strict sense. However, it does not include moist evergreen forest (rainforest), grasslands and dwarf shrublands, such as heathlands and fynbos. Where more specific vegetation types are being described, the appropriate term (e.g. dense woodland, wooded grassland) is used.

Forest Structure at Landscape Level

The terminology used in describing African vegetation is often confusing, with many different approaches. In the 1950s, a continent-wide framework was established at a meeting in Yangambi, Belgian Congo (now Democratic Republic of Congo) (CCTA/CSA, 1956), but this was considered by many as too biased towards the moist forest and dense woodland formations of humid western Africa. Workers more concerned with rangeland and land-use planning developed a well-structured system based on percent canopy cover and canopy height (Greenway, 1973; Pratt et al, 1966), while the worldwide UNESCO vegetation mapping legend (UNESCO, 1973) provided a middle ground, but was only really usable at very broad scales. The most recent, and perhaps most thorough overview of African vegetation is that by White (1983), who made a point of avoiding the use of terms such as steppe and savanna owing to difficulties in their definition and usage. White himself, for example, uses the term 'dry forest' very specifically to mean vegetation dominated by a continuous stand of trees over 10m in height which experience a dry season of a few months during which atmospheric humidity is low – a quite restricted vegetation type. In this book we use the term 'dry forest' for the tall closed-canopy seasonal or deciduous forests that separate tropical rainforest from the mesic woodlands to the immediate south and north of the equator. The term woodland is used to include vegetation types ranging from open woodlands and wooded grassland with greater than 10 per cent woody cover, sometimes loosely termed savanna. Also included are shrublands with a canopy at only 2m or so high. Dry forests and woodlands therefore incorporate seven of White's structural vegetation formations including forest, woodland, transition woodland, bushland and thicket (see Box 2.1), scrub woodland, shrubland and wooded grassland.

Box 2.1 Itigi-Sumbu thicket

A thicket is a low forest consisting of a closed stand of bushes and climbers between 3 and 7m tall (White,1983). Itigi-Sumbu thicket is a unique vegetation form surrounded by almost unrelated vegetation types (Almond, 2000). The thicket occurs in Tanzania (central parts of the country) and Zambia (between Lakes Mweru Mweru-Wantipa and Tanganyika (Wild and Fernandes, 1967) and covers a total area of about 7800km^2. Little is known about the ecology of the Itigi-Sumbu thicket but it has a woody flora of about 100 species (Fanshawe, 1971) and characteristic species include *Baphia burttii*, *B. massaiensis*, *Bussea massaiensis*, *Burttia prunoides*, *Combretrum celastroides*, *Grewia burttii*, *Pseudoprosopsis fischeri* and *Tapiphyllum floribunda*. The vegetation type is considered endangered with about 50 per cent of it in Tanzania and as much as 71 per cent in Zambia having been cleared; apparently clearing takes place even in protected areas (Almond, 2000). In Tanzania none of this vegetation is in protected areas. However, protection status does not appear to prevent clearing of Itigi-Sumbu thicket.

Source: Based on World Wide Fund for Nature (2001)

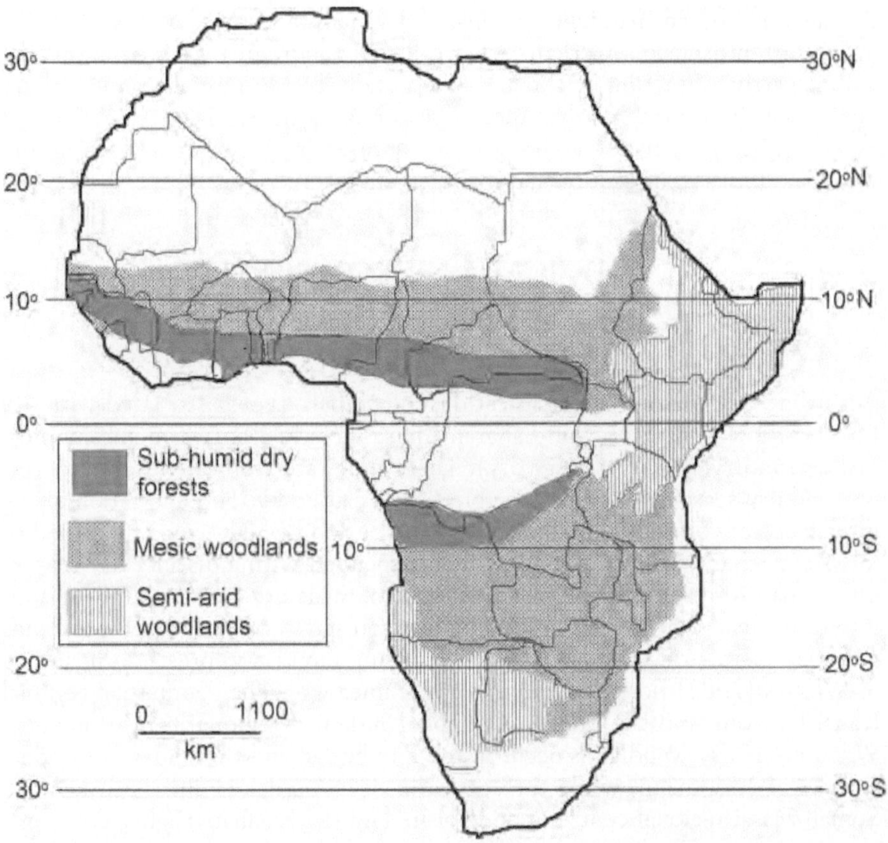

Figure 2.1 *Distribution of dry forest and woodlands across sub-Saharan Africa*

Source: Based on White (1983)

Another important consideration is the difference – in composition, structure and ecological processes – between woodlands dominated by broader-leaved species, such as *Brachystegia*, *Isoberlinia* and *Combretum*, and those dominated by fine-leaved or microphyllous species such as *Acacia* (Frost, 1996). Most of the dry forests and woodlands discussed here are dominated by trees of the legume family (Leguminosae or Fabaceae), but broad-leaved woodlands are mostly composed of species from the Caesalpinoid subfamily of the legumes, while microphyllous savannas are dominated by species from the Mimosoideae subfamily. The difference is not just in species composition, but also in ecology, and hence response to management. Broad-leaved woodlands, such as miombo, teak woodland and *Isoberlinia* woodlands in western Africa, tend to occur on nutrient-poor soils, are slower-growing when regenerating from seedlings and also comprise species that do not fix nitrogen. Whereas microphyllous savannas, such as the Kalahari *Acacia* woodlands of southern Africa, the eastern African *Acacia–Commiphora* savannas, and those across the Sahel region, often occur on nutrient-richer soils with more nitrogen-fixing species. This is partly a function of climate – soils under higher rainfall conditions are more leached of nutrients than those in much drier areas. These features are of great significance in determining the responses of different dry forest and woodland types to utilization and management interventions.

PHYSICAL ENVIRONMENT

Climate

According to the Köppen classification, the climate types associated with dry forests and woodlands in sub-Saharan Africa include 3Af (warm sub-humid), 4Af (warm dry) and 5Af (very dry). These types are tropical with alternating wet and dry seasons in which precipitation is caused by the penetration of the inter-tropical convergence zone during the period of high sun. The period of low sun is characterized by trade winds associated with a distinct dry season. On the basis of these climatic features, we can divide dry forests and woodlands into three main types: warm sub-humid dry forests, warm mesic dry woodlands and warm semi-arid woodlands. The warm sub-humid dry forests occur in two of White's (1983) floristic regions – the Guinea-Congolia/Zambezian regional transition zone and the Guinea-Congolia/Sudanian regional transition zone. Warm mesic dry woodlands occur in the Zambezian and the Sudanian regional centres of endemism, while the semi-arid dry woodlands cover most of the Somali-Masai regional centre of endemism, and the Kalahari-Highveld regional transition zone. These vegetation formations, which occur between 20°N and 30°S, are shown in Figure 2.1.

In western Africa, the sub-humid dry forests and warm mesic dry woodlands correspond to the Guinea savanna and the Sudan savanna of Keay (1959), respectively. In eastern Africa the semi-arid dry woodlands correspond to the *Acacia–Commiphora* region, and in southern Africa the warm mesic dry forests correspond predominantly to the *Brachystegia–Julbernardia* (miombo) woodlands while the semi-arid dry woodlands correspond to *Acacia–Combtretum* formations. Within the miombo woodlands are embedded other vegetation types such as undifferentiated woodlands and mopane woodlands.

The broad climatic features associated with dry forests and woodlands are summarized in Figure 2.2. The sub-humid dry (Guinea) forests of western Africa have a shorter rainy season of about 7 months compared to 8 months in the southern sub-humid dry forests; mean annual rainfall is similar and ranges from 1200 to 2000mm. In contrast, the warm mesic dry (Sudanian) woodlands of western Africa experience a longer wet period of about 4.5 months compared to 4 months in the warm mesic dry (Zambezian) woodlands of southern Africa, but the range in mean annual rainfall of 600 to 1200mm is again similar in the two woodland regions. In the semi-arid dry woodlands the rainy season is shorter (2 months) in eastern Africa and 4.5 months in southern Africa. Another

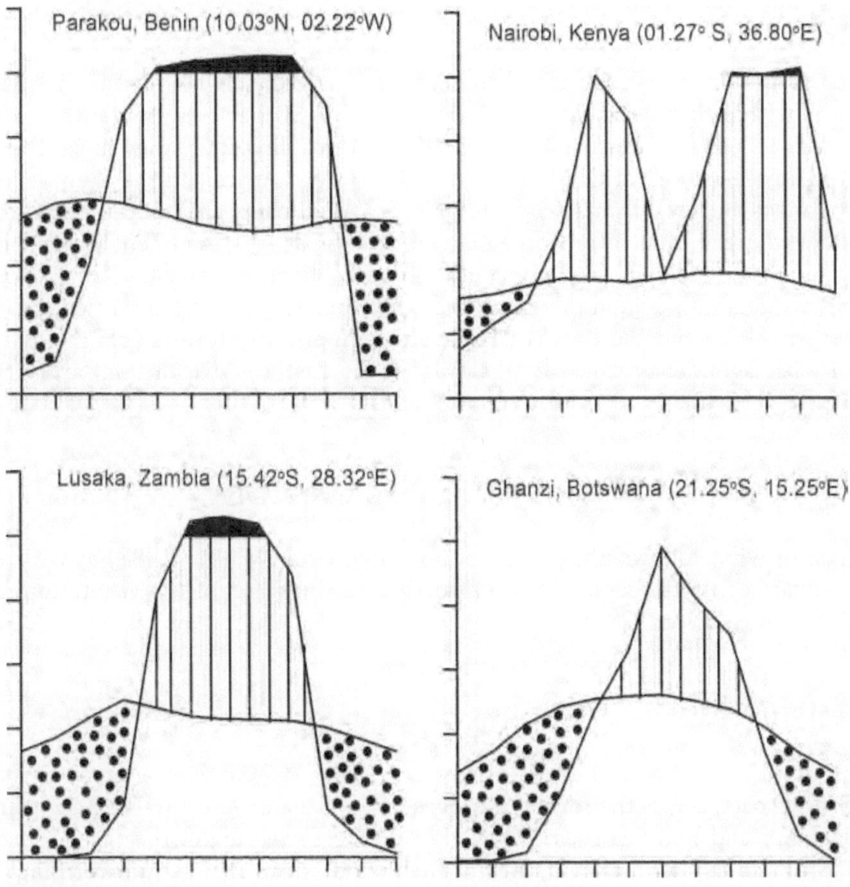

Figure 2.2 *Climate diagrams for stations in the warm dry woodland areas of western Africa (Parouku) and southern Africa (Lusaka), and in semi-arid areas of eastern Africa (Nairobi) and southern Africa (Ghanzi)*

Note: The drought period is shown by dots while the humid period is vertical hatched
Source: Chapter authors

characteristic feature of semi-arid formations of eastern Africa is the bimodal rainfall pattern, which contrasts with other semi-arid formations in which the rainfall pattern is basically unimodal (see Figure 2.2). Annual precipitation is lower (around 250mm) in the semi-arid areas of eastern Africa and higher (around 450mm) in southern Africa.

Soils

Soils in the dry forest and woodland areas are usually shallow (less than 1.5m deep) with a high sand content in the range 30–90 per cent, while the clay content is 5–20 per cent and silt is 5–45 per cent. Texturally they range from silt loam in eastern Africa to sandy loam in western Africa and the Zambezian dry forests and woodlands of southern Africa, and loamy sand in the Kalahari-Highveld area. Total soil nitrogen is very low (0.02–0.10 per cent), as is the organic matter content (1–3 per cent) and available phosphorus (2–30mg kg^{-1}). Soils are generally acid in the Zambezian dry forests, (pH 4–5), intermediate in western Africa and the Kalahari-Highveld semi-arid dry forests (pH 6.0–6.5) and with a tendency towards alkalinity in the eastern African semi-arid dry forests (pH 7–8).

Floristic Composition

Much of what follows under this section is based on White (1983) as we are not aware of more recent work on floristic composition of the vegetation in sub-Saharan Africa.

Western Africa

The dry forests and woodlands of western Africa stretch west–east from the Atlantic coast, across the southern fringes of the Sahara Desert to the Ethiopian Highlands and Red Sea coast. They can be categorized under the three broad climatic zones – sub-humid, warm mesic and warm dry – each forming a relatively narrow band. Much of the area is low (under 750m altitude) and gently undulating, with a few higher areas such as the Jos Plateau in northern Nigeria. It has also been extensively modified by human activities such as cultivation and fire so that the natural vegetation is often not easy to determine – most vegetation now present is secondary wooded grassland and grassland.

The Guinea sub-humid dry forest stretches from the Atlantic coast in Guinea across the middle regions of Ivory Coast, Ghana, Nigeria and Cameroon to the Central African Republic and northern Democratic Republic of Congo (DRC), covering an extent of around 1.4 million km^2. Locally it verges into Guineo-Congolian rainforest in both structure and species composition, while elsewhere, to the north, it can be difficult to separate from the warm mesic dry Isoberlinia woodlands. Rainfall ranges from 1200–1500mm per year in the east to 1500–2000mm per year in the west. A characteristic species is *Parinari excelsa*, sometimes forming a forest 18–20m high, along with *Erythrophleum suaveolens*, *Detarium senegalense*, *Khaya senegalensis* and *Afzelia africana*. At the western extent *Berlinia grandiflora*, *Cynometra vogelii* and others are typical, while in the eastern parts in the Central African Republic *Isoberlinia doka*, *Afzelia africana*, *Anogeissus leiocarpus*, *Burkea africana*, *Borassus*

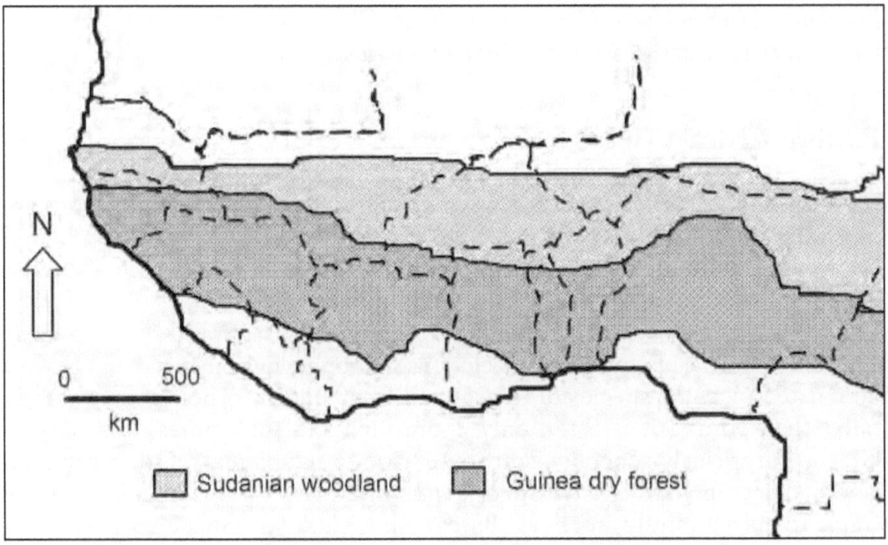

Figure 2.3 *Distribution of dry forest and woodland in western Africa*

Note: Broken lines represent national boundaries.
Source: Based on Keay (1959)

aethiopicum and *Terminalia* species are found. The oil palm *Elaeias* and *Parkia* are commonly found across transformed landscapes.

The Sudanian warm mesic dry woodland lies in a parallel band, perhaps 500–700km wide, to the north of the Guinea dry forest belt, and covers around 2.6 million km^2. Stretching from the Atlantic coast in Senegal, it includes much of Mali, northern Ghana and northern Nigeria, to the southern Sudan and southwest Ethiopia. Soils are predominantly recent and sandy. The major part of the vegetation is disturbed, consisting of open broad-leaved woodland to wooded grassland and bush fallow, often maintained by frequent fire. The moister southern portion, from Guinea to Nigeria, is typified by low woodland, rarely more than 5m high, of *Isoberlinia doka, I. angolensis, Burkea africana, Daniellia diveri* and *Erythrophleum africanum,* with tall *Hyparrhenia* grass. It has been likened to a depauperate miombo woodland, but without the *Brachystegia* and *Julbernardia.*

To the drier north, the environment is marginal for agriculture and the microphyllous vegetation types are naturally more open, characterized by species of *Acacia* (especially *Acacia laeta, A. nilotica var. adansonii, A. senegal, A. seyal, A. macrostachya*), *Balanites aegyptiaca, Bauhinia rufescens, Boscia salicifolia, Capparis tomentosa, Commiphora africana, Dalbergia melanoxylon, Grewia flavescens, G. vilosa, Pterocarpus lucens, Saba senegalensis.* The combretaceae family is well represented with C*ombretum micranthum,* C. *glutinosum,* C. *nigricans.* Scattered remnant trees, such as baobab (*Adansonia digitata*) and Shea tree (*Vitellaria paradoxa*) are common in the landscape. The most common grass species are *Aristida hordeacea, Schoenefeldia gracilis,*

Brachiaria xantholeuca, *Cenchrus biflorus*, *Chloris lamproparia*, *C. prieurii*, *Ctenium elegans* and scattered taller *Andropogon gayanus*.

Eastern Africa

Compared to western and southern Africa, the woodlands of eastern Africa are much less varied and it is difficult to separate them out. The typical sub-humid dry forests and mesic woodlands, so well represented in western and southern Africa, are very restricted in extent in eastern Africa.

By far the most extensive woodland type is in the semi-arid zone, covering 1.6 million km^2. This comprises deciduous microphyllous bushland and thicket dominated by spiny species of *Acacia* and *Commiphora*. The canopy is rarely taller than 10m, but is often only 3–5m high and sometimes quite dense. Grasses surprisingly generally contribute little to the biomass. Other common woody plants include *Grewia* species, *Balanites*, and various members of the Capparidaceae family such as *Boscia* and *Cadaba*. Succulents such as *Euphorbia* and *Adenium* are not uncommon. The baobab tree (*Adansonia*) is also characteristic at lower altitudes towards the coast. Some of the most diverse vegetation is found in a narrow band along perennial or seasonal rivers, and on rocky outcrops and hills. True moist forest, outside the scope of this book, is found above 2000m on larger mountains, or lower down closer to the coast.

Southern Africa

The sub-humid dry forests of southern Africa mostly comprise forest or dense woodlands, much of which has now been reduced to wooded grassland or secondary grassland with scattered forest trees after destruction by humans and fire. Total extent is around 980,000km^2. The larger forest remnants have canopies about 25m high, but elsewhere this is only around 10m. In these remnants characteristic species are *Marquesia macroura*, *M. acuminata*, *Berlinia giorgii*, *Daniellia alsteeniana*, *Brachystegia spiciformis*, *B. wangermeeana* and *Parinari curatellifiolia*. However, in the more common secondary vegetation comprising grassland and wooded grassland ('mikwati') the most frequent fire-resistant trees are *Erythrophleum africanum*, *Dialium englerianum*, *Burkea africana*, *Diplorhynchus condylocarpon*, *Pterocarpus angolensis* with an under storey comprising species of *Protea*, *Combretum* and *Strychnos* with tall *Hyparrhenia* grasses.

Within the warm dry forest and woodland area of southern Africa there are four extensive vegetation types: undifferentiated woodland, miombo woodland, mopane woodland and semi-arid shrubland (White, 1983; Figure 2.4).

Undifferentiated woodlands

The most extensive undifferentiated woodlands are teak and acacia woodlands. Teak woodland is a deciduous broad-leaved dry forest or woodland type dominated by *Baikiaea plurijuga*, often called Zambezi Teak that occurs on Kalahari sands. This species, much valued for its hard heavy timber, is common, along with other valued timbers such as *Pterocarpus angolensis*, *Guibourtia coleosperma* and *Schinziophyton rautanenii*. *Baikiaea* woodland areas cover around 265,000km^2 on the Kalahari sands of northwestern Zimbabwe, northeast Botswana, southwest Zambia, northeast Namibia and southeast Angola (Figure 2.4) at an elevation of around 900–1100m on the gently undulating plateau. This area is a mosaic of *Baikiaea*-dominant dry forest and woodland on the deeper sands, *Burkea africana* on shallower sands and a wide network of shallow valleys with seasonally wet grasslands called dambos or vleis. Many of these are associated with the headwaters of the Upper Zambezi and Kavango rivers. To the north and to the southeast in western Zimbabwe, *Brachystegia spiciformis* comes in, forming a transition to more typical miombo woodland. *Baikiaea* woodland is generally 6–10m high, with few emergents. Although deciduous, *Baikiaea* itself can remain green long into the dry season owing to its ability to access moisture from deep in the soil profile. In some regards,

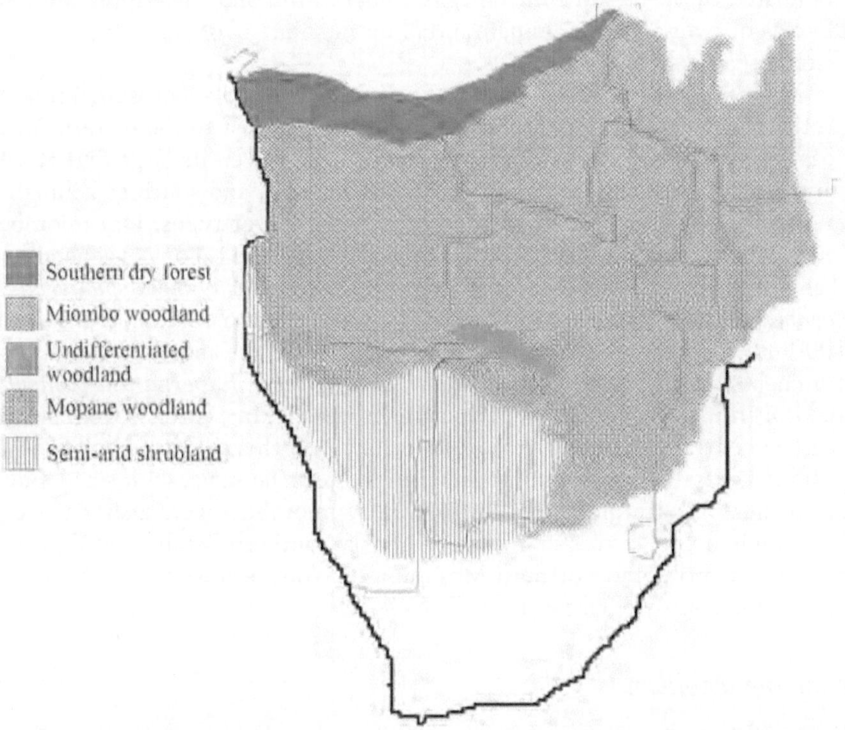

Figure 2.4 *Distribution of dry forest and woodland types in southern Africa*

Source: Based on White (1983)

Baikiaea woodland, occurring as it does on deep sandy soils that are not prone to erosion, is resilient in the face of major perturbations such as drought, frost, herbivory, fire and, more recently, logging and clearance.

In the drier southeastern parts of the warm dry forest region open mixed (microphyllous and broad-leaved) acacia woodlands, dominated by *Acacia* (especially *A. nigrescens*, *A. nilotica* and *A. gerrardii*) and *Combretum* (especially *C. apiculatum*), are found. Other characteristic and useful species include *Burkea africana*, *Terminalia sericea*, *Kirkia acuminata*, *Pseudolachnostylis maprouneifolia*, *Sclerocarya birrea* and *Ziziphus mucronata*. Fruit trees and species with useful woods are more commonly found here than in the miombo or mopane woodlands. The grass growing on these more fertile soils generally has a high grazing value. Fire is an important ecological feature, and a number of species are tolerant of it.

Miombo woodland

This is by far the most extensive warm dry forest type in southern Africa (Frost, 1996). Miombo (or muombo) is a common local name for various species of *Brachystegia* in parts of Malawi, Tanzania and Zambia, which are the woodland type's most characteristic species. These broad-leaved deciduous woodlands, dominated by species of *Brachystegia*, *Julbernardia* and *Isoberlinia*, are found in areas of over 700mm annual rainfall on nutrient-poor soils where there is distinct seasonality.

Miombo is found from the Atlantic coast in Angola, across Zambia, the southern DRC, Malawi and much of Zimbabwe, almost to the eastern African coast in Mozambique and Tanzania (Figure 2.4). Given its large extent, with surprisingly little variation from Angola to Tanzania and south to Zimbabwe, miombo is often arbitrarily divided into dry and wet types. Dry miombo is found where rainfall is less than 1000mm per year, the canopy is generally less than 15m high, and the dominant tree species are *Brachystegia spiciformis*, *B. boehmii* and *Julbernardia globiflora*. In wet miombo rainfall is more than 1000mm per year, the canopy exceeds 15m in height and the characteristic dominants are *Brachystegia floribunda*, *B. longiflora*, *Julbernardia paniculata* and *Isoberlinia*. Wet miombo, covering 1.36 million km^2, extends from near the coast in central and northern Angola, through northern Zambia and southern DRC to central and northern Malawi and western Tanzania, with small extents in northern Mozambique and Burundi. Dry miombo covers a similar extent (1.21 million km^2) across southeastern Angola, southern Zambia and Zimbabwe to south, central and northern Mozambique, southern Malawi and much of central and southern Tanzania.

Mopane woodland

Whereas miombo woodland is generally found on lighter-textured, nutrient-poor, well-drained soils on the African Plateau, mopane woodland is mostly confined to lower-lying areas with clay- and nutrient-rich soils. The dominant tree, often to the exclusion of many others, is *Colophospermum mopane*, commonly known as mopane or chanate.

Mopane woodland, covering 380,000km^2, is a dry broad-leaved woodland type found on heavier-textured soils in the wide, flat valley bottoms of lower-altitude river valleys of southern Africa, such as the Limpopo, Zambezi, Okavango, Cunene, Shire and Luangwa (Mapaure, 1994; Timberlake, 1999) in southern Angola, northern Namibia, northern and eastern Botswana, Zimbabwe, northern South Africa, southern Zambia, southern Malawi and south and central Mozambique (Figure 2.4), at an elevation of 200–1200m, but normally from 300–900m. Rainfall in these areas ranges from 400 to 700mm per year, although the species itself can be found in drier areas in northwestern Namibia.

Semi-arid shrubland

Semi-arid shrubland types in southern Africa cover over 900,000km^2 and comprise either microphyllous wooded grassland or shrubland in which there is a more-or-less continuous grass layer. Due to water stress, the sparse tree canopy is around 5–8m high. Characteristic species in the drier Kalahari sands area, at around 1000m altitude, include various *Acacia* (*A. erioloba, A. luederitzii, A. fleckii, A. hebeclada, A. mellifera, A. tortilis*), *Boscia albitrunca, Dichrostachys cinerea* and *Terminalia sericea*. In the northern, less arid parts, broad-leaved trees such as *Combretum collinum, Lonchocarpus nelsii* and *Ziziphus mucronata* are common, while on more stony soils towards the South Africa Highveld these are replaced by *Acacia karroo* and numerous shrub species, with an increasing dominance of the grass *Themeda triandra*.

FOREST DYNAMICS

Spatial integrity

Large areas of African dry forests have been subjected to clearing for agriculture, which has greatly affected their spatial integrity (Table 2.1). In addition, logging and wood harvesting for fuel have exacerbated this fragmentation process. The greatest impact of agriculture has been in western and eastern Africa, where more than 50 per cent of dry forest land has been converted to agricultural use. Because of the nature of shifting cultivation that is widely practiced in Africa, land under agriculture represents a number of cover types, including cropland, abandoned fields and fallows at various stages of recovery. For example, Lanly and Clement (1982) estimated that 20 per cent of open woodland in Africa is made up of fallows. When converting woodland to agriculture not all trees are necessarily cleared and, as a result, cultivated land is often dotted with trees: *Adansonia digitata, Vitellaria paradoxa, Sclerocarya, Borassus, Faidherbia albida*. Such landscapes are called farmed parklands in western Africa. Population pressure on dry forests is growing. For example, land per person in the Zambezian dry forest zone decreased from 1.45ha in 1975 to

Table 2.1 *Woodland and conversion to agriculture in sub-Saharan Africa in 2000*

White's phytoregion	*Woodland area (km²)*	*Extent under agriculture (km²)*	*(%)*
Sudanian	3,233,160	1,900,850	58.79
Somali-Masai	875,880	706,880	80.70
Zambezian	4,483,380	679,700	15.16

Source: Based on Mayaux et al (2004)

1.05ha in 2000, and corresponding values for the Sudanian zone of western Africa are 1.58ha in 1975 and 1.28ha in 2000 (Eva et al, 2006).

Plant cover and phenology

The level of vegetation cover in dry forests depends on its structure and phenology, but generally tree cover is highest in sub-humid dry forests (over 70 per cent) and lowest in semi-arid woodlands (10–40 per cent). Because of its deciduous nature, vegetation cover also varies seasonally (Figure 2.5). The increase in greenness during the late dry season is due to the woody component, while that in the early wet season is due to the grass component (Chidumayo, 2001).

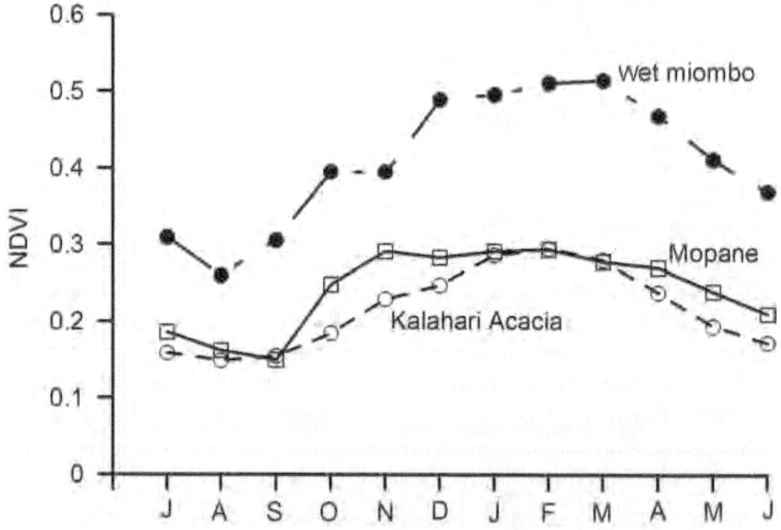

Figure 2.5 *Monthly changes in vegetation greenness as determined by the Normalized Difference Vegetation Index (NDVI) in wet miombo at Mpika (Zambia), mopane woodland at Francistown (Botswana) and Kalahari Acacia savanna at Ghanzi (Botswana) in southern Africa*

Source: Based on Chidumayo (2001)

Changes in the tree component caused by fire, timber and wood harvesting, drought or browsing/grazing by wildlife or livestock, tend to reduce dominance of the woody component in favour of the grass component. This grass dominance can then be maintained by fire or grazing.

Climate change and land use

There will undoubtedly be some marked changes in distribution and extent of African vegetation types over this century, probably resulting from an increased tendency in many areas to lower mean annual precipitation. But such effects have also got to be seen in the context of equally – or more – rapid changes in land use resulting from intensification and increased population pressures. However, climate change and increased human impacts on vegetation are likely to be intertwined, and it is difficult to predict the consequences. There are also the changes in land-use practices caused by concentration of people and control of fire that lead to increased woody cover in some places, and such changes may be more profound over short periods than those predicted from the increases in carbon dioxide (CO_2) or temperature or rainfall.

One approach is to model climate change and determine what the impacts on vegetation would be 50 years hence of, for example, decreased rainfall. Such a study has been done for southern Africa (Hulme, 1996), but also recognizes that global models of climate change, whether resulting from higher CO_2 levels, higher temperatures or decreased precipitation, are inherently weaker in such seasonal environments where there are naturally large annual fluctuations in temperature and rainfall. The models can only predict what might happen at a very coarse, sub-continental level, but not the local detail. Indeed, the study points out that local vegetation patterns are determined more by local differences in moisture and nutrient availability than global changes in annual means, and the response of vegetation in such areas to extreme wet or dry years is possibly more important than the response to long-term averages.

The southern African study suggested an increase in mean annual temperature (based on the 1961–1990 mean) of 1.5°C over the period 1990 to 2050, a figure supported by the observed increase in southern Africa over the last century of 0.05°C per decade. Inherent uncertainties in the study suggest that this figure could be reached by 2030, or as late as the end of the 21st century. Such a warming will result in lower annual rainfall. Using the BIOME model (Prentice et al, 1992), which predicts the distribution of broad vegetation types based on climate, soils and CO_2 concentrations, the study went on to predict changes in vegetation distribution across the region. Under the 'dry' (20 per cent less moisture) scenario, there would be a 10 per cent increase in what we term sub-humid dry forest, which would be derived from moist evergreen forest, a 30 per cent increase in acacia semi-arid bushland and wooded grassland at the expense of warm dry forest, and a significant change in the distribution of warm dry woodland types and those in the semi-arid region. Increasing temperatures in southern Africa are also predicted to either extend the growing season in some ecosystems or shorten it in others (Rutherford et al, 1999).

Elsewhere in Africa, retraction of sub-humid and warm dry climate species to areas of higher rainfall and lower temperature has been recorded as a result of desertification in the latter part of the 20th century in Senegal (Gonzalez, 2001), while modelling of the distribution of forest species has projected changes from warm dry forests to semi-arid forests in Tanzania and Gambia (Jallow and Danso, 1997).

Added to these impacts of climate change will be the reduced productivity that comes with reduced rainfall, and the effects of reduced and perhaps much more variable rainfall on regeneration and reproductive success. With the increase in bushland vegetation, cattle-raising may become less significant, to be replaced by increased numbers of small stock, which may in turn also impact on both vegetation structure and regeneration. Climate change is also likely to alter the frequency, intensity and seasonality of fires that are so characteristic of dry forests and woodlands in Africa, which may also change both in structure and composition.

However, under these predicted changes in climate, it must be remembered that species will respond individually, perhaps with substantial time lags, rather than there being major shifts in vegetation formations and their associated fauna.

Herbivory

There are various types of impact of herbivory and herbivores on dry forests and woodlands. The most obvious are vertebrate herbivores, which can have marked effects on both structure and composition of woodlands through physical damage, through browsing and indirectly through grazing. Invertebrates are also important herbivores, and can influence woodlands through defoliation and seed predation. The impacts of herbivores, both direct and indirect, are often larger than realized.

Elephant impacts on vegetation are very marked, but tend to be confined to protected areas or adjacent lands. Principally confined to the warm dry woodlands of eastern and southern Africa, in particular miombo and mopane woodland types, these impacts are also seen in semi-arid acacia woodlands such as in Tsavo and Amboseli in Kenya. In Tarangire National Park in Tanzania, for example, the impacts varied significantly between tree species. The density of *Balanites aegyptiaca*, *Maerua triphylla* and *Commiphora* and *Combretum* species declined while that of *Cordia* and *Acacia tortilis* increased (Figure 2.6). Although elephants affected the tree size distribution, the decline in density of small trees in both *Acacia* and *Combretum* woodlands was attributed to a severe drought in 1993 (Van de Vijver et al, 1999). Under certain circumstances, elephant can reduce large areas of woodland to coppice shrubland by pushing over and breaking larger trees, particularly with *Colophospermum mopane*. This results in a 'lawn' of young coppice foliage about 2m high, which the elephant

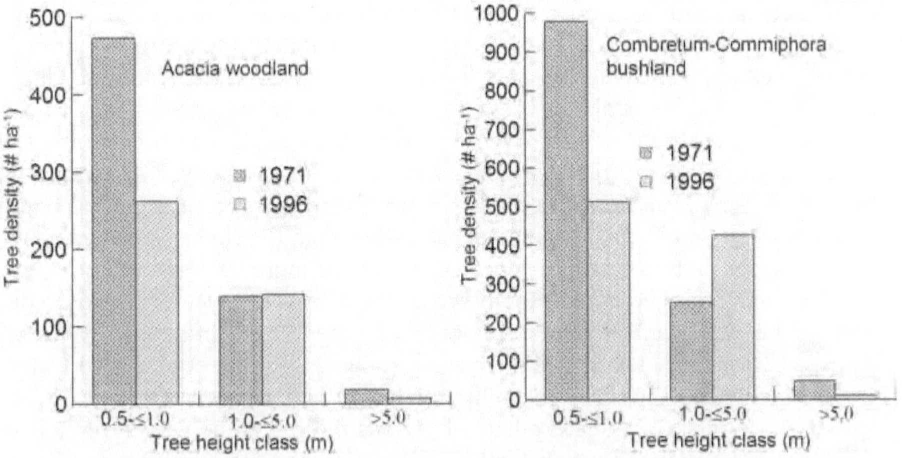

Figure 2.6 *Changes in tree density caused by elephants and drought over 25 years in Tarangire National Park, Tanzania*

Source: Based on data from Van de Vijver et al (1999)

maintain by browsing (Conybeare, 2004). In miombo woodland such destruction can also result in larger quantities and a greater depth of grass growth, leading to greater fire risk and fuel loads. The more frequent and more intense fires, coupled with increased frost risk and higher grazing pressures, help maintain the areas as open woodland or wooded grassland, rather than the dense woodland they once were, as can be seen in some protected areas in Zimbabwe and southern Tanzania.

Impacts by other browsers such as giraffe, kudu and eland, are more local, although they can be marked. For example, many *Acacia* trees in the eastern African savannas are shaped by giraffe, and acacias generally have developed thorns as a defence against such herbivory (Pellew, 1983). Browse lines are common in areas with many livestock – these are at 1.5–2m high in areas with goats, and 2–2.5m high in areas with many cattle, although the latter are primarily grazers except during the dry season. Such pressures can maintain the grass–shrub–tree balance in areas that, without browsing, would revert to woodland.

Many browsers help disperse woody plants by ingesting seeds that are then deposited elsewhere in the faeces. Species such as *Acacia nilotica* and *A. erioloba* have hard impervious seeds adapted to ingestion and dispersal by mammals, while elephants are known to disperse palms such as *Hyphaene* and *Borassus*. It is likely that the baobab (*Adansonia digitata*), through its useful fruits, has a wider range across Africa owing to its dispersal by humans, and likewise with the tamarind (*Tamarindus indica* and *Sclerocarya birrea*).

The impacts on woody vegetation by grazing mammalian herbivores are more indirect as these animals primarily eat grasses. They also depend on the intensity, duration and season of grazing. At low grazing levels, grass growth can be encouraged, which tends to shade and kill off woody seedlings inhibiting recruitment and regeneration. But when an area is more heavily grazed, the woody seedlings can be eaten. When grazing pressure is reduced, seedlings and suppressed saplings can grow rapidly to form what is often termed bush encroachment – the beginnings of woodland regeneration. However, the species involved tend to be those that grow rapidly from seed and with invasive properties, such as many *Acacia* species and *Dichrostachys cinerea*, rather than the slower growing broad-leaved species such as *Brachystegia*, that reproduce more by resprouting.

Although relatively minor compared to the effects of browsers, seed predation can be significant. A large proportion of a seed crop can be lost by weevil damage (Ernst et al, 1989) and from small rodents (Walters et al, 2005).

Normally insect herbivory is minor, perhaps 10–20 per cent of the leaf crop. However, defoliation by caterpillars of various moths can be very severe, as in the case of the mopane worm (*Imbrasia*) in eastern Botswana that consumes 19.5 tonnes per hectare of mopane leaves in just a few weeks (Styles, 1994). Outbreaks of defoliating behaviour can also be seen with other moths in broad-leaved or nutrient-poor semi-arid and warm dry forests, such as with *Julbernardia paniculata* (Malaisse, 1997) and *Burkea africana*, but rarely in microphyllous savannas. It is not clear if this phenomenon significantly changes woodland composition or structure.

Fire

Although some fires originate from lightning around the start of the rainy season, most are caused by humans. Some originate accidentally from burning undertaken during land preparation, from making charcoal or from honey collecting, but others are set deliberately to flush out mammals or birds for hunting or to stimulate new grass growth for livestock (Frost, 1996). Some have suggested that savannas (here taken to be the more open types of dry forest, including woodland) are anthropogenic and not natural. It must be recognized that fire is a natural feature of dry forests and woodlands in Africa, although its frequency is probably now much higher than historically. Nevertheless, in some areas there has been a deliberate control of fire with widespread increase in woody growth and a total change in woodland dominance. Fire has a composite nature (Frost and Robertson, 1987), its behaviour, timing, intensity, frequency all vary independently, and all affect vegetation structure and composition differently. Fire can kill individual organisms, such as trees, but it can also change the competitive status, allowing vegetation to change over time through differential recruitment and survival of seedlings and saplings.

Fire is one of the few dry forest and woodland determinants that can be readily manipulated, and so has great importance to managers. In most cases the plants that burn – the fuel – predominantly comprise grass and other herbaceous material, and not the woody material. This fuel load can be from a few centimetres to 2m in height. However, woody material does burn and often continues to smoulder long after the main fire has passed.

Most fires occur during the dry season or at the start of the rains – from May to October in the southern hemisphere and from November to February north of the equator. The impact depends on the intensity and timing in relation to plant phenology. For example, late dry season fires are more intense and destructive, particularly if they occur after leaf flush. Other important determinants of impact are the amount of moisture in the fuel load, the atmospheric temperature and relative humidity, and the wind speed.

Although hot intense fires can cause quite spectacular changes to vegetation, it is the frequency – the period between fires in any one area – which causes many of the changes in vegetation structure and composition, changes that can be both subtle and only become apparent over a decade or so. Fire frequency is often higher in vegetation types having higher fuel loads, such as moist miombo woodland (warm dry woodland), with fire return intervals of between 1 and 5 years (Frost and Robertson, 1987). In Zambia, Chidumayo (1995) reports a fire-return interval of 1.6 years, while remote-sensing across a much larger area of miombo woodland shows that only 37 per cent of land is burnt in any one year, suggesting an average fire-return interval closer to 3 years (Frost, 1996). In more arid savannas (semi-arid shrubland), a much longer fire return interval of 5–50 years is suggested, depending on fuel loads.

Frequent late dry season ('hot') fires can transform woodland into open tall grass savanna with only isolated fire-tolerant canopy trees and scattered under storey trees and shrubs. In Africa a number of long-term fire exclusion experiments have shown that tree cover in sub-humid and warm dry forests is limited by fire, notably the analysis by Trapnell (1959) of the fire plots in moist miombo woodland in northern Zambia and that by Geldenhuys (1977) in the drier forest types in northern Namibia. The major structural change that occurs under early burning and fire protection relates to the greater thickening of the shrub layer (Figure 2.7) and occasionally the colonization by fire-sensitive sub-humid forest trees (Trapnell, 1959; Louppe et al, 1995). Trapnell recognized four categories of species with regard to fire:

- fire-intolerant mostly evergreen species which occur only where there is protection from fire (e.g. *Parinari excelsa*, *Entandrophragma delevoyi*, *Syzygium guineense*);
- fire-tender species which decline under regular burning and increase under protection, with higher mortality rates under late dry season burning than under early dry season burning (e.g. *Brachystegia spiciformis*, *Julbernardia paniculata*);
- semi-tolerant species that are relatively unaffected by early dry season fires but decline under late dry season fires (e.g. *Uapaca kirkiana*, *Pseudolachnostylis maprouneifolia*, *Strychnos pungens*);
- fire-tolerant species that are able to survive late dry season fires (e.g. *Pterocarpus angolensis*, *Parinari curatellifolia*, *Erythrophleum africanum*, *Anisophylla boehmii*, *Diplorhynchus condylocarpon*, *Detarium microcarpum*, *Combretum glutinosum*, *Combretum nigricans*).

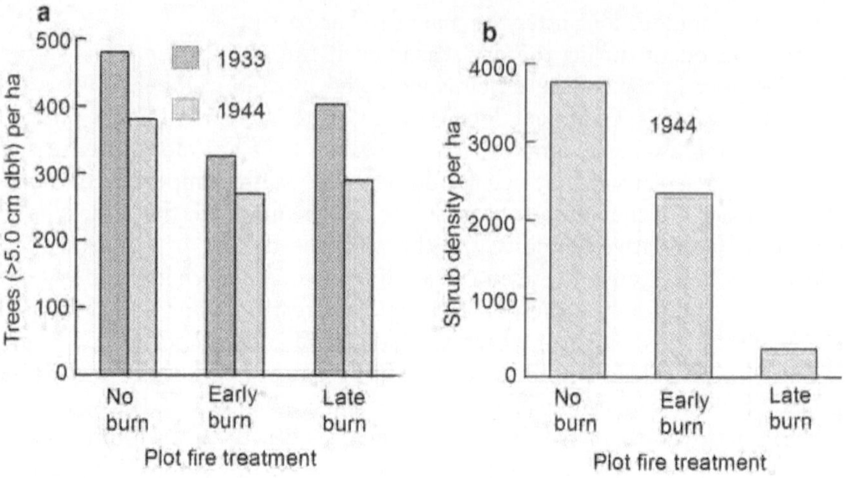

Figure 2.7 *Changes in tree density and shrub density, only enumerated in 1944, after 11 years of fire protection, early burning and late burning in old-growth miombo woodland at Ndola, Zambia*

Source: Based on data from Trapnell (1959)

However, even early dry season ('cool') fires can have a significant impact on vegetation structure and composition, as shown by the long-term fire plots in *Baikiaea* woodland in western Zimbabwe (Calvert and Timberlake, 1993). Plots protected from fire for 35 years reverted to dense *Baikiaea* woodland, but annual or biennial late dry season burns did not eliminate, only suppress, the woody component, shifting the regrowth underground. A resilient reservoir of underground rootstocks remained, which could rapidly grow up with protection.

Although often regarded as a negative feature, with care, fire can be used as a management tool to increase desirable attributes. Ranchers use it both to stimulate new grass growth and to control encroachment or recruitment of woody plants in rangeland, while foresters have used cool fires in mature stands to clear undergrowth or to reduce the risk of hotter, more damaging fires late in the dry season.

Stem density, basal area and wood biomass

The analysis of data from literature (as cited in next paragraph) for 164 dry forest sample plots revealed that stem density does not vary significantly in African dry forests and woodlands. Density ranged from 300 to 900 per hectare with a tendency for higher stem densities in semi-arid dry forests (800–900 per hectare) and lower densities in warm dry forests (700–800 per hectare). However, data for 92 sample plots indicated that basal area varies significantly

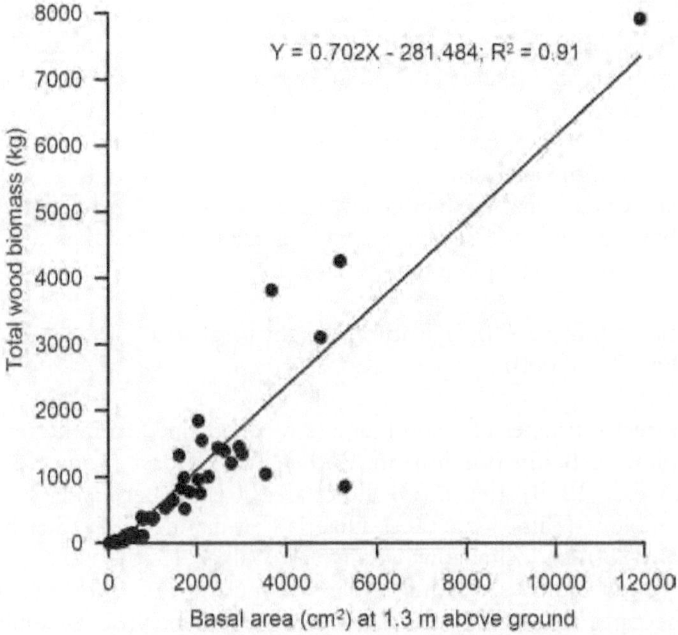

Figure 2.8 *Linear equation for determining total above ground wood biomass (oven-dry basis) from basal area of individual and stand trees in African dry forests and woodlands*

Source: Based on data from Chapter authors

among African dry forests. The highest basal area at breast height occurs in sub-humid dry forests (24–28m^2 per hectare) and this declined to a range of 11.5–14.1m^2 per hectare in warm dry forests in southern Africa and 8.4–11.0m^2 per hectare in Kalahari-Highveld semi-arid dry forests. The basal area in the Sudanian warm dry forests of western Africa ranges from 3.6 to 10.1m^2 per hectare.

The determination of above ground wood biomass of trees is generally achieved using the relationship between tree diameter and/or height and biomass derived from felled sample trees. For the purposes of this chapter, a combination of tree- and stand-based basal area (derived from diameter measurements) was used to develop a common equation (Figure 2.8) for estimating wood biomass from basal area data. This approach is particularly useful because most forest standing stocks are reported as basal area per hectare and an equation based on basal area could have a wider application in estimating wood biomass in dry forests and woodlands of Africa. The equation in Figure 2.8 was applied to basal area data from different dry forests in Africa to compute wood biomass. Data for a total of 92 sample plots in Benin (Schreckenberg, 1999), Burkina Faso (Nikiema, 2005; Zida et al, 2007), DRC (Malaisse, 1978, 1984; Malaisse and Binzangi, 1985; Malaisse et al, 1970,

1975), Mali (Picard et al, 2006), Mozambique (Musanhane et al, 2000; Sambane, 2005; Williams et al, 2008), South Africa (Shackleton et al, 1994; Scholes, 1990), Tanzania (Banda et al, 2006; Byström et al, 1987; Isango, 2007; Mugasha and Chamshama, 2002), Zambia (Araki, 1992; Chidumayo, 1985, 1986, 1987, 1997a, 1997b; Endean, 1968) and Zimbabwe (Ward and Cleghorn, 1970) were used in the analysis.

The basal area data that were originally based on stump height (between 15 and 30cm above ground) measurements were converted to basal area at breast height using the following equation:

$$\text{Basal area} = -0.0019 + 0.71 \times \text{stump height basal area},$$
$$r^2 = 0.999, P < 0.0001$$

In addition direct estimates of wood biomass were obtained from literature for 73 sample plots in Benin (Orthmann, 2005), Ivory Coast (Lamotte, 1979; Menaut and Cesar, 1979), Kenya (Okello et al, 2001; Western and Ssemakula, 1981), Mozambique (Sambane, 2005), Nigeria (Fatubarin, 1984), South Africa (Higgins et al, 1990) and Zimbabwe (Kelly and Walker, 1976).

Above ground biomass was highest in sub-humid dry forests (160–209 tonnes per hectare) followed by the Zambezian woodlands (88–97 tonnes per hectare), Sudanian woodlands (56–78 tonnes per hectare), Kalahari scrubland (22–34 tonnes per hectare) and lastly Somali-Masai bushland (13–18 tonnes per hectare).

Dry Forest and Woodland Regeneration

Regeneration in dry forest and woodland trees occurs through either sexual or vegetative means. Sexual regeneration is achieved through seed germination and establishment of seedlings and their recruitment into the tree phase. Vegetative regeneration occurs through the recruitment of sprouts or resprouts into the tree phase from pre-existing trees that get cut or damaged, sometimes termed coppice. Sprouting is the production of secondary trunks as an induced response to injury or to profound changes in growing conditions. There are four types of sprouts: sprouts from the trunk, sprouts from specialized underground stems (lignotubers and rhizomes), sprouts from roots (root suckers), and opportunistic sprouts from layered branches. Seed dispersal, predation, desiccation and seedling mortality can act as strong constraints that impede dry forest and woodland recovery after disturbance. The ability to sprout after severe injury from disturbances such as herbivory, fire, floods, logging or drought overcomes these barriers, as these individuals bypass the seed stage and tend to have more vigorous shoots than seedlings presumably because vegetative shoots may take advantage of the extensive root system and the substantial food storage in the remaining parts of the parent plant. However, sprouting ability varies with the age or size of a plant and also with the type and severity of injury.

Figure 2.9 *Common regeneration mechanisms in dry forests: true seedling, seedling sprout, stump sprouts or coppice and root sucker*

Source: Chapter authors

The common mechanisms of regeneration in dry forests and woodlands include current-year seedlings, seedling sprouts that occur after seedling shoot dieback (including sprouts from underground rootstocks), coppice or sprouts from stumps of mature trees and root suckers that arise from lateral roots (Figure 2.9). The importance of each of these regeneration mechanisms depends on the floristic composition of the forest/woodland and the type of disturbance. Ky-Dembele et al (2007) found that the regeneration pool in a Sudanian savanna dominated by *Detarium microcarpum*, *Terminalia avicennioides* and *Acacia* species in Burkina Faso was dominated by seedlings and seedling sprouts that made up 88 per cent of the regeneration stock, while these sources of regeneration constituted nearly 95 per cent of the regeneration pool in a Zambian miombo woodland (Chidumayo, 1993; Table 2.2). However, true seedlings made up only 3.5 per cent of the regeneration pool in the Sudanian savanna compared with 56 per cent of the regeneration pool in miombo woodland. Although root suckers and water sprouts (plantlets developed at the base of live mature trees) were observed in the Sudanian woodland, these were absent in miombo woodland.

Seedlings and sprouts face frequent and severe fire damage that can retard their recruitment into the tree layer (Figure 2.10). Between fires, seeds have to germinate and build enough root reserves to survive the next fire. However, given that fires usually occur once in every 2–4 years (Chidumayo, 2004; Sankaran et al, 2007; Jacobs and Schloeder, 2002), sprouts would need to grow rapidly to escape damage. Once dry forest has been cleared and the land abandoned, regeneration often takes place but the speed of forest/woodland recovery depends on the methods used in clearing, the sources available for regeneration and site history (type, frequency and intensity of stress and/or disturbance). Initially regeneration is mostly from trunk and root sprouts, hence tree species composition in regrowth stands tends to remain similar to that of the preceding old-growth stand. Consequently, the regrowth forest consists of

Table 2.2 *Structure of the tree regeneration pool one year after clear cutting miombo woodland in Zambia and two years after cutting a Sudanian woodland in Burkina Faso*

Regeneration type	*Miombo woodland*		*Sudanian savanna*	
	Plants per ha	*Percent*	*Plants per ha*	*Percent*
True seedlings	14,575	56.4	1154	3.5
Seedling sprouts	9850	38.1	28,269	84.5
Stump sprouts	1425	5.5	2000	6.0
Root suckers	0	0.0	1308	3.9
Water sprouts	0	0.0	718	2.2

Source: Based on Chidumayo (1993) for miombo woodland and Ky-Dembele et al, (2007) for Sudanian savanna

Table 2.3 *Stem density in old-growth and regrowth stands in woodlands in sub-Saharan Africa*

Woodland type	*Stem density (number per hectare)*	
	Old-growth stands	*Regrowth stands*
Kalahari	837–1131	7264–9700
Sudanian	425–590	1280–2250
Zambezian	710–820	2446–2830

Source: Calculated by Chapter authors from data in various literature sources (see text)

multi-stemmed trees often with a stem density higher than in the previous old-growth woodland it replaced (Table 2.3).

Much work has been done on the effects of fire on woodland recovery following clearing. In Mozambique the number of sprouts produced per stump declined with increasing fire frequency. For example, the average number of *Diplorhynchus condylocarpon* and *Julbernardia globiflora* sprouts per plant decreased from 14.7 under fire protection to 7.7 when burnt once and 6.4 and 4.6 when burnt twice and three times, respectively (Zolho, 2005). However, in Burkina Faso fire protection or burning had no significant effect on stump mortality, although mortality was highest (15 per cent) in the first year after cutting and only increased marginally by 1 per cent in subsequent years (Sawadogo et al, 2002). In Zambia, stem density in an 11-year-old regrowth miombo was significantly higher under fire protection and lower under late dry season burning (Table 2.4). Similar results have been reported for Sudanian woodland in Ghana (Figure 2.11). The density of trees (>0.5cm dbh, i.e. diameter at 'breast height', which is deemed to be 1.3m) was also lower under late dry season burning (214/ha) in a 65-year-old fallow located in the transitional zone between the Guinea-Sudan woodland and Guinea dry forest in Ivory Coast, compared to 2244/ha and 6877/ha under early dry season burning and fire protection (Louppe et al, 1995). Clearly, some degree of fire control is important in enhancing dry forest recovery after clearing.

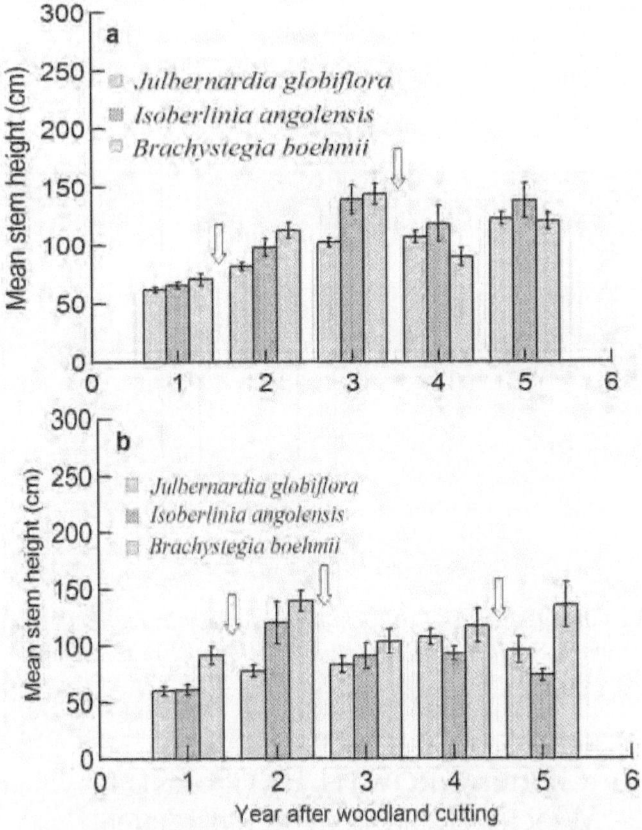

Figure 2.10 *Mean height development of sprout stems of miombo trees at sites and during five years following clear cutting of forest*

Note: Vertical lines on bars indicate 1SE and inverted arrows indicate fire occurrence at each site.
Source: Based on Chidumayo (unpublished data)

Table 2.4 *Stem density in wet miombo woodland plots before clear-cutting in 1933 (stems >6.4cm dbh) and in regrowth (stems >0.9m tall) under three fire treatments at Ndola in Zambia*

Fire treatment	*Woodland type*	*Stems per hectare*
Fire protection	Before cutting	205
	11-year-old regrowth	2963
	49-year-old regrowth	393
Early dry season fire	Before cutting	550
	11-year-old regrowth	2080
	49-year-old regrowth	520
Late dry season fire	Before cutting	455
	11-year-old regrowth	475
	49-year-old regrowth	no data

Source: Based on Chidumayo (1997b)

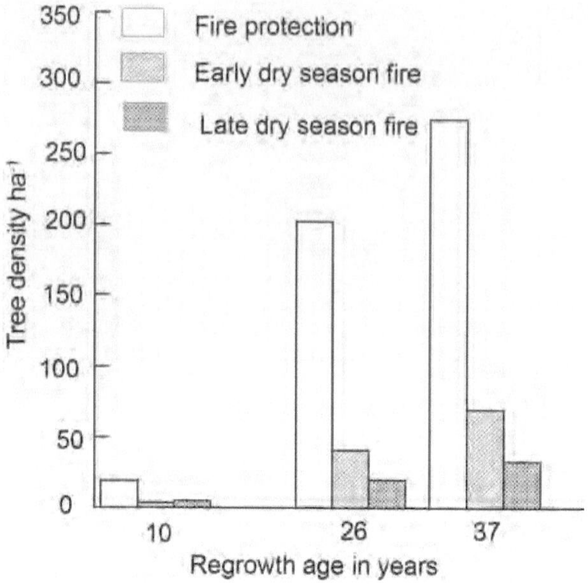

Figure 2.11 *Changes in tree (>9.0cm dbh) density in regrowth Sudanian savanna in northeast Ghana under different fire treatments*

Source: Based on Swaine and Brookman-Amissah (1987)

Tree Growth Rates and Wood Biomass Production

Tree growth is influenced by many factors, including genetic, climate and soils, as well as levels of disturbance such as fire and disease. The majority of tree species studied have diameter increments ranging from 0.03 to 2.6cm per annum. Among the most studied species in the Zambezian savannas are *Brachystegia spiciformis* (Grundy, 2006; Holdo, 2006; Trouet et al, 2006) and *Pterocarpus angolensis* (Boaler, 1966; Holdo, 2006; Stahle et al, 1999). *Brachystegia spiciformis* in western Zambia grew by 0.24 to 0.33cm diameter per annum while in Zimbabwe the species grew by 0.03 to 0.27cm per annum (Grundy, 2006; Trouet et al, 2006). A study in Tanzania (Jeffers and Boaler, 1966) revealed that humidity and minimum temperature were the most significant factors affecting the seasonal growth of *Pterocarpus angolensis* and, in a similar study, Boaler (1966) found that the annual diameter increment ranged from 0.08 to 0.48cm. However, growth ring bands indicated that the species also exhibited inter-decadal variations in growth rate. The annual diameter growth of *Isoberlinia doka* and *Anogeisus leiocarpus* in western Africa is estimated at 0.44cm and 0.45cm (Schöngart et al, 2006) while that of *Acacia drepanolobium* in eastern Africa is estimated at 0.7cm (Okello et al, 2006).

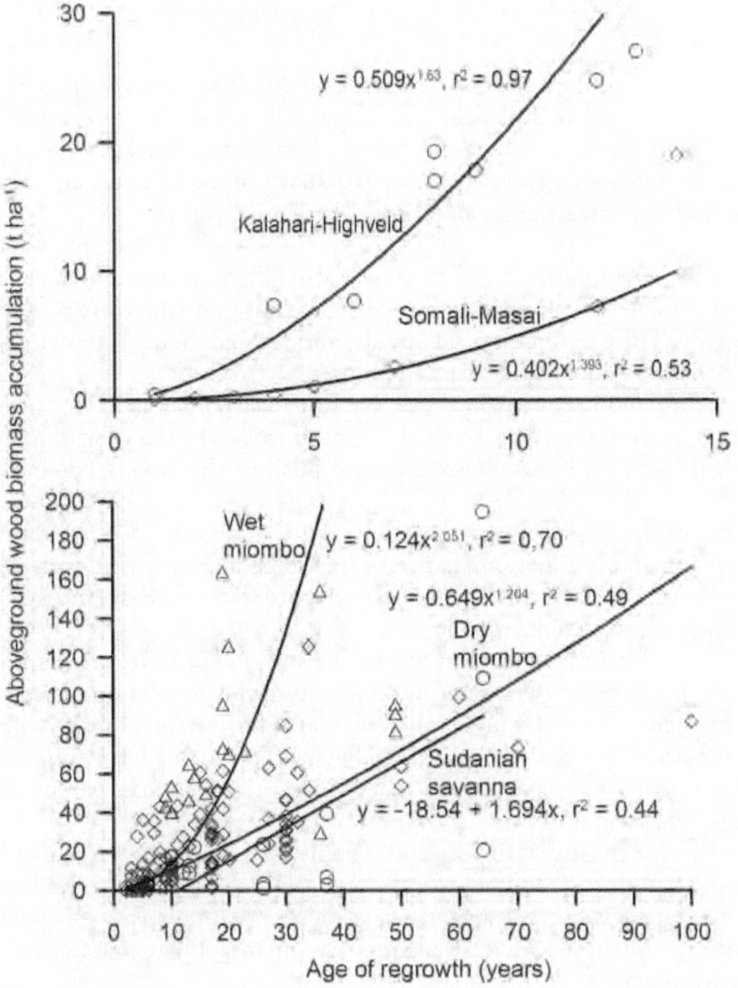

Figure 2.12 *Above ground wood biomass accumulation in Kalahari and Somali-Masai woodlands and wet miombo, dry miombo and Sudanian woodland*

Source: Based on data from literature (see text)

In most dry forests and woodlands woody biomass accumulation with increasing age was best explained by a steep rise in biomass with increasing age of regrowth (Figure 2.12). The steepness of the slope of the line relating biomass to age was highest in wet miombo followed by dry miombo and semi-arid savannas of the Kalahari region of southern Africa suggesting that these woodlands may be more productive than those in eastern and western Africa. The lowest biomass accumulation was observed in the Somali-Masai region. Generally, however, the age of regrowth was more important in determining woody biomass than phytoregion.

Key Issues and Questions

The chapter has described the distribution and general characteristics of dry forests and woodlands in sub-Saharan Africa and raised a number of issues concerning the management of these forests. Among the management issues raised, six key questions stand out as particularly important to sustainable management of dry forests and woodlands. These are:

- What will the response of dry forests/woodlands to predicted changes in regional climate be? We need to develop scenarios that will assist continued management of existing natural stands, and also guide us in re-establishment practices, where these are called for.
- What ability and capacity do these dry forests/woodlands have, through coppicing, to respond to harvesting? We need to determine what the limits are to this, and find out whether we are pushing the woodlands beyond them.
- What is the long-term impact of repeated fire on dry forest/woodland regeneration and production? Is fire protection a necessity, and is it worth the high cost. What is the regeneration potential of these forests and woodlands in the face of frequent fire.
- Although we have many scattered data, we still do not have a comprehensive idea of dry forest/woodland productivity levels in relation to stage of growth, moisture availability, nutrient status, forest/woodland type or composition. We need to understand what the main determinants of productivity are for each type, so that management practices can be more strategically applied.
- It appears that most dry forests and woodlands regenerate and spread by means of root suckers, and only locally through seed. We need to find out if this observation is correct and, if so, is there a particular impediment to regeneration of these species that improved management practices could readily overcome.

References

Almond, S. (2000) 'Itigi thicket monitoring using Landsat™ Imagery', MSc dissertation, University College, London

Araki, S. (1992) 'The role of miombo woodland ecosystem in chitemene shifting cultivation in northern Zambia', *Japan InforMAB*, vol 11, pp8–15

Banda, T., Schwartz, M. W. and Caro, T. (2006) 'Woody vegetation structure and composition along a protection gradient in a miombo ecosystem of western Tanzania', *Forest Ecology and Management*, vol 230, pp179–185

Boaler S. B. (1966) *The Ecology of* Pterocarpus angolensis *in Tanzania*, Ministry of Overseas Development, London

Byström, M., Stern, M., Kikula, I. S. and Shishira, E. K. (1987) 'Assessment of woodfuel resources in Tanzania using satellite remote sensing', A pre-study report to joint UNDP-World Bank Energy Sector Management Assistance Programme,

Institute of Resource Assessment, University of Dar es Salaam
Calvert, G. M. and Timberlake, J. R. (1993) 'Burning trials in Baikiaea woodland, Gwayi Forest, Zimbabwe', in G. D. Piearce and D. J. Gumbo (eds), *The Ecology and Management of Indigenous Forests in Southern Africa*, pp347–365, Proceedings of an international symposium, Victoria Falls, Zimbabwe, 27–29 July 1992, Zimbabwe Forestry Commission and SAREC, Harare
CCTA/CSA (1956) 'Phytogéographie/Phytogeography', CCTA/CSA Publication no 53
Chidumayo, E. N. (1985) 'Structural differentiation of contiguous savanna woodland types in Zambia', *Tropical Ecology and Geography*, vol 9, pp51–66
Chidumayo, E. N. (1986) 'Species diversity in some Zambian forests', *Tropical Ecology and Geography*, vol 10, pp71–78
Chidumayo, E. N. (1987) 'A survey of wood stocks for charcoal production in the miombo woodlands of Zambia', *Forest Ecology and Management*, vol 20, pp105–115
Chidumayo, E. N. (1993) *Responses of Miombo to Harvesting: Ecology and Management*, Stockholm Environment Institute, Stockholm
Chidumayo, E. N. (1995) *Handbook of Miombo Ecology and Management*, Stockholm Environment Institute, Stockholm
Chidumayo, E. N. (1997a) 'Effects of accidental and prescribed fires on miombo woodland, Zambia', *Commonwealth Forestry Review*, vol 76, pp268–272
Chidumayo, E. N. (1997b) *Miombo Ecology and Management: An introduction*, Intermediate Technology Publishers, London
Chidumayo, E. N. (2001) 'Climate and phenology of savanna vegetation in southern Africa', *Journal of Vegetation Science*, vol 12, pp347–354
Chidumayo, E. N. (2004) 'Development of *Brachystegia-Julbernardia* woodland after clear-felling in central Zambia: Evidence for high resilience', *Applied Vegetation Science*, vol 7, pp237–242
Conybeare, A. M. (2004) 'Elephant impacts on vegetation and other biodiversity in the broadleaved woodlands of S-Central Africa', in J. R. Timberlake and S. L. Childes (eds), *Biodiversity of the Four Corners Area: Technical Reviews*, vol 2, pp477–508. Occasional Publications in Biodiversity no 15, Biodiversity Foundation for Africa/The Zambezi Society, Bulawayo and Harare
Endean, F. (1968) 'The productivity of miombo woodland in Zambia', *Forest Research Bulletin*, no 14, Government Printer, Lusaka
Ernst, W. H. O., Tolsma, D. J. and Decelle, J. E. (1989) 'Predation of seeds of *Acacia tortilis* by insects', *Oikos*, vol 54, pp294–300
Eva, H. D., Brink, A. and Simonetti, D. (2006) *Monitoring Land Cover Dynamics in Sub-Saharan Africa: A Pilot Study Using Earth Observing Satellite Data from 1975 and 2000*, European Commission, Directorate-General Joint Research Centre, Institute for Environment and Sustainability, Luxembourg
Fanshawe, D. B. (1971) *The Vegetation of Zambia*, Government Printer, Lusaka
Fatubarin, A. (1984) 'Biomass estimates for the woody plants in a savanna ecosystem in Nigeria', *Tropical Ecology*, vol 25, pp208–213
Frost, P. (1996) 'The ecology of miombo woodlands', in B. Campbell (ed), *The Miombo in Transition: Woodlands and Welfare in Africa*, Center for International Forestry Research, Bagor
Frost, P. G. H. and Robertson, F. (1987) 'The ecological effects of fire in savannas', in B. H. Walker (ed), *Determinants of tropical savannas*, IRL Press, Oxford
Geldenhuys, C. J. (1977) 'The effect of different regimes of annual burning on two woodland communities in Kavango', *South African Forestry Journal*, vol 103, pp34–42

Gonzalez, P. (2001) 'Desertification and a shift of forest species in the west African Sahel', *Climate Research*, vol 17, pp217–228

Greenway, P. J. (1973) 'A classification of the vegetation of East Africa', *Kirkia*, vol 9, pp1–68

Grundy, I. M. (2006) 'Age determination of miombo species *Brachystegia spiciformis* (Leguminosae-Ceasalpinadeae) in Zimbabwe using growth rings', *South African Forestry Journal*, vol 206, pp5–12

Higgins, S. I., Shackleton, C. M. and Robbie, E. (1990) 'Changes in woody community structure and composition under contrasting landuse systems in a semi-arid savanna, South Africa', *Journal of Biogeography*, vol 26, pp619–627

Holdo, R. M. (2006) 'Tree growth in an African woodland savanna affected by disturbance', *Journal of Vegetation Science*, vol 17, pp369–378

Hulme, M. (ed) (1996) *Climate Change in Southern Africa: An Exploration of Some Potential Impacts and Implications in the SADC Region*, Climatic Research Unit, University of East Anglia, Norwich, UK

Isango, J. A. (2007) 'Stand structure and tree species composition of Tanzanian miombo woodlands: A case study from miombo woodlands of community based forest management in Iringa District', *Proceedings of the First MITMIOMBO Project workshop held in Morogoro, Tanzania, 6– 12 February 2007*

Jacobs, M. J. and Schloeder, C. A. (2002) 'Fire frequency and species associations in perennial grasslands of southwest Ethiopia', *African Journal of Ecology*, vol 40, pp1–9

Jallow, B. P. and Danso, A. A. (1997) 'Assessment of the vulnerability of the forest resources of The Gambia to climate change' in *Republic of The Gambia: Final Report of The Gambia/U.S. Country Study Program Project on Assessment of the Vulnerability of the Major Economic Sectors of The Gambia to the Projected Climate Change*, Banjul, Gambia

Jeffers, J. N. R. and Boaler, S. B. (1966) 'Ecology of a miombo site. Lupa North Forest Reserve, Tanzania. I. Weather and plant growth, 1962–64', *Journal of Ecology*, vol 54, pp447–463

Keay, R. W. J. (1959) *Vegetation Map of Africa*, Oxford University Press, London

Kelly, R. D. and Walker, B. H. (1976) 'The effects of different forms of land use on the ecology of a semi-arid region in south-eastern Rhodesia', *Journal of Ecology*, vol 64, pp553–576

Ky-Dembele, C., Tigabu, M., Bayala, J., Ouedraogo, S. J. and Oden, P. C. (2007) 'The relative importance of different regeneration mechanisms in a selectively cut savanna-woodland in Burkina Faso, West Africa', *Forest Ecology and Management*, vol 243, pp28–38

Lamotte, M. (1979) 'Structure and functioning of the savanna ecosystem of Lamto (Ivory Coast)', in Unesco (ed), *Tropical Grazing Land Ecosystems*, Unesco, Paris

Lanly, J. P. and Clement, J. (1982) *Forest Resources of Tropical Africa*, FAO, Rome

Louppe, D., Oattara, N. and Coulibaly, A. (1995) 'The effects of brush fires on vegetation: The Aubreville fire plots after 60 years', *Commonwealth Forestry Review*, vol 74, pp288–292

Malaisse, F. (1978) 'The miombo ecosystem', in UNESCO (ed), *Tropical Forest Ecosystems*, UNESCO/UNEP/FAO, Paris

Malaisse, F. (1997) *Se nourrir en forêt Claire africaine. Approche écologique et nutritionnelle*, Centre Technique de Coopération Agricole et Rurale, CTA, Gembloux, Belgium

Malaisse, F. (1984) 'Structure of a Zambezian dry evergreen forest of the Lubumbashi surroundings (Zaire)', *Bulletin de la Société Royale de Botanique de Belgique*, vol 177, pp428–458

Malaisse, F. and Binzangi, K. (1985) 'Wood as a source of fuel in Upper Shaba (Zaire)', *Commonwealth Forestry Review*, vol 64, pp227–239

Malaisse, F., Malaisse-Mousset, M. and Bulaimu, J. (1970) 'Contribution à l'étude de l'ecosystème forêt dense seche (Muhulu). Note 1: Phenologie de la defoliation', *Trav. Serv. Sylv. Pisc.*, vol 9, pp1—11, University of Congo, Lumbumbashi

Malaisse, F., Freson, R., Goffinet, G. and Malaisse-Mousset, M. (1975) 'Litter fall and litter breakdown in miombo', in F. B. Golley and E. Medina (eds), *Ecological Systems: Trends in Terrestrial and Aquatic Research (Ecological Studies)*, vol 11, pp137—152

Mapaure, I. (1994) 'The distribution of mopane', *Kirkia*, vol 15, pp1–5

Mayaux, P., Bartholome, E., Fritz, S. and Belward, A. (2004) 'A new land-cover map of Africa for the year 2000', *Journal of Biogeography*, vol 31, pp861–877

Menaut, J. C. and Cesar, J. (1979) 'Structure and primary productivity of Lamto savannas, Ivory Coast', *Ecology*, vol 60, pp1197–1210

Mugasha, A. G. and Chamshama, S. A. O. (2002) 'Trees biomass and volume estimation for miombo woodlands at Kitulangalo, Morogoro, Tanzania. Indicators and Tools for Restoration and Sustainable Management of Forests in East Africa', Working Paper no 9

Musanhane, J., Nhamuco, L. and Virtanen, P. (2000) 'A traditionally protected forest as a conservation area: A case study from Mozambique', *Silva Carelica*, vol 34, pp89–115

Nikiema, A. (2005) 'Agroforestry parkland species diversity: Uses and management in semi-arid West Africa (Burkina Faso)', PhD thesis, Wageningen University, Netherlands.

Okello, B. D., O'Connor, T. G. and Young, T. P. (2001) 'Growth, biomass estimates, and charcoal production of *Acacia drepanolobium* in Laikipia, Kenya', *Forest Ecology and Management*, vol 142, pp143–153

Okello, B. D., Young, T. P., Riginos, C., Kelly, D. and O'Connor, T. G. (2006) 'Short-term survival and long-term mortality of *Acacia drepanolobium* after a controlled burn', *African Journal of Ecology*, vol 46, pp395–401

Orthmann, B. (2005) 'Vegetation ecology of a woodland-savanna mosaic in central Benin (West Africa): Ecosystem analysis with focus on the impact of selective logging', PhD thesis, Rostock University, Germany

Pellew, R. A. (1983) 'The impacts of elephant, giraffe and fire upon the *Acacia tortilis* woodlands of the Serengeti', *African Journal of Ecology*, vol 21, pp41–74

Picard, N., Ouattara, S., Diarisso, D., Ballo, M. and Gautier, D. (2006) 'Defining units for savanna management in Sudano-Sahelian areas', *Forest Ecology and Management*, vol 236, pp403–411

Pratt, D. J., Greenway, P. J. and Gwynne, M. D. (1966) 'A classification of East African rangeland', *Journal of Applied Ecology*, vol 3, pp369–382

Prentice, I. C., Cramer, W., Harrison, S. P., Leemans, R., Monserud, R. A. and Solomon, A. M. (1992) 'A global biome model based on plant physiology and dominance, soil properties and climate', *Journal of Biogeography*, vol 19, pp117–134

Rutherford, M. C., Powrie, L. W. and Schulze, R. E. (1999) 'Climate change in conservation areas of South Africa and its potential impact on floristic composition: A first assessment', *Diversity and Distributions*, vol 5, pp253–262

Sambane, E. C. C. (2005) 'Aboveground biomass accumulation in fallow fields at the Nhambita community – Mozambique', MSc Dissertation, University of Edinburgh

Sankaran, M., Hanan, N. P. and Scholes, R. J. (2007) 'Characteristics of African savanna biomes for determining woody cover', www.daac.ornl.gov

Sawadogo, L., Nygard, R. and Pallo, F. (2002) 'Effects of livestock and prescribed fire on coppice growth after selective cutting of Sudanian savannah in Burkina Faso', *Annuals of Forest Science*, vol 59, pp185–195

Scholes, R. J. (1990) 'The regrowth of *Colophospermum mopane* following clearing', *Journal of Grassland Society of South Africa*, vol 7, pp147–151

Schöngart, J., Ortrhmann, B., Hennenberg, K. J., Porembaski, S. and Worbes, M. (2006) 'Climate-growth relationships of tropical species in West Africa and their potential for climate reconstruction', *Global Change Biology*, vol 12, pp1139–1150

Schreckenberg, K. (1999) 'Products of a managed landscape: Non-timber forest products in the parklands of the Bassila region, Benin', *Global Ecology and Biogeography*, vol 8, pp279–289

Shackleton, C. M., Griffin, N. J., Banks, D. I., Mavrandonis, J. M. and Shackleton, S. E. (1994) 'Community structure and species composition along a disturbance gradient in a community managed South African savanna', *Vegetation*, vol 115, pp57–167

Stahle, D. W., Mushove, P. T., Cleaveland, M. K., Roig, F. and Haynes, G. A. (1999) 'Management implications of annual growth rings in *Pterocarpus angolensis* from Zimbabwe', *Forest Ecology and Management*, vol 124, pp217–229

Styles, C. V. (1994) 'Mopane worms: More important than elephants?', *Farmer's Weekly*, 29 July, pp14–16

Swaine, M. D. and Brookman-Amissah, J. (1987) 'Effects of fire on trees in savanna in N. E. Ghana', *Tropical Biology Newsletter*, vol 52, p1

Timberlake, J. R. (1999) '*Colophospermum mopane*: An overview of current knowledge', in J. R. Timberlake and S. Kativu (eds), *African Plants: Biodiversity, Taxonomy and Uses*. Proceedings of the 1997 AETFAT Congress, Harare, Zimbabwe and Royal Botanic Gardens, Kew, London

Trapnell, C. G. (1959) 'Ecological results of woodland burning experiments in Northern Rhodesia', *Journal of Ecology*, vol 47, pp129–168

Trouet, V. Coppin, P. and Beechman, H. (2006) 'Annual growth ring patterns in *Brachystegia spiciformis* reveal influence of precipitation on tree growth', *Biotropica*, vol 38, pp375–382

UNESCO (1973) 'International classification and mapping of vegetation', *Ecology and* Conservation, vol 6, UNESCO, Paris

Van de Vijver, C. A. D. M., Foley, C. A. and Olff, H. (1999) 'Changes in the woody component of an East African savanna during 25 years', *Journal of Tropical Ecology*, vol 15, pp545–564

Walters, M. W., Milton, S. J, Somers, M. J. and Midgley, J. J. (2005) 'Post-dispersal fate of *Acacia* seeds in an African savanna', *South African Journal of Wildlife Research*, vol 35, pp191–199

Ward, H. K. and Cleghorn, W. B. (1970) 'Effects of grazing practices on tree regrowth after clearing indigenous woodland', *Rhodesia Journal of Agricultural Research*, vol 8, pp57–65

Western, D. and Ssemakula, J. (1981) *A Survey of Natural Wood Supplies in Kenya and an Assessment of the Ecological Impact of its Usage*, Beijer Institute, Stockholm

White, F. (1983) *The Vegetation of Africa: A Descriptive Memoir to Accompany the UNESCO/AETFAT/UNSO Vegetation Map of Africa*, UNESCO, Paris

Wild, H. and Fernandes, A. (eds) (1967) 'Vegetation map of the *Flora Zambesiaca* area', *Flora Zambesiaca*, Supplement, Salisbury

Williams, M., Ryan, C. M., Rees, R. M., Simbane, E., Fernado, J. and Grace, J. (2008) 'Carbon sequestration and biodiversity of re-growing miombo woodlands in Mozambique', *Forest Ecology and Management*, vol 254, pp145–155

World Wide Fund for Nature (2001) *Itigi-Sumbu thicket* (AT0708), World Wide Fund for Nature, Washington, DC

Zida, D., Sawadogo, L., Tigabu, M. and Oden, P. C. (2007) 'Dynamics of sapling population in savanna woodlands of Burkina Faso subjected to grazing, early fire and selective tree cutting for a decade', *Forest Ecology and Management*, vol 243, pp102–115

Zolho, R. (2005) 'Effect of fire frequency on the regeneration of miombo woodland in Nhambita, Mozambique', MSc Dissertation, University of Edinburgh

3

Biodiversity of Plants

Enos Shumba, Emmanuel Chidumayo, Davison Gumbo, Cynthia Kambole and Mwale Chishaleshale

Defining Biodiversity of African Dry Forests and Woodlands

The Convention on Biological Diversity (CBD) defines biodiversity as the variation between ecosystems and habitats; the variation between different species; and the genetic variation within individual species. Biodiversity can therefore be described in terms of the diversity of ecosystems, species and genes. This chapter describes the floristic and ecosystem diversities in dry forests and woodlands of sub-Saharan Africa and their current status and management.

Floristic Diversity and Endemism

Sub-Saharan Africa has a wide range of dry forest and woodland formations, each with diverse flora. Some of these formations have been described in Chapter 2 but many more subtypes were described by White (1983) and Table 3.1 lists a selection of these.

Species richness (total number of species in a given area) and endemism (proportion of species restricted to a particular area) are often used to describe biodiversity. Endemic taxa are species or genera or families that have at least 75 per cent of their geographical range within one ecoregion. An ecoregion is characterized by a suite of plant taxa that respond to distinct patterns of landform, geology, soils and climate. Centers of endemism are areas of high

Table 3.1 *Diversity of vegetation types in African dry forests and woodlands*

Phytoregion	*Main vegetation types*
Guineo-Congolian/Zambezian Regional Transition Zone	Southern dry evergreen forest and transitional woodland Wooded grassland
Guineo-Congolian/Sudania Regional Transition Zone	Guinea dry forest
Zambezian Region	Dry deciduous forest and scrub forest Zambezian wooded grassland Itigi deciduous thicket Miombo woodland Mopane woodland Undifferentiated woodland
Sudanian Region	Sudanian *Isoberlinia* woodland Undifferentiated woodland *Acacia* wooded grassland
Kalahari-Highveld Regional Transition Zone	*Acacia* woodland Wooded grassland Semi-arid shrubland
Somali-Masai Region	*Acacia-Commiphora* bushland and thicket Evergreen bushland and secondary wooded grassland Semi-arid shrubland

Source: Based on White (1983)

concentrations of taxa that are endemic to an ecoregion. Because of taxonomic revisions, variable sampling effort and differences in delineating phytoregions, there are often large differences in estimates of species richness and levels of endemism among different workers. This problem is particularly acute in the case of dry forests and woodlands of Africa that are diverse and their delineations vary considerably among workers.

Floristic diversity in African dry forests and woodlands was assessed by White (1983) and has recently been re-evaluated by Linder et al (2005) (Table 3.2). Both the assessments by White (1983) and Linder et al (2005) indicate that the Zambezian Regional Centre of Endemism has the highest floristic diversity of dry forests and woodland types. Mittermeier et al (2003) focusing on the Zambezian woodlands also identified the miombo-mopane woodlands as one of the five ecozones (together with Amazonia, Congo, New Guinea and the North American deserts) needing to be prioritized for biodiversity conservation because of their irreplaceability in terms of species endemism. The Zambezian Regional Centre of Endemism has eight endemic genera compared to four in the Sudanian Regional Centre of Endemism; however, endemic genera in the Somali-Masai Regional Centre of Endemism are even higher at 50 (White, 1983). The Zambezian phytoregion is also a centre of diversity for the *Brachystegia* and *Monotes*. There are also considerable similarities in the flora of the different phytoregions; some flora in the Guineo-Congolian/Sudanian and the Guineo-Congolian/Zambazian are also found in the Sudanian and Zambezian woodlands. Similarly, about a quarter of the species in the Zambezian phytoregion are also found in the Sudanian phytoregion.

Table 3.2 *Floristic diversity and levels of endemism in phytoregions in which dry forests and woodlands are dominant formations in sub-Saharan Africa*

Phytoregion	*Plant species*		*Endemic species*		*Percent endemic species*	
	White (1983)	*Linder et al (2005)*	*White (1983)*	*Linder et al (2005)*	*White (1983)*	*Linder et al (2005)*
Guineo-Congolian/ Sudania RTZ[1]	2000	711	50	5	3	1
Guineo-Congolian/ Zambezian RTZ[1]	2000	571	50	28	3	5
Sudanian RCE[2]	2750	684	960	6	35	1
Zambezian RCE[2]	8500	1725	4590	377	54	22
Somali-Masai RCE[2]	2500	931	1250	103	50	11
Kalahari/Highveld RTZ[1]	3000	583	50	10	20	2

Notes: 1. RTZ is Regional Transition Zone.
2. RCE is Regional Centre of Endemism.
Source: White (1983)

Protection of Biodiversity

The protection of biodiversity is closely linked to protected area (World Park Congress, 2003) and these can be divided into two broad categories, those meant for conservation and the other for resource utilization. IUCN, the International Union for Conservation of Nature (IUCN) defines the former as 'protected areas' (Chape et al, 2003) while those established as sites for controlled resource utilization in forests and woodlands are termed 'forest reserves' (Burgess et al, 2005, 2007). The early forest reserves established in Africa were not for conservation purposes (Lovett, 2003) but largely for timber extraction and at times for water harvesting but, ultimately, the two categories have been at the forefront of biodiversity conservation in Africa. In the dry forest and woodland countries, Burgess et al (2007) report that there are close to 4604km^2 of protected areas and 2027km^2 of forest reserves and the latter is made up of classified forests, reserved or designated forests, national forests, state forests and state reserved forests. It is critical to note that both protected areas and forest reserves have effectively conserved forests and woodland, but more so in the protected areas and those specialized forest reserves such as botanical gardens and sanctuaries. A noticeable development is the fact that over 70 per cent of the protected areas and forest reserves lies across international boundaries (Olson and Dinerstein, 1998; Brooks et al, 2004). These provide opportunities for trans-frontier conservation area initiatives on the continent.

The distribution of protected areas in the dry forest and woodland zones in sub-Saharan Africa is shown in Figure 3.1. The Guinea and southern dry forests are poorly covered by protected areas. Protected areas (6390 in all) of all categories cover about 2.4 million km^2 (World Resources Institute, 2003).

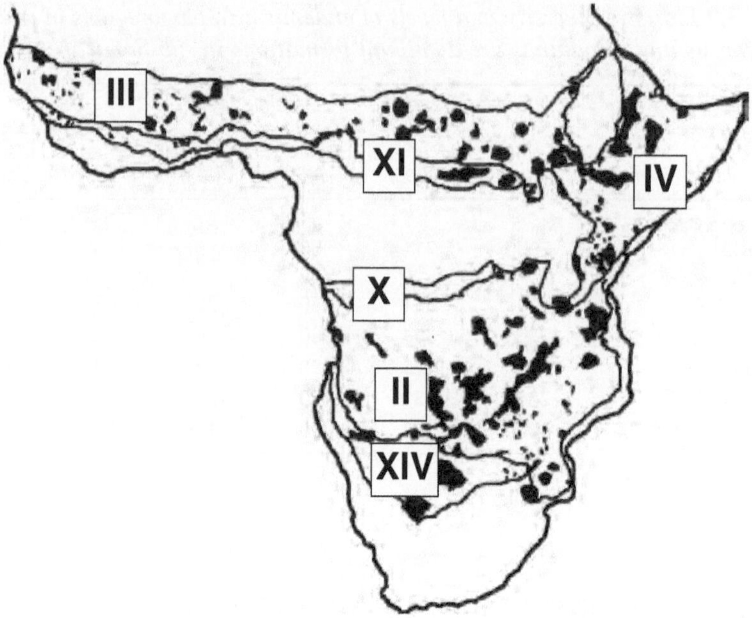

Figure 3.1 *Distribution of protected areas in the dry forest and woodland phytoregions of sub-Saharan Africa*

Note: (II) Zambezian Regional Center of Endemism, (III) Sudanian Regional Center of Endemism, (IV) Somali-Masai Regional Centre of Endemism, (X) Guineo-Congolian/Zambezian Regional Transition Zone, (XI) Guineo-Congolian/Sudanian Regional Transition Zone and (XIV) Kalahari Region of the Kalahari-Highveld Regional Transition Zone.
Source: Based on World Conservation Monitoring Centre (1997)

Figure 3.2 shows the extent and number of protected areas; the area under protection represents about 9 per cent of total land area in the Sudanian zone, 11 per cent in the Somali-Masai zone, 14 per cent in the Zambezian zone and 16 per cent in the Kalahari zone. The Conference of Parties (CoP7) of the Convention on Biological Diversity required that at least 10 per cent of each of the world's ecological regions be protected (Chape et al, 2005). This would imply that other than the dry forests and Sudanian woodlands, there is adequate coverage of woodland phytoregions in protected areas in sub-Saharan Africa. The average size of a protected area ranges from 260km^2 in the Sudanian zone to 430km^2 in the Zambezian zone and 670km^2 and 830km^2 in the Somali-Masai and Kalahari zones, respectively.

Species Conservation Status

Species are declining to critical population levels, important habitats are being destroyed, and ecosystems are being destabilized through climate change, pollution, alien invasive species and direct human impacts. Thus, the conservation

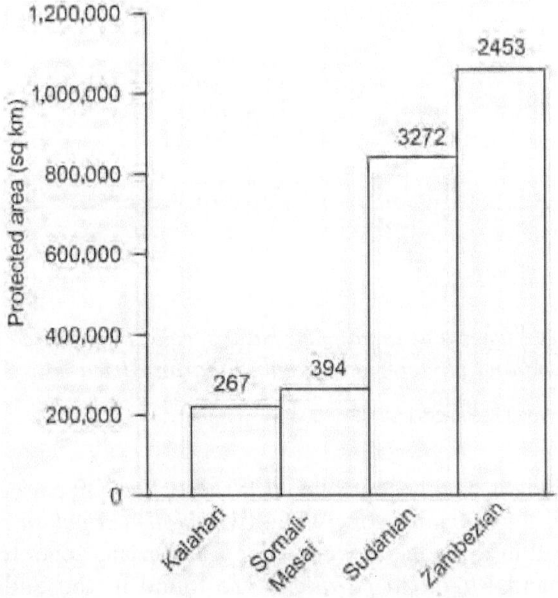

Figure 3.2 *Protected area in dry woodland phytoregions in sub-Saharan Africa*

Note: Numbers on top of bars indicate number of protected areas.
Source: Based on World Resources Institute (2003)

status of a species is a good indicator of the impact of threats as the likelihood of a species remaining extant either in the present day or the near future has a bearing on planning and management (Hamilton and Hamilton, 2006). An assessment of the conservation status of a species should not however be limited to the number remaining, but the overall increase or decrease in the population over time, breeding success rates, known threats, and so on. This means that even a species with high levels of regeneration, both sexually and vegetatively as is the case with many dry forest and woodland species (see Chapter 2), must be evaluated as threatened on the basis of reproductive adults.

Figure 3.3 shows threatened higher plant species in African dry woodland phytoregions. It is difficult to determine the number of threatened plant species by phytoregion from data that are often presented by country and, in addition, some countries contain vegetation formations that are not dry forest and woodland. Nevertheless, the data in Figure 3.3 indicate that the number of threatened plant species per country increases from the Kalahari zone to the Somali-Masai and Sudanian zones and is highest in the Zambezian zone.

Loss of some tree species has been largely through trade – an aspect that the Convention on International Trade in Endangered Species (CITES), signed by 164 countries, has been trying to address by controlling (Appendix II of CITES) or curtailing (Appendix I of CITES) trade. In Table 3.3 we list 13 tree species from dry forests and woodlands of Africa that are on Appendix II and we note that these are not necessarily threatened with extinction now but may

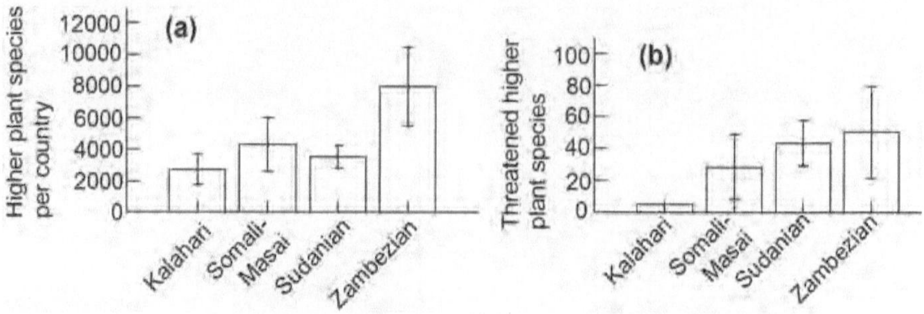

Figure 3.3 *Average higher and threatened plant species in protected areas in woodland phytoregions in sub-Saharan Africa*

Source: Based on data in World Resources Institute (2003)

become so unless trade is closely controlled. Of these, five (39 per cent) of the species are found in the Sudanian zone while two, *Hallea stipulosa* and *Khaya anthotheca*, are found in this zone as well as the Zambeziana zone. In addition, *Pouteria altissima* and *Vitellaria paradoxa* are found in the Sudanian and Somali-Masai zones and *Pterocarpus angolensis* in zones II and XIV. The Zambeziana zone holds two trees species that are exclusive to this regional centre of endemism and these are *Baikiaea plurijuga* and *Entandrophragma caudatum* while *Cordeauxia edulis* and *Pericopsis elata* are exclusive to the Somali-Masai zone. This suggests that the majority of threatened tree species are in the Sudanian and Zambezian zones, which also share 25 per cent of flora (White, 1983).

Table 3.3 *Tree species on the CITES list occurring in dry forests and woodlands of sub-Saharan Africa*

Species	*Conservation status*	*Threats*	*Distribution II*	*III*	*IV*	*XIV*
Afzelia africana	Vulnerable	Exploitation		X		
Baikiaea plurijuga	Lower risk	Exploitation	X			
Cordeauxia edulis	Vulnerable	Local use/ browsing			X	
Entandrophragma caudatum	Lower risk	Local use	X			
Hallea stipulosa	Vulnerable	Habitat loss	X	X		
Khaya anthotheca	Vulnerable	Exploitation	X	X		
Khaya grandifolia	Vulnerable	Exploitation/ Habitat loss		X		
Khaya senegalensis	Vulnerable	Exploitation/ Habitat loss		X		
Pouteria altissima	Lower risk	Exploitation		X	X	
Pterocarpus angolensis	Lower risk	Exploitation	X			X
Pericopsis elata					X	
Vitellaria paradoxa	Vulnerable	Local use		X	X	
Warburgia salutaris	Endangered	Exploitation	X			

Note: II Zambezian regional centre of endemism, III Sudanian regional centre of endemism, IV Somali-Masai regional centre of endemism, XIV Kalahari-Highveld regional transition zone
Source: IUCN, 2009

THREATS TO PLANT BIODIVERSITY

The demand for certain species lies at the very base of their survival (Stedman-Edwards, 1998). Hamilton and Hamilton (2006) place the blame for loss of plant materials on increasing human populations, increased demand for these resources, destruction and modification of habitats, expansion and intensification of agriculture. These threats do not necessarily operate alone but in combination with others. There are many threats to plant biodiversity in dry forests and woodlands of sub-Saharan Africa (Burgess et al, 2005; IUCN, 2009) but the important ones are those cited by Hamilton and Hamilton (2006) above and include poor management of protected areas, population and land-use pressures, climate change, over-harvesting of plant resources and proliferation of invasive species.

Poor management of protected areas

In the majority of sub-Saharan African countries investment in protected areas and forest reserves is chronically low. As a consequence of this, there is poor infrastructure and inadequate personnel, equipment and law enforcement and research: important components of good and effective management of protected areas. The minimum budgetary requirements for effective law enforcement for African protected areas is estimated at US$200–230 per km^2 (Lindberg, 2001) but expenditure on protected areas in most southern African countries, for example, is below the minimum requirements (Figure 3.4). In addition, the history of the establishment of protected areas is dominated by opportunistic acquisitions of land often at the expense of rural people (Siegfried, 1989) who over time have sought to reclaim their rights and often do so through encroachment and counter claims (Palmer, 2001). After the Rio Earth Summit of 1992, there has been an upsurge in the number of new protected areas and these are having negative impacts on the livelihoods of local communities through a loss of rights, exclusion from natural resources and displacement from traditional lands (Wittemeyer et al, 2008). These encroachments have huge implications for the management of these protected areas as well as the status of the biodiversity found in them.

Population and land-use pressure

The population inhabiting dry forests and woodlands in sub-Saharan Africa was estimated at 320 million people in 2000 (Eva et al, 2006). Despite the adverse impacts of the HIV/AIDS pandemic, the continent's population is growing at an average rate of 2.4 per cent per annum and the highest human footprint in dry forests and woodlands in sub-Saharan Africa is in western Africa where population densities are in the range of 30–45 per km^2 (Table 3.4). One of the challenges facing the continent is how to increase agricultural

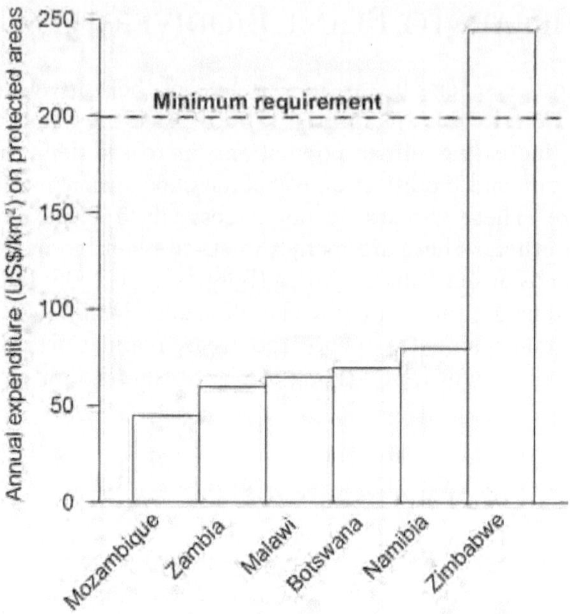

Figure 3.4 *Expenditure in government protected areas in dry forest and woodland countries in southern Africa*

Source: Based on Lindberg (2001) and Cumming (2004)

output in order to adequately feed the growing population. Given the limited availability of suitable agricultural land, there is increasing pressure to convert remaining dry forests and woodlands to agriculture. This is contributing to loss of biodiversity.

Given the difficulties of modelling deforestation and degradation of tropical open woodlands (Grainger, 1999), estimates of woodland cover loss in Africa tend to vary greatly depending on the methodology used to estimate deforesta-

Table 3.4 *Human population density in dry forest and woodland regions of sub-Saharan Africa in 2000*

Phytoregion	*Area (km²)*	*Population size*	*Population density per km²*
Guinea-Congolia-Zambezia	779,911	1,516,8813	19.45
Somalia-Masai	1,974,420	3,366,6625	17.05
Guinea-Congolia-Sudania	1,225,983	52,659,006	42.95
Kalahari-Highveld	1,277,340	13,298,317	10.41
Zambezian	3,924,240	70,158,185	17.88
Sudanian	3,641,240	112,929,909	31.01
Sahel	2,570,970	21,557,690	8.39

Source: Based on Eva et al (2006)

tion and degradation. Estimates of woodland loss therefore can only be indicative of the extent of the problem of deforestation in woodland areas.

During 1990 to 2000 it was estimated that dry forest and woodland countries in sub-Saharan Africa lost nearly 5 million ha of forest cover annually or nearly 1 per cent of the forest cover in 2000 (FAO, 2005). Much of this loss occurred in the Sudanian zone (2.5 million ha) and southern Africa (2.3 million ha). According to Kigomo (2003) the causes of woodland cover degradation and loss in semi-arid Africa are overgrazing, agricultural expansion and overexploitation of forest resources. Mayaux et al (2004) estimated that nearly 15 per cent of the Zambezian woodlands has been converted to agriculture while similar values for the Sudanian and Somali-Masai woodlands are 60 per cent and 80 per cent, respectively.

It is therefore not surprising that sub-Saharan Africa is experiencing human-induced biodiversity decline. The trend continues unabated as human activities (e.g. agriculture, exotic timber plantations, mining and urban development) transform habitats and replace indigenous biota. The loss of biodiversity results in the loss of ecosystem goods and services and translates into reduced economic opportunities for present and future generations.

Climate change

Dry forest and woodland vulnerability to climate change refers to the degree to which these vegetation types are susceptible to or unable to cope with adverse effects of climate change, its variability and extreme events. Some of the possible impacts of climate change on African dry forests and woodlands have been mentioned in Chapter 2. Therefore only a few additional examples are given in this chapter.

Observations made in acacia woodland in central Zambia involving five species revealed that temperature significantly affected seedling emergence in 80 per cent of the species and germination rate under a 1°C warmer climate was predicted to decline in three of the species while an increase was predicted in one species (Chidumayo, 2008). Temperature also significantly affected seedling mortality in all the five species such that under a warmer climate, mortality was predicted to increase in two of the species and decrease in the other three species. The conclusion was that woodland trees would respond to climate warming in different but predictable ways. Results of tree growth monitoring at the same woodland site showed that the radial growth of the majority of trees declined due to additive effects of temperature factors, suggesting that different species will respond differently to climate change (Chidumayo, in preparation; Table 3.5).

The study by McClean et al (2005) revealed shifts in ranges of individual species in African woodlands as a result of climate change. These authors through modelling have predicted that 25–75 per cent of the plant species in African woodlands might lose all their currently climatically suitable ranges under a future warmer and drier climate.

Table 3.5 *Climate factors affecting radial growth of woodland trees at a Zambezian woodland site in central Zambia, 1998–2008*

Species	*Significant predictor factors*	*Growth variation caused by climatic factor(s) (%)*	*Predicted annual growth rate under a 1°C warmer climate*
Acacia polyacantha	Rainfall	16	No change
Acacia sieberiana	Average temperature	20	Decrease
Combretum molle	Rainfall and minimum temperature and average temperature	39	Increase
Piliostigma thonningii	Average and maximum temperature	30	Decrease

Source: Based on Chidumayo (in preparation)

A recent study by Maranz (2009) in western Africa has also shown that the high mortality of mesic woodland tree species in the northern portion of the Sudanian zone and their apparent retreat southwards has been due to the return of arid conditions during the latter half of the 20th century (Figure 3.5). The disappearance of tree species has been particularly noticeable in parkland landscapes where *Vitellaria paradoxa* and *Parkia biglobosa* are either disappearing or retreating to more mesic habitats. In addition, the savanna areas of northern Nigeria are reported to be losing plant species as a result of increasing desertification due to inadequate rainfall, excessive drought and sand dune encroachment.

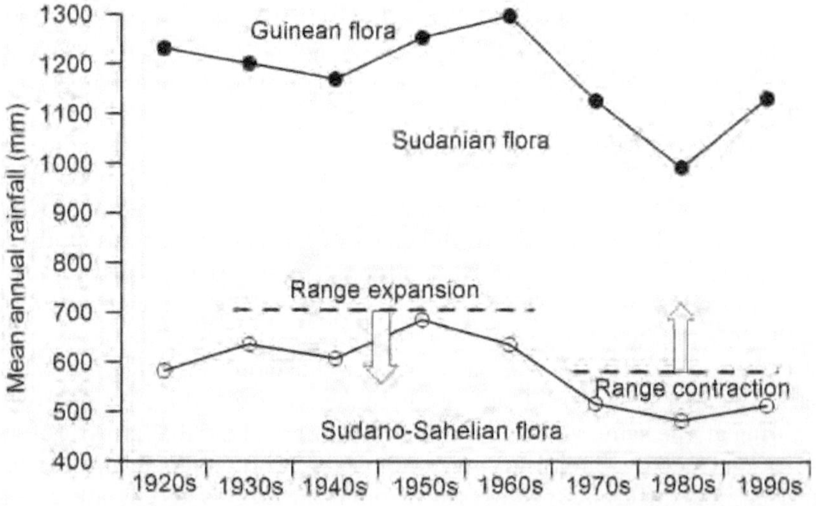

Figure 3.5 *Decadal rainfall pattern averaged for seven western African weather stations representing the Sudano-Sahelian zone (low mean rainfall, bottom line) and seven stations representing the southern Sudanian zone (high mean rainfall, top line)*

Note: Expansion in the range of some Sudanian tree species occurred from the 1930s to 1960s while range contraction occurred from the 1970s to 1990s.
Source: Based on Maranz (2009)

Overharvesting of plant resources

Overexploitation of plant resources is a growing threat to biodiversity in dry forest and woodland countries in sub-Saharan Africa. For example, of the 13 tree species on the CITES list (see Table 3.3), nearly 90 per cent of them are threatened by overexploitation and 11 per cent are threatened by habitat loss. Over-reliance on traditional medicinal plants for primary health care by the majority of the sub-Saharan population has contributed to the overexploitation of some other species, such as *Walburgia salutaris* in Zimbabwe and *Albizia brevifolia* in Namibia and many others that are now threatened. Similarly, the commercialization of crafts, like baskets and wood curios, has led to a decline in tree species such as *Berchemia discolor* which is used as a palm leaf fibre dye in Botswana and Namibia. There has also been overharvesting of *Afzelia quanzensis* and *Pterocarpus angolensis* in a number of woodland countries in response to the flourishing woodcraft industry (Cunningham et al, 2005; Shackleton, 2005). Some of these shortages and losses can be at local level (site specific) while this may not be the case at the regional level, e.g. *Berchemia discolor* which is under threat in Namibia but actually spreads from Ethiopia to northern parts of South Africa. Some tree species may be facing acute pressure at a country level, but because of their abundance at regional level may not qualify to be placed on the IUCN Red Data List or CITES Appendix II.

Management and conservation measures in the past had always been influenced by taboos that restricted people from destructive harvesting (Osemeobo, 1994). But, these have become largely dysfunctional under increasing pressures and have not been replaced by alternatives. Indeed, management services provided by the government are weak and ill equipped. Alternative lesser known substitutes need to be brought to light so as to reduce the pressure on over-sourced species. The importance of some species for multiple uses should also be highlighted. As a management strategy, proper records of plant status must be kept and abundance and collection rates monitored. The perception and orientation of harvesters must also be changed for they believe that plants can never be overexploited.

Proliferation of invasive alien species

Invasive alien species are species introduced deliberately or unintentionally outside their natural habitats where they have the ability to establish themselves, invade, out-compete natives and take over the new environments (IUCN, 2000). The problem and impact of invasives is likely to increase as more plants move across borders and destabilize natural vegetation (Hamilton and Hamilton, 2006) especially in areas where phytosanitary regulations are lax. Such species are found in all categories of organisms and all types of ecosystems. Some of them have significant environmental and economic impacts. In its compilation of the Red Data List of threatened species, IUCN cited alien species as directly affecting 15 per cent of all threatened plants (Carlton, 1998). Alien species disturb nutrient recycling, pollination and the regeneration of soils and energy, among other things; they also threaten the integrity of natural systems. For example, the 'fixing' or sequestration of carbon is becoming a major consideration regarding global warming and where fire-promoting alien species have replaced indigenous vegetation, the release of carbon has accelerated.

Opportunities for Conserving Biodiversity

Preserving trees in transformed landscapes and tree domestication

Under subsistence farming some tree species, such as indigenous fruit trees, are left in the field and may contribute to biodiversity conservation. For example, Dean et al (1999) found that large *Acacia erioloba* trees in semi-arid scrubland in the Kalahari-Highveld phytoregion of southern Africa increased biodiversity through provision of habitat for fleshy-fruited plants, frugivores, nectivorous and tree-nesting birds, raptors, weaver birds, tree rats and shade-seeking large mammals. Similarly the parklands of western Africa contain a high number of fruit trees, thereby contributing to the maintenance of tree biodiversity in transformed landscapes (Maranz, 2009; Figure 3.6). In addition there is a growing interest in the domestication of fruit trees throughout the dry forests and woodlands of sub-Saharan Africa (Akinnifesi et al, 2006; Schreckenberg et al, 2006).

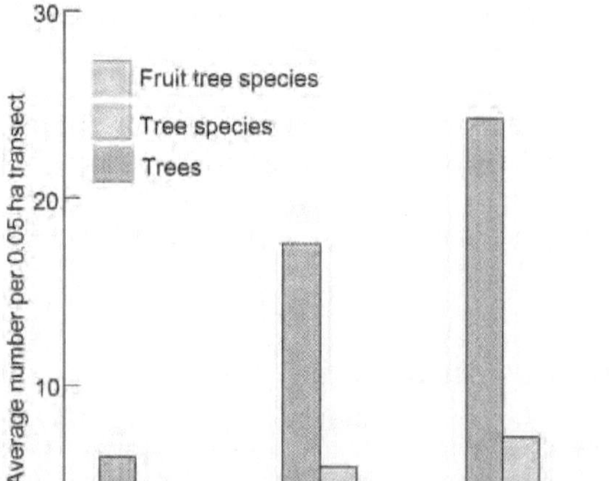

Figure 3.6 *Density of trees, tree species and fruit tree species in three landscapes in western African woodlands*

Source: Based on Maranz (2009)

Community and private sector involvement in biodiversity management

Sustainably harvested and fairly traded indigenous products offer a significant opportunity to improve the livelihoods of poor communities living in rural areas with access to natural resources. This industry has the potential to benefit both the natural environment and those in the supply chain – including the rural poor primary producer groups, buyers, processors and exporters (PhytoTrade Africa, 2005). The use of resources, which are accessible to and owned by the poor rural people and are ecologically and culturally adapted to local conditions, underpin a 'biodiversity-friendly' industry, with low barriers to entry. As will be shown under Chapters 4 and 5, forests and woodlands are important sources of income from wild foods, fuel, fodder and thatch grass (Vedeld et al, 2004). Use and access are often based on local rules as well as resource tenure and property rights (Bruce, 1989; Ostrom, 1999) but recent trends have shown a greater interest from private capital and calls for resource concessions. At the same time, through various decentralization schemes, local and community level institutions have become more assertive in the management of local forests and woodland resources and importance of resource tenure. Thus for any efforts targeted at biodiversity management to be successful, local communities must be involved and these can be linked to private capital in a number of innovative ways (FAO, 2002b).

For more than two decades, some African countries have been implementing strategies that support human livelihoods through the sustainable use of biological resources within the context of community-based natural resource management (CBNRM). In this approach, communities are given rights of access to wild resources and legal entitlements to benefits that accrue from managing the resources (Kellert et al, 2000; Child, 2004). This creates positive social and economic incentives for the people to invest their time and energy in natural resource conservation (Crook and Clapp, 1998). Typically, CBNRM initiatives have been implemented in ecologically marginal areas, with limited potential for agriculture.

Operationally, CBNRM involves the following:

- the devolution of control and management responsibilities for natural resources from the state to local people through appropriate legislative and policy changes;
- building the technical, organizational and institutional capacity of local communities to assume management responsibilities over natural resources.

The success of CBNRM has depended on the level of devolution, donor commitment and policy changes; and links with tourism and hunting. The key economic driver of CBNRM has been wildlife (large mammals), mostly through trophy hunting and eco-tourism outside protected areas.

The potential role of natural products is only beginning to be realized through value addition and commercialization (PhytoTrade Africa, 2005). Such

Box 3.1 Public–private sector partnerships to commercialize plant resources in Africa

1. Makoni Tea in Zimbabwe

Fadogia ancyalantha is used to produce the herbal Makoni tea. The Southern Alliance for Indigenous Resources (SAFIRE), an NGO (non-governmental organization), has facilitated the establishment of a community-based enterprise by encouraging members of Ward 23 of Nyanga district in Zimbabwe to form an indigenous tea producers association. The association consists of 200 members who collect leaves of the herb and pre-process them for the production of Makoni tea. This is done in partnership with private companies, Katiyo, Tanganda and Speciality Foods of Africa. The companies are involved in the final processing and packaging of the leaves and marketing and selling the tea. The association earns revenue from the sale of the pre-processed leaves and receives dividends based on returns from tea sales locally and abroad. However, revenue receipts have been limited due to competition with established herbal teas and inadequate promotion and marketing. It is however noted that some *Fadogia* species have phytochemicals that are known to be toxic to livestock and potentially to people, but that marketing has occurred without health and safety checks.

2. The Swazi Secrets project in Swaziland

The Swazi Secrets project harvests marula fruits (in the wild) for processing into a variety of products. The project is working with 14 producer/collector communities (comprising 2500 individual suppliers) who sell marula kernels to Swazi Indigenous Products Pvt. Ltd. The project has a strong capacity building component that trains communities on appropriate harvesting techniques.

Since the community derives direct economic benefits from harvesting wild Marula fruits and since fruit harvesting is non-destructive, the sustainable harvesting of the tree species can be guaranteed. This demonstrates how economic incentives can promote biodiversity conservation.

Source: SADC (in press)

products have potential for nutritional, pharmaceutical and industrial use, as well as for generating income and a number of initiatives involving rural local people; local and international companies have been set up in a number of dry forest and woodland countries (Grote, 2003; Hailwa, 1998; Moyo and Epulani, 2002; Sola, 2005; Shackleton, 2005) and these new initiatives are broadening the economic viability of CBNRM initiatives through their wider distribution when compared to wildlife (Machena et al, 2005).

There has been limited investment in bio-prospecting and natural product value addition by national governments in Africa. This is partly because most development models on the continent consider biological resources as a source of sustenance and not as a source of wealth. There is, however, growing interest in adding value and commercializing biological resources on the continent. For example, the Southern African Natural Products Trade Association (PhytoTrade

Africa) is developing commercial opportunities from natural products (products derived from indigenous plants) for the benefit of rural communities in the sub-region. It does this through investment in research and development (R&D) and market development, whilst facilitating linkages between rural producers and private sector processors and manufacturers. Through the creative use of public funds, PhytoTrade Africa has been able to leverage significant private sector investment into R&D. However, it remains one of the very few cases in which favourable conditions for private sector investment have been successfully created (Le Breton, personal communication).

CHALLENGES

Sub-Saharan Africa is experiencing increased pressure and demand on agricultural land and biodiversity due to limited alternative livelihood opportunities. The need to explore other livelihood opportunities and to refocus national policy development models beyond the primary sectors of production cannot therefore be over-emphasized (Frost et al, 2007). In fact, this is the development route that was followed by the currently developed nations and highlights the fact that natural resources alone are not a panacea to Africa's development problems. How can forest biodiversity contribute more to livelihoods on a sustainable basis?

There have been limited national, sub-regional and regional level inventories of various biodiversity components on the continent as illustrated by the following:

- Only large and commercial species of wildlife are regularly monitored (because of their importance in national economies). Similarly, regular inventory and monitoring programmes are usually in place for commercial indigenous timber species and exotic timber plantations. Other species that provide a range of timber and non-timber forest products to local communities have not been catered for.
- The monitoring of biodiversity habitats, some of which are under extreme pressure, is often lacking. However, such information is critical for the effective management of protected areas, including trans-frontier conservation areas (TFCAs).

The inadequacy of up-to-date information on biodiversity and limited ability to handle the available information makes it difficult to effectively plan, manage and monitor biodiversity conservation and its sustainable use in Africa. It also makes it difficult to demonstrate the value and impact of biodiversity losses to national, sub-regional and regional economies. There is therefore need to develop and implement comprehensive but simple biodiversity inventory and monitoring programmes covering key species and habitats. Skills to handle and package the information are needed to improve knowledge and management of biodiversity.

Africa's protected areas have been a cornerstone of biodiversity conservation. However, existing legislation precludes neighbouring communities from accessing goods and services from them. This has created 'islands of green' surrounded by degraded communally owned landscapes. The result has been increased illegal timber and game harvesting and illegal settlements in some protected areas. How can community participation and the development of appropriate access and benefit sharing arrangements be advanced to facilitate sustainable management of protected areas?

TFCAs offer opportunities to raise funds for biodiversity conservation in protected areas through tourism. However, their success depends on the creation of a conducive environment for public-private sector partnerships through targeted incentives and appropriate legislation that ensures that part of the generated revenue is ploughed into biodiversity management and promotion of TFCAs and trans-boundary tourism through appropriate national policies and legislation and capacity building at various levels.

The bulk of the continent's biodiversity lies outside protected areas and is under extreme pressure from various threats. In spite of efforts to conserve biodiversity outside protected areas through joint management, especially in community forestry, these efforts to improve the management of biodiversity in off-reserve areas through CBNRM initiatives are being hampered by the inadequacy of incentives to local communities. There is therefore need for:

- concerted R&D efforts that unleash the economic potential locked up in the region's biological resources through bio-prospecting and value addition and finding innovative ways to equitably share benefits there from;
- building the capacity of local communities that live with the biological resources in the management of common property resources;
- formulating policies, legislation and bye-laws that regulate access to and use of biological resources.

REFERENCES

Akinnifesi, F. K., Kwesiga, F., Mhango, J., Chilanga, T., Mkonda, A., Kadu, C. A. C., Kadzere, I., Mithofer, D., Saka, J. D. K., Sileshi, G., Dhliwayo, P. and Swai, R. (2006) 'Towards developing the miombo indigenous fruit trees as commercial tree crops in southern Africa', *Forests, Trees and Livelihoods*, vol 16, pp103–121

Brooks, T. M., Bakarr, M. I., Boucher, T., Da Fonseca, G. A. B., Hilton-Taylor, C., Hoekstra, J. M., Moritz, T., Olivieri, S., Parrish, J., Pressey, R. L., Rodrigues, A. S. L., Sechrest, W., Stattersfield, A., Strahm, W. and Stuart, S. N. (2004) 'Coverage provided by the Global Protected Area system: Is it enough?' *BioScience*, vol 54, pp1081–1091

Bruce, J. W. (1989) 'Rapid Appraisal of Tree and Land Tenure', *Community Forestry Note 5. Rome, Italy: Food and Agriculture Organization of the United States*

Burgess, N., Kuper, W., Mutke, J., Brown, J., Westaway, S., Turpie, S., Meshack, C., Taplin, J., McClean, C. and Lovett, J. C. (2005) 'Major gaps in the distribution of

protected areas for threatened and narrow range Afrotropical plants', *Biodiversity and Conservation*, vol 14, pp1877–1894
Burgess, N., Loucks, C., Solton, S. and Didley, N. (2007) 'The potential of forest reserves for augmenting the protected area network in Africa', *Oryx*, vol 41, no 2, pp151–159
Carlton, J. T. (1998) 'Bioinvaders in the sea: Reducing the flow of ballast water', *World Conservation*, vol 4/97, pp9–10
Chape, S., Blyth, S., Fish, L., Fox, P. and Spalding, M. (2003) *2003 United Nations List of Protected Areas*, UNEP-World Conservation Monitoring Centre, Cambridge
Chape, S., Harrison, J., Spalding, M. and Lysenko, I. (2005) 'Measuring the extent and effectiveness of protected areas as an indicator for meeting global biodiversity targets', *Philosophical Transactions of The Royal Society*, vol B360, pp443–455
Chidumayo, E. N. (2008) 'Implications of climate warming on seedling emergence and mortality of African savanna woody plants', *Plant Ecology*, vol 198, pp61–71
Chidumayo, E. N. (in preparation) 'Tree growth and climate at a savanna site in cental Zambia over a 10-year period, 2000 to 2010', Makeni Savanna Research Papers, Lusaka, Zambia
Child, B. A. (2004) 'Principles, practice and results of CBNRM in Southern Africa', in Lyman, M. W. and Child, B. (eds) *Natural Resources as Community Assets: Lessons from Two Continents*, Sand County Foundation, Madison , WI, pp19–50
Crook, C. and Clapp, R. A. (1998) 'Is market-oriented forest conservation a contradiction in terms?', *Environmental Conservation*, vol 25, no 2, pp131–145
Cumming, D. (2004) 'Performance of parks in a century of change', in *Parks in Transition: Biodiversity, Rural Development and the Bottom-line*, EarthScan, London
Cunningham, A., Campbell, B. and Belcher, B. (eds) (2005) *Carving Out a Future: Forests, Livelihoods and the International Woodcarving Trade*, Earthscan, London
Dean, W. R. J., Milton, S. J. and Jeltsch, F. (1999) 'Large trees, fertile islands, and birds in arid savanna', *Journal of Arid Environments*, vol 41, pp61–78
Eva, H. D., Brink, A. and Simonetti, D. (2006) *Monitoring Land Cover Dynamics in Sub-Saharan Africa: A Pilot Study Using Earth Observing Satellite Data from 1975 and 2000*, European Commission, Directorate-General Joint Research Centre, Institute for Environment and Sustainability, Luxembourg
FAO (Food and Agriculture Organization) (2002b) *Law and Sustainable Development Since Rio: Legal Trends in Agriculture and Natural Resources Management*, FAO Legislative Study 73, Rome, Italy
FAO (Food and Agriculture Organization) (2005) *State of the World's Forests*, FAO, Rome
Frost, P. G. H., Campbell, B. M., Luckert, M. L., Mutamba, M., Mandondo, A. and Kozanayi, W. (2007) 'In search of improved rural livelihoods in semi-arid regions through local management of natural resources: Lessons from case studies in Zimbabwe', *World Development*, vol 35, pp1961–1974
Grainger, A. (1999) 'Constraints on modeling the deforestation and degradation of tropical open woodlands', *Global Ecology and Biogeography*, vol 8, pp179–190
Grote, K. (2003) *The Increased Harvest and Trade of Devil's Claw (*Harpagophytum procumbens*) and Its Impacts on the Peoples and Environment of Namibia, Botswana and South Africa*, Global Facilitation Unit for Underutilized Species, Maccarese, Italy
Hailwa, J. (1998), *Non-wood Forest Products in Namibia, EC-FAO Partneship Programme (1998-2000)*, Directorate of Forestry, Ministry of Environment and Tourism, Windhoek, Namibia

Hamilton, A. and Hamilton, P. (2006) *Plant Conservation: An Ecosystem Approach*, Earthscan, London

IUCN (2000) *Guidelines for the Prevention of Biological Invasions*, IUCN, Gland

IUCN (2009) *IUCN Red List of Threatened Species*, Version 2009.2, downloaded on 11 November 2009, http://www.iucnredlist.org/?sess=e23c06fe4478bb296ebb683ea145df7

Kellert, S. R., Mehta, J. N., Ebbin, S. A. and Lichtenfeld, L. L. (2000) 'Community natural resource management: Promise, rhetoric and reality', *Society and Natural Resources*, vol 13, no 8, pp705–715

Kigomo, B. (2003) 'Forests and woodlands degradation in dryland Africa: A case for urgent global attention', Paper presented at the XII World Forestry Congress, Quebec, Canada

Lindberg, K. (2001) 'Financial aspects of tourism in protected areas', www.ecotourism.org/retiesself.html, Cooperative Research Centre for Sustainable Tourism, Australia

Linder, H. P., Lovett, J. C., Mutke, J., Barthlott, W., Jurgens, N., Rebelo, T. and Kuper, W. (2005) 'A numerical re-evaluation of the sub-Saharan phytochoria', in I. Friis and H. Balslev (eds) *Plant Diversity and Complexity Patterns – Local, Regional and Global Dimensions*, Royal Danish Academy of Sciences and Letters, Copenhagen

Lovett, J. C. (2003) 'Tanzania forest law', in B. Chaytor and K. Gray (eds) *International Environmental Law and Policy in Africa*, Kluwer Law International, The Hague

Machena, C., Mpofu, B., Chitsike, L. T. and Shumba, E. M. (2005) 'Status of access and benefit sharing of genetic resources in Zimbabwe', in E. M. Shumba (ed) *The Status of Biodiversity in Zimbabwe*, IUCN, Harare

Maranz, S. (2009) 'Tree mortality in the African sahel indicates an anthropogenic ecosystem displaced by climate change', *Journal of Biogeography*, vol 36, pp1181–1193

Mayaux, P., Bartholome, E., Fritz, S. and Belward, A. (2004) 'A new land-cover map of Africa for the year 2000', *Journal of Biogeography*, vol 31, pp861–877

McClean, C. J., Lovett, J. C., Kuper, W., Hannah, L., Sommer, J. H., Barthlott, W., Termansen, M., Smith, G. F., Tokumine, S. and Taplin, J. R. D. (2005) 'African plant diversity and climate change', *Annals of the Missouri Botanical Garden*, vol 92, pp139–152

Mittermeier, R. A., Mittermeier, C. G., Brooks, T. M., Pilgrim, J. D., Konstant, W. R., da Fonseca, G. A. B. and Kormos, C. (2003) 'Wilderness and biodiversity conservation', *Proceedings of the National Academy of Science*, vol 100, pp10309–10313

Moyo, N. and Epulani, F. (2002), *Community Partnerships for Sustainable Resource Management in Malawi: Examples of CBNRM Best-Practices in Malawi*, Development Alternatives, Blantyre, Malawi

Olson, D. M. and Dinerstein, E. (1998) 'The global 2000: A representation approach to conserving the Earth's most biologically valuable ecoregions', *Conservation Biology*, vol 12, pp502–515

Osemeobo, G. J. (1994) 'The role of folklore in environmental conservation: Evidence from Edo State, Nigeria', *International Journal of Sustainable Development and World Ecology*, vol 1, pp48–55

Ostrom, E. (1999) *Self-governance and Forest Resources*, Occasional Paper no 20, CIFOR, Bogor

Palmer, R. (2001) *Land Tenure and Insecurity on the Zambian Copperbelt, 1998: Can Anyone Go Back to the Land*, Oxfam, UK

PhytoTrade Africa (2005) www.phytotradeafrica.com/
SADC (in press) *Regional Guidelines on Innovative Financing Mechanisms for Biodiversity Management and Sustainable Use in Southern Africa*, SADC Secretariat, Gaborone
Schreckenberg, K., Awono, K., Degrande, A., Mbosso, C., Ndoye, O. and Tchoundjeu, Z. (2006) 'Domesticating indigenous fruit trees as a contribution to poverty reduction', *Forests, Trees and Livelihoods*, vol 16, pp35–51
Shackleton, S. E. (2005) 'The significance of the local trade in natural resource products for livelihoods and poverty alleviation in South Africa', PhD thesis, Rhodes University, Grahamstown, South Africa
Siegfried, W. R. (1989) 'The protected area network', in B. J. Huntely (ed) *Biotic Diversity in Southern Africa*, Oxford University Press, Cape Town, pp186–201
Sola, P. (2005) 'Impacts and outcomes of commercialization of non-timber forest products (NTFPs) on human well being and ecosystem health', PhD thesis, Bangor, UK
Stedman-Edwards, P. (1998) *Root Causes of Biodiversity Loss*, WWF (World Wide Fund for Nature), Gland, Switzerland
Vedeld, P., Angelsen, A., Sjaastad, E. and Kobugabe Berg, G. (2004) 'Counting on the environment. Forest incomes and the rural poor', *World Bank Environmental Economics Series* Paper 98, Washington, DC
White, F. (1983) *The Vegetation of Africa: A Descriptive Memoir to Accompany the UNESCO/AETFAT/UNSO Vegetation Map of Africa*, UNESCO, Paris
Wittemeyer, G., Elsen, P., Bean, W. T., Coleman, A., Burton, O. and Brashares, J. S. (2008) 'Accelerated human population growth at protected area edges', *Science*, vol 321, pp123–126
World Conservation Monitoring Centre (1997) *Forest and Protected Area Distribution within Ecological Zones: Africa*, World Conservation Monitoring Centre and CIFOR, London
World Resources Institute (2003) *EarthTrends: The Environmental Information Portal*, World Resources Institute, Washington, DC

4

Contribution of Non-wood Forest Products to Livelihoods and Poverty Alleviation

Sheona Shackleton and Davison Gumbo

DEFINING NON-WOOD FOREST PRODUCTS: DIVERSITY AND CONTEXT

The term non-timber forest product (NTFP) was first popularized by de Beer and McDermott (1989) in an attempt to raise awareness of the importance of forests for uses other than commercial logging. They drew attention to the fact that many forest resources, other than timber, held significant value for local people and their economies and certainly did not deserve the label 'minor' forest products. They defined NTFPs to encompass 'all biological materials other than timber, which are extracted from forests for human use'. Since then there have been various refinements of this definition and much debate over what should and should not be included as an NTFP (Belcher, 2003). Wickens (1991) defined 'timber' more explicitly as industrial roundwood and derived sawn timber, woodchips, wood panel and pulp, and added plantations as well as natural forests as a source of products. With this clarification, woodcarvings, fuelwood, charcoal and other locally manufactured wood products such as furniture all fall under the NTFP banner. This contrasts with the Food and Agriculture Organization of the UN (FAO) term – non-wood forest products (NWFP) – that purposely excludes wood in all its forms. Recently, the use of the term NWFP has become so widespread, that it now often encompasses products from ecosystems other than forests (Belcher, 2003).

Drawing on the above, an NWFP has been defined for the purposes of this book as any raw or processed product, excluding commercial timber, that is produced from an indigenous or wild biological resource found within the dry forest zone and that is harvested for either domestic consumption or trade. In some instances, the resource may be cultivated or sourced from modified or non-natural systems (as for some edible leafy plants), but cannot be regarded as a conventional agricultural crop.

Non-wood forest products are an extremely diverse and complex category (Figure 4.1). As is evident from the above, they include a diversity of biological groups or 'types of resource' (herbaceous plants, climbers, trees, insects, birds, mammals, fungi, etc.) as well as thousands of species, all with differing biological and ecological characteristics. Many species have multiple and sometimes even competing uses. The types of raw and processed products are also vast, including wood products, fruits, seeds, exudates, leaves, bark, roots, bulbs, stems, fibres, whole organisms (e.g. insects, mammals and fungi), honey, oils, juices and various extracts and derivatives, some requiring sophisticated processing technology. Some products (e.g. wild fruits) may be extremely seasonal, only available for very short periods in the year. The uses NWFPs can be put to are just as varied, and may include, for example, for food, medicines, craft, building and fencing, energy, as inputs into farming systems (e.g. wooden tools and implements, organic fertilizers, veterinary medicines), and as new organic products for the cosmetic, botanical, pharmaceutical and health food industries.

Categories of NWFPs thus include a mix of products and species with very different ecological, social, livelihood and market niches, and equally diverse management and trade practices, and end products and consumers (Wynberg and Laird, 2007). This multidimensionality is often bewildering, creating numerous challenges for policy-makers in designing policy, legislation and programmes for these products, as well as for their development and sustainable management. It also makes it tremendously difficult to touch on all aspects of these products in any substantial way in one book chapter, particularly given the scope of countries and forest and woodland types involved.

Superimposed on this complexity related to the biological and physical characteristics of the resource and product, are a whole range of social, institutional and, for traded products, market factors (Alexiades and Shanley, 2004). As already mentioned, NWFPs are not only used to meet subsistence needs, but are also economic resources that may be traded between different kinds of actors in different types and scales of markets. This trade may be locally driven, as in the majority of situations, or externally facilitated, and may involve government, non-government and private sector stakeholders working in partnership with local communities. Sometimes competition between uses and different markets may arise, with possible deleterious impacts on the resource base and on local social systems. Non-wood forest products are also embedded in the political and cultural life of the people involved in their collection and consump-

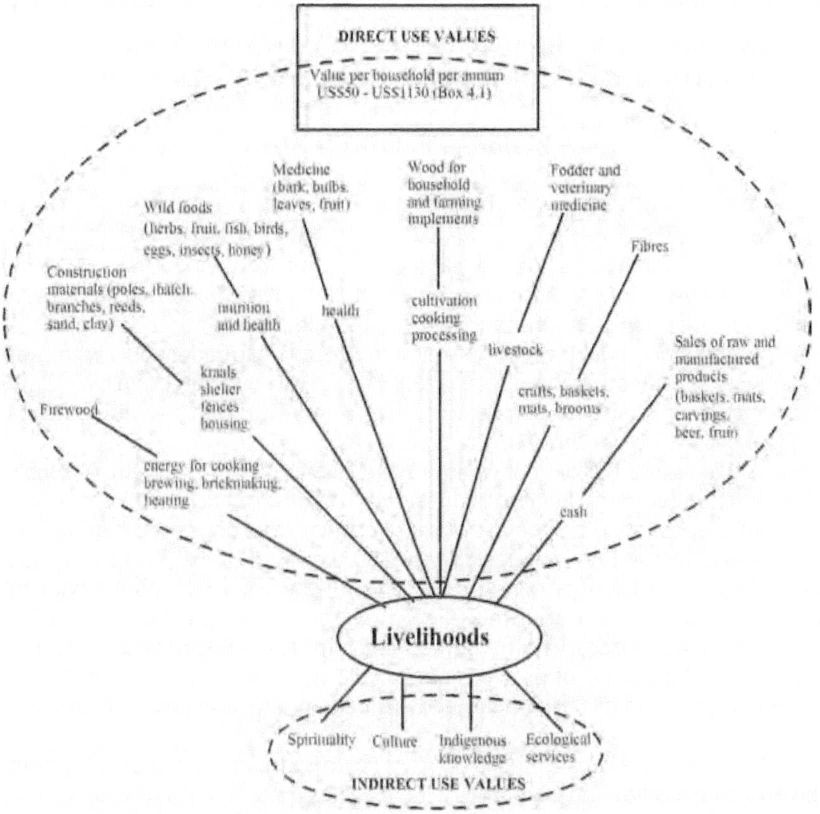

Figure 4.1 *Diversity of products and services that contribute to livelihoods in sub-Saharan dry forests and woodlands*

Source: Chapter authors

tion, with their 'multidimensionality evident in the myriad of processes, actors and factors that shape their access, management and commercialisation' (Alexiades and Shanley, 2004). The fact that NWFPs may be sourced from land under different types of ownership and management, including communal, private and state protected areas, is just one important institutional dimension affecting access and management. Furthermore, the way in which NWFPs are integrated into users' livelihoods, and their significance for the household economy, varies amongst different social actors and groupings, and according to the types of assets households have.

Despite this, and the fact that many millions of poor people benefit daily from NWFPs, their crucial importance for livelihood security and significant economic contribution, primarily in the informal sector, is generally poorly recognized and appreciated and sometimes even ignored in terms of national policy and forest management (Bird and Dickson, 2005; Petheram et al, 2006). Such neglect may undermine the potential of these products to deliver benefits

Box 4.1 Tapping new markets: PhytoTrade Africa's approach to natural product commercialization

Lucy Welford, PhytoTrade Africa

PhytoTrade Africa (www.phytotradeafrica.com) is the Southern African Natural Products Trade Association. Since its inception in 2001, PhytoTrade Africa has been committed to its objective of improving rural livelihoods through developing a sustainable natural products sector in southern Africa. PhytoTrade works with over 50 members in southern Africa (Botswana, Malawi, Mozambique, Namibia, South Africa, Swaziland, Zambia and Zimbabwe), who in turn work with tens of thousands of natural products producers in the region.

PhytoTrade Africa has developed environmentally sustainable and ethical supply chains for natural cosmetic and food ingredients that are wild harvested from indigenous plant species found predominantly in woodlands from across southern Africa. The association is currently researching over 300 species of useful plants, but focal species include manketti/mongongo (*Schinziophyton rautanenii*), baobab (*Adansonia digitata*), sausage tree (*Kigelia africana*), kalahari melon (*Citrullus lanatus*), marula (*Sclerocarya birrea*), mobola plum (*Parinari* spp) and sour plum (*Ximenia* spp). Categories of products produced include herbal teas, essential oils, gums and resins, lipid oils, fruit pulps and a variety of botanical raw materials and extracts.

With training and capacity building from PhytoTrade Africa, and utilizing both internal audit measures and external Fair Trade and environmental and organic certification schemes, association members are able to assure industry of reliable supply chain management and adherence to strict quality control measures. PhytoTrade's members supply industry with products for the nutraceutical, phyto-medicinal, botanical, flavour and fragrance, herbal remedy, dietary supplement, functional food, cosmeceutical and personal care industries. In order to do this, the association develops commercial opportunities on behalf of its members based on partnerships with commercial companies in key natural products markets. This involves not only developing long-term trusting partnerships with international commercial companies, but also ensuring that strong legal and technical agreements are in place. Commercial partnerships are based on a sound approach to both market and product development that demonstrates meaningful financial and technical commitment by both parties.

PhytoTrade Africa works in four key areas:

1. Institutional development
   PhytoTrade Africa acts as a service provider to the natural products industry and a stakeholder within the industry and realizes that:
   - industry will always be driven by private sector;
   - for primary producers to make money, all actors in the market chain must also make money;
   - intervention points for industry tend to be higher up the chain.

2. Product development
   - PhytoTrade Africa's strategy is to pass the responsibility and cost for product development to commercial partners;
   - this maximizes the association's R&D funding;
   - and ensures that products are developed that the market wants.
3. Market development: negotiating commercial partnerships
   - market development is closely linked with PhytoTrade Africa's R&D approach;
   - there is recognition that it is better to develop a relationship with one reliable client than many one-off buyers;
   - ensures that strong legal agreements are in place;
   - the Association has a marketing office in its primary export market and maintains a presence at key trade fairs and industry events;
   - the Association maintains a strong corporate identity.
4. Supply chain development
   - PhytoTrade is not part of the supply chain;
   - PhytoTrade ensures quality standards are maintained through a 'pre-qualified supplier' audit;
   - the Association provides guidance on production and processing technologies and quality standards;
   - ensures volumes of production meet demand;
   - provides assistance with certification;
   - provides assistance with logistics and paperwork.

in the future, erode vital safety nets, and exacerbate the already persistent poverty situation so endemic in Africa.

Daily, millions of men, women and children across the dry forest and woodland countries of sub-Saharan Africa enter the forests and woodlands surrounding their villages to procure an array of wild natural resources, or NWFPs, for home consumption and sale. Non-wood forest products have long formed a vital component of people's everyday livelihood needs, providing energy, food, medicines and raw materials for building, crafts, tools and implements (Campbell and Luckert, 2002; see Figure 4.1). In addition to subsistence use, NWFPs are also sold, in raw or processed form, in local and regional markets, offering an important means for poor individuals and households to generate cash income. In other instances, these products may form the basis of commercial enterprises, with some commodities reaching high-value international markets. In the last 10–15 years, facilitated projects aimed at the dual goals of biodiversity conservation and rural development have further stimulated and developed the trade in NWFPs, often in new external markets (Neumann and Hirsch, 2000; Box 4.1).

There is a distinct geographical variation in the importance of NWFPs across sub-Saharan Africa (Figure 4.2). Exudates (gum Arabic, olibanum, myrrh, tannins, opopanax, gum karaya) are more important in the Sudanian woodlands while edible plant foods are equally important in both the Sudanian and Zambezian woodlands. The importance of insect foods appears to be restricted

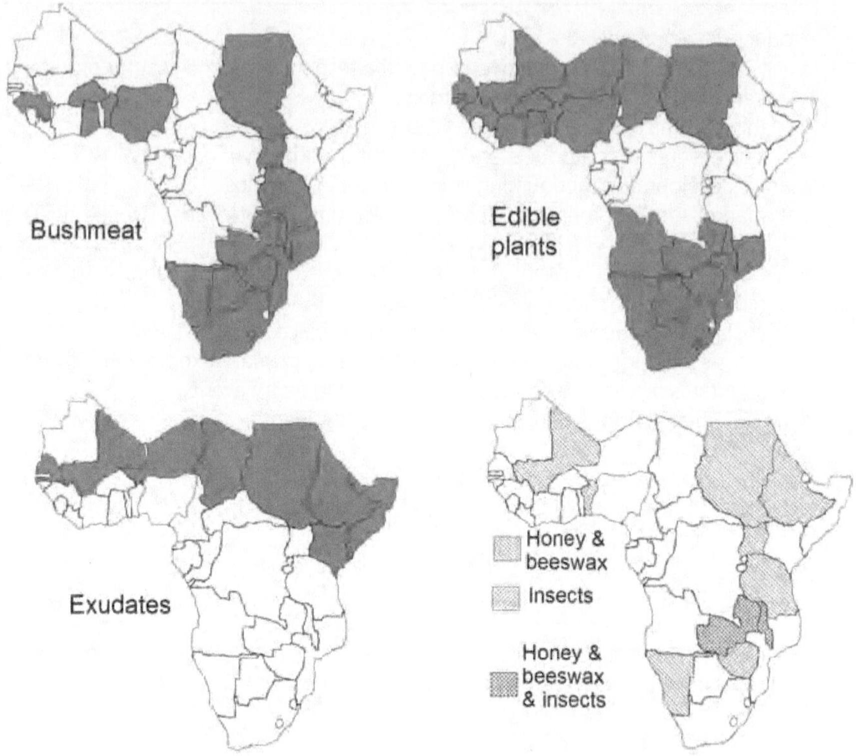

Figure 4.2 *Important non-wood forest products across sub-Saharan Africa*

Source: Walter (2001)

to Zambezian woodlands in Malawi, Zambia and Zimbabwe while the importance of honey and beeswax is widespread in the Kalahari, Zambezian and Somali-Masai phytoregions, although it is also important in a few countries in the Sudanian phytoregion. Unlike plant-based NWFPs, bushmeat is important in most countries across all the dry forest and woodland regions of sub-Saharan Africa. Some of these NWFPs, such as exudates are discussed in Chapter 9.

It is within this context that this chapter seeks to document the uses, livelihood significance, value and key challenges for a number of important categories of NWFPs, drawing from across the dry forest and woodlands of Africa. The chapter is structured as follows. Firstly, a definition of the term NWFP is provided (see above) and some of the intricacies and complexities of the sector highlighted. This is followed by synopses of livelihood, poverty and safety net roles of dry forests and woodlands. Additional synopses cover NWFPs' value and importance for subsistence, culture and commercial trade. This is followed by a section that summarizes some of the main challenges.

Poverty in Dry Forest and Woodland Countries: The Place of NWFPs

Africa has the redoubtable reputation of being home to some of the poorest countries in the world. Poverty in Africa is rife; almost 60 per cent of rural Africans live on less than US$1 per day (Kaimowitz, 2003; World Bank, 2005). In many of the continent's rural areas, poverty appears entrenched and intractable with few opportunities for relief, especially in the context of the huge and devastating impacts of HIV/AIDs (Bryceson and Fonseca, 2006; Shackleton, 2006; Wiegers et al, 2006). Moreover, within the dry forest and woodland regions, low and erratic rainfall, frequent droughts, and generally poor soils render farming activities, particularly those based on arable agriculture, exceptionally risky, contributing further to the hardships rural people face (Frost and Mandondo, 1999; Mortimore, 1998). All of this is aggravated by socio-economic conditions typical of underdeveloped areas, including inadequate infrastructure and services, poor exposure and access to markets, weak political power, low human capacity due to poor education and health, and often struggling local institutions (Belcher, 2005; Hedge, 2007).

It therefore comes as no surprise that some 320 million people depend on Africa's dry forests and woodlands to meet many of their basic needs (Petheram et al, 2006). The collection, and in some cases the sale, of NWFPs provides one of the most widely accessible livelihood opportunities available to poor rural people in these regions. Most NWFPs can be accessed 'freely' by households, and low barriers to entry mean people can trade in these products with little capital investment. In Zimbabwe, up to a third of household income has been estimated to come from NWFPs, with the proportion consistently higher in poorer households (Cavendish, 2000; Campbell et al, 2002). Mutamba (2007) established a similar pattern in Zambia. Such households typically engage in low return activities but often fail to accumulate capital from such activities (Kamanga et al, 2009). As will be shown later poor households focus on using forests and woodlands for consumption and as safety nets (Angelsen and Wunder, 2003). Further, some activities (e.g. bushmeat hunting) are predominantly carried out by males, while the increased marketing of plant materials is carried out by women, thereby increasing their opportunities to earn incomes. Notwithstanding all this, the rational extraction of NWFP contributes towards poverty reduction in the dry forests and woodlands of sub-Saharan Africa.

Valuation studies undertaken in a number of dry forest and woodland countries, mainly within southern Africa, have demonstrated the significant economic value of NWFPs for rural households, with the income share from these products reaching as much as one third of total household income (Table 4.1). In most cases, the largest proportion of this value can be attributed to firewood consumption, followed by wild foods and construction materials.

Table 4.1 *Value of dry forest and woodland NWFPs to rural households*

Region/Country	*Value (US$ per household per year)*	*Percentage contribution to total household income*	*Source*
Botswana	335	20.1	Zitzmann (2000) in Chipeta and Kowera (2004)
Zimbabwe	436	28.4	Cavendish (2000)
Zimbabwe	50–85	–	Campbell et al (1996)
Zimbabwe	120	–	Clarke et al (1996)
Zimbabwe	578	–	Campbell et al (1997)
Zimbabwe	320	–	FAO (1999)
Zimbabwe, Chivi	99	15.0	Campbell et al (2002)
South Africa, Bushbuckridge, Limpopo Province	572	19.4	Dovie (2004)
South Africa, Mogano, Limpopo Province	1130	–	Shackleton et al (2002)
South Africa, Ha-Gondo, Limpopo Province	565	–	Shackleton et al (2002)
South Africa, Mametja, Limpopo Province	620	–	Twine et al (2003)
Eritrea, Dighe, Gash-Barka Administrative Zone	386	–	Araia (2005)

Note: Values are not directly comparable as different studies have varying criteria regarding what to include/exclude from the analysis, for example the Cavendish study includes livestock browse while the others do not. Local currencies have been converted to US$ at the exchange rate for the year that fieldwork was completed.

Subsistence and Cultural Uses

In terms of subsistence use, NWFPs are critical for health, food, nutrition, shelter and energy. Considering food security and nutrition alone, poor people depend on NWFPs for many regularly utilized foods, for crisis or famine foods, for firewood to cook, for nutrients and vitamins, for grazing, for genetic resources, for inputs into agricultural production, such as implement handles and ploughs, and for the raw material for manufacturing such items as canoes for fishing (Bass et al, 2001). Non-wood forest products make significant contributions to livelihoods for the poor and more so for remote area dwellers. These are areas that are physically or frictionally distant from locations of strong economic activity and may lie behind ecological barriers such as mountains or disease prone areas (McCall, 1985; Smith, 1992; Barrett et al, 2005).

Recent work to place an economic value on this auto-consumption of wild products has shown it to be worth several hundreds of dollars per annum to user households (Clarke et al, 1996; Cavendish, 2000; Campbell et al, 2002; Chipeta and Kowero, 2004; Shackleton and Shackleton, 2004a, 2004b; Vedeld et al, 2004; Box 4.1). Moreover, this daily subsistence use of 'free' NWFPs allows households to enjoy a significant saving of scarce cash resources that can be redirected to meet other needs (Shackleton and Shackleton, 2004b). Non-wood forest products are also frequently used to barter for other goods or as a source of exchange for labour. For example, female-headed or elderly-headed households needing the help of a strong man to assist with, for instance, building repairs often brew traditional beers (Shackleton and Shackleton, 2005) or collect wild fruits as 'payment'.

In addition to the practical uses described, many NWFPs also have important social and cultural functions providing indirect livelihood benefits. Non-wood forest products such as medicinal plants can be symbolically and culturally important, providing livelihood benefits through social significance (Coad et al, 2008). Medicinal plants may be held in special religious, nationalistic or ideological esteem, thus providing a culturally based support for the value of flora and fauna (Hamilton, 2004) in the dry forests and woodlands of sub-Saharan Africa. Further, the sharing of marula (*Sclerocarya birrea*) beer (a widespread woodland product) plays a key role in building and maintaining vital social support systems, allowing people to draw on these networks in times of need (Shackleton and Shackleton, 2005). Marula beer could, thus, be viewed as having an indirect safety net or risk insurance function. Many NWFPs are collected and purchased for their spiritual and cultural significance and value (Cocks, 2006). These products may be used as traditional gifts, as cultural symbols in rituals, as charms and talismans against external agents like witches, as 'protectors' against events such as lightning strikes, and to build friendships and reciprocity. Often the process of gathering and processing NWFPs results in important social benefits as groups of people cooperate in these activities and can be a basis for community-based systems for conservation (den Adel, 2002). Furthermore, local knowledge of traditional practices centred on NWFPs, e.g. medicinal plants can be a source of employment opportunities and local identity (Coad et al, 2008). In this way, customs, taboos and superstitions associated with the use of NWFPs can assist in their sustainable management. These important traditional and cultural values of NWFPs are often forgotten in the pursuit of demonstrating economic value and in maximizing financial outcomes.

Non-wood forest products are a key resource for many poor communities (Sunderlin et al, 2005; Mutamba, 2007; Vedeld et al, 2007). In western Africa, for example, bushmeat provides 25 per cent of protein requirements, and can be the principal source for some indigenous groups (Bennett and Robinson, 2000; Fusari and Carpaneto, 2006). While there is some truth in the above statement, increased poverty and food insecurity are leading many people to turn to wildlife as a source of food. De Merode et al (2003) noted that bushmeat is more important for income than food and hunters often sell their catch to buy cheaper alternative foods (Juste et al, 1995). In addition, poor households may not benefit as often they may not even have males to carry out the hunting. In eastern and southern Africa, where the sale of wildlife products is commonly outside the formal economy the contribution of bushment goes unnoticed (de Merode et al, 2003; Fusari and Carpaneto, 2006).

Safety Net Role

Non-wood forest products derived from dry forest and woodlands of Africa play different roles in the livelihood strategies of different types of users ranging from being a source of food for subsistence, materials, medicines and equipment to safety nets (Belcher and Kusters, 2004). Depending on circumstances, dry forest and woodland products can be treated as 'daily net' and a 'safety net'. The 'daily net' describes everyday use, with products meeting current household needs, offering a reliable source of income to purchase agricultural inputs (Shackleton and Shackleton, 2004b), or fodder for livestock herds. A 'safety net' comes into play when other sources of household income (e.g. agriculture) fail to meet dietary shortfalls, or whenever a quick cash option is required (McSweeney, 2003). Studies have shown that NWFPs provide a 'safety-net' function in terms of quick, easy access to goods for household consumption or sale to earn cash in unpredictable misfortunes such as illness, death or natural disasters. In addition, Africa's dry forest and woodland countries experience frequent crop failure often resulting in poor nutrition of local people. In such cases, particular forest or woodland products will act as a form of 'natural insurance' (Arnold and Ruiz-Pérez, 2001; McSweeny, 2004; Takasaki et al, 2004), at critical times of the year to bridge income gaps, and/or to meet specific needs such as school fees or the costs of a celebration (Wunder, 2001; Belcher and Kusters, 2004). This safety net, buffering, and gap filling role of NWFPs also extends to the use of goods for own consumption during droughts, floods or other lean times, and as substitutes for purchased products during cash flow crises (Byron and Arnold, 1999; Kaimowitz, 2003). It is noted that certain livelihood activities (e.g. the sale of NWFPs) that may have originated as a response to misfortune, can also be upgraded to a permanent strategy (Angelsen and Wunder, 2003). 'Income smoothing' is another widely mentioned benefit of NWFP trading, especially at times when on-farm labour is in low demand (Fereday et al, 1997). In all of these situations, NWFPs serve the function of reducing household risk and vulnerability (Arnold, 2002), often helping to prevent households, particularly the poorest (Takasaki et al, 2004), from sinking lower into poverty during difficult times.

Trade and Cash Income

Millions of people in sub-Saharan dry forest and woodland countries trade in a diverse range of NWFPs and this trade appears to be growing worldwide, with numerous examples of supportive evidence from Africa (Campbell et al, 2002; Lowore, 2003; Clarke and Grundy, 2004; Box 4.2). Increasingly, rural dwellers are selling products previously used only for subsistence and cultural purposes. This growth is driven, at the local level, by a greater need for cash income as people become more integrated into the market economy, and by economic hardship and shock due to, amongst other factors, unemployment, retrenchment, withdrawal of agricultural subsidies and HIV/AIDS (Devereux, 1999; Monela et al, 1999; Rogerson and Sithole, 2001; Campbell et al, 2002; Kepe, 2002). In sub-Saharan Africa alone, it is estimated that several million people earn their primary cash income from the sale of NWFPs (Kaimowitz, 2003). In South Africa, some 3–14 per cent of rural households within the savanna biome are trading in at least one natural resource product, albeit often on an irregular basis. Some products such as medicinal plants form part of a multi-million dollar industry in South Africa, providing income-earning opportunities for many gatherers and traders, mainly poor rural and peri-urban women, and some 300,000 traditional healers (Mander, 1998, also see Box 4.2). Others, such as everyday items like traditional brooms and mats, provide more localized benefits, but are nonetheless still critical for the households involved (Shackleton, 2005).

Generally the returns from the sales of NWFPs are modest for the majority of participants, although extremely variable both within and across products and between households. Cash incomes earned may range from a few hundred to thousands of dollars per annum, even amongst those producing and selling the same product (Shackleton, 2005), and may contribute all or only a small proportion of total household income, with the average being between 10 and 25 per cent (Ndoye et al, 1997; Cavendish, 2000; Vedeld et al, 2007). For individuals and households specializing in niche markets, incomes may be more significant, but, on the whole, returns are limited by the conditions characteristic of rural sub-Saharan Africa mentioned above, as well as isolation, poor roads, high transport costs, a lack of markets or limited markets, insecure property rights and poor education and levels of organization amongst small-scale entrepreneurs (Sunderland and Ndoye, 2004).

Participation in forest product commercialization represents different types of livelihood strategy for different categories of poor households (Ruiz-Pérez et al, 2004; Shackleton, 2005; Marshall et al, 2006; Wiersum and Ros-Tonen, 2006). The sale of NWFPs may form part of an income diversification or risk reduction strategy, as households or individuals seek ways to supplement other sources of income or smooth their earnings throughout the year. For example, the NWFP trade often complements agricultural production in many regions of the world and in large parts of Africa (Byron and Arnold, 1999; Campbell et al, 2002). Alternatively, the trade may be the primary source of income for households, resulting in high levels of specialization (Ruiz-Pérez et al, 2004). Such a scenario is most likely for high value-added products, often with external markets; in situations that are conducive to trading such as close proximity to a major urban centre; and where intensified production is taking place. In these cases, if the value of the product increases significantly then the danger exists that the trade and benefits may be captured by the better off (Dove, 1993; see above). At the other end of the spectrum, it is not uncommon to find that individuals and households turn to trading in NWFPs in the absence of any alternative income-earning opportunities, particularly after experiencing hardship or shock (e.g. Shackleton et al, 2000; Rogerson and Sithole, 2001). The opportunity to sell NWFPs thus provides a safety net for people desperately looking for

Box 4.2 Regional, national and export value of trade in selected dry forest and woodland NWFPs

The value of the commercial trade in NWFPs to regional and national economies can be substantial, although in general data are extremely scarce or unreliable. Often, where statistics are available, these only capture the value of formally traded export goods. Thus the considerable worth of domestic markets and the extensive informal trade is generally underreported and underestimated. The following provides some illustration of the potential economic value of a range of dry forest and woodland products.

- The mopane worm industry in Botswana was valued at £4.42 million (then US$7 million) in 1995 and employed as many as 10,000 local people (Styles, 1995).
- The informal trade in medicinal plants in southern Africa is valued at US$75–150 million per annum with some 35,000–70,000 tonnes of plant material traded each year (Mander and Le Breton, 2006).
- The value of gum and resin exports from Ethiopia from 2001–2003 amounted to US$2.8 million, 3.3 million and 4.1 million respectively. Natural gum tapping and collection activities create seasonal employment opportunities for 20,000–30,000 people.
- In 2005, some 256 tonnes of beeswax worth US$1 million were exported from Ethiopia, while the corresponding figures for Tanzania were 277 tonnes and US$1.3 million, respectively (ITC, 2006).
- The woodcarving industry in Kenya is worth over US$20 million annually in export products and employs some 60,000–80,000 carvers supporting over 400,000 dependants (Choge, 2004).
- Zambia and Tanzania are the two woodland countries exporting the largest volumes of honey. In Zambia in 2005, 219 tonnes of honey were exported with a value of US$491,000 while Tanzania exported 466 tonnes with a value of US$674,000. In both countries, volumes exported have risen by 20–30% since 2001 (ITC, 2006).
- Shea butter is the third most important crop in Burkina Faso and provides income to about 300,000–400,000 women (Schreckenberg, 2004; Harsch, 2001). Imports of shea butter to Europe from Sahelian countries were estimated at US$13 million in 1999 (Schreckenberg, 2004).

some source of income. Unlike the short-term safety net functions described earlier, this may evolve into a long-term or permanent source of livelihood or coping strategy if the conditions that initially forced the individual into the trade prevail, or if the producer subsequently chooses to make their living from the trade. It is not unusual to find all these livelihood strategies represented for the same product, but in different types of households (Lowore, 2003; Stack et al, 2003; Shackleton, 2005). Furthermore, in many instances, households sell more than one NWFP, depending on a portfolio of different products to ensure a more reliable income and reduce risk. These subtleties are often missed in studies that are product rather than livelihood focused.

IMPORTANCE FOR THE MARGINALIZED AND VULNERABLE

Most work to date suggests that it is the poorest members of society who are most dependent on NWFPs (Fereday et al, 1997; Cavendish, 2000; Neumann and Hirsch, 2000; Beck and Nesmith, 2001; Fisher, 2004; Shackleton and Shackleton, 2006), although recent findings from the humid regions of Cameroon reveal that NWFPs in that country contribute most significantly to middle-income households (Ambrose-Oji, 2003). The collection of NWFPs is an activity that is generally available to all households, but that is more likely to be exploited by poorer groups with limited land resources and other assets, minimal education and skills, and few other income sources, contributing a greater proportion of total income to these households (Arnold, 2002; Fisher, 2004). In addition, households living in remote areas are also likely to be vulnerable. Wild foods are extremely important for the nutrition and food security of children, especially those from poor and HIV/AIDS affected households (Kaschula in Shackleton, 2006, also see Chapter 5). The low barriers to entry to the trade in many NWFPs means that this activity provides an important option for poor and marginalized households who would have difficulty accessing other employment opportunities, or who are less able to cope with or insure against risk than better-off households (Fisher, 2004).

It is noted that vulnerability affects both the rich and poor households (Devereux et al, 2006) as this is a dynamic concept that is constantly changing and forward looking. They further argue that poverty is a static concept that measures proxies for well-being at a point in time. For the poor, however, it can be noted that resilience and the ability to cope with risk will be subject to the extent to which their asset base is 'degraded'. Poor and risk-prone households are likely to prevent neighbouring households from coming out of poverty and thus remain vulnerable (Corcoran, 1995).

Women in particular benefit widely from the use and sale of these products, as do older and less educated people who cannot compete effectively in the formal job market (Falconer, 1996; Terry, 1999; Kaimowitz, 2003). Non-wood forest products are often of particular importance to women, but the context can lead to radically different situations for them. In situations where women are the traditional harvesters/producers of NWFPs as is the case in much of sub-Sahara Africa they may be able to use their skills and knowledge to improve their status and increase their contribution to decision-making processes. In some cases returns may be too low to such an extent that the earnings realized may not help in any way to contribute to their economic and political emancipation. Conversely, the returns may be very high and they will be edged out by the men. In the worst cases, women may even become excluded from their traditional role in NWFPs because rights and/or benefits are captured by men in case of increasingly attractive benefits.

Role in Poverty Alleviation

Non-wood forest products are essentially a niche for the poor (Arnold and Ruiz-Pérez, 1998). That is the reason why any effort aimed at developing the sector will be very important for poverty reduction. There has been considerable debate recently regarding the ability of NWFPs to contribute significantly to the Millennium Development Goal of poverty alleviation. Much depends on how the notion of poverty is defined and understood (Angelsen and Wunder, 2003). Poverty has over time been an outcome-based measure of livelihood performance (Sunderlin et al, 2005). In this respect, poverty has often been measured in terms of absolute income, with a common indicator defining the 'poor' as those who earn less than US$1 per day (Anglesen and Wunder, 2003). Lately, the concept of a Human Development Index (HDI) was developed by the United Nations Development Programme (UNDP) and this concept includes health and education parameters. Thus, poverty assessment frameworks now no longer view poverty as a matter of income alone. They tend to incorporate natural, human, social and physical capital, using indicators ranging from income, access to resources and basic infrastructure, to the vulnerability of populations to shock, and level of community organization (Dubouis, 2002).

Thus in accepting to combat poverty, it is necessary to consider, inter alia, aspects of income, income distribution, access to assets, human capital, empowerment and rights, vulnerability and risk, food security, alternatives and choice, health and well-being, and the ability of the poor to devise appropriate coping strategies when faced with shocks and crises (Sen, 2003; Ashley and Maxwell, 2001), then NWFPs clearly have a vital role to play. As outlined above, extensive evidence exists to support their importance in reducing vulnerability, in ensuring food security, in providing cash income to some of the poorest sectors of society, and in contributing more generally to improved rural welfare, livelihood security and diversification (e.g. Alexiades and Shanley, 2004; Kusters and Belcher, 2004; Sunderland and Ndoye, 2004). However, the picture is somewhat less unambiguous regarding how these products may assist poor people to accumulate assets, improve their standards of living and move out of poverty, certainly in any enduring way. Non-wood forest products, thus, tend to be more central to poverty mitigation, that is, preventing the deepening of poverty, than to poverty reduction or elimination, or lifting people out of poverty (FAO, 2003). Further, existing policy frameworks sometimes prevent this essential contribution of NWFPs being realized (Michon, 2005). With respect to the former, the essential poverty prevention role of NWFPs assumes magnified significance in the context of increased exposure of the poor to risk due to, amongst other factors, climate change (a significant concern in the dry forest and woodland regions), HIV/AIDS, changes in trade, globalization, and increasing violence and crime (Aliber, 2003; World Bank, 2000).

In summary, it is important to note that poverty reduction refers to a successful improvement of livelihoods (Sunderlin et al, 2003). When discussing forest-based poverty alleviation, it must be recognized that commercialization

Table 4.1 *Interactions between well-being factors and social organization levels with regards to poverty*

	Individual	*Community*	*National*	*International*
Income and growth	• Labour/non labour income • Enhanced land rights • Access to subsistence products • Small enterprise development • Skills and knowledge	• Infrastructure improvements • Local spending • Improved public services • Skills and knowledge	• Infrastructure improvements • Skills and knowledge • MDGs attained • Small and medium enterprise development	• Development and biodiversity conservation targets addressed
Equity	• Level and distribution of income and resource access from NWFP within households	• Level and distribution of income and resource access from NWFP within community	• Level and distribution of income and resource access from NWFP across regions	• Mechanism for international trade in NWFP
Voice and choice	• Effective participation in community discussions of NWFP	• Effective participation indecision making over NWFPs • More viable representative local level institutions	• Effective participation in national discussions over NWFP	• Effective participation in global discussions over NWFP

Note: Millennium Development Goals (MDGs) are internationally agreed time-bound goals that provide concrete, numerical benchmarks for addressing extreme poverty in the world. They are eight in number and include targets on income poverty, hunger, maternal and child mortality, disease, inadequate shelter, gender inequality, environmental degradation and a Global Partnership for Development.

Source: Authors

and increased market access to forest resources does not necessarily provide opportunities for the poor, and may shift access and use towards the richer sections of a community (Arnold and Ruiz-Pérez, 2001). In Table 4.1 the potential benefits and risks of these systems for the poor are looked at with respect to dimensions of poverty, including income and growth, equity, and voice and choice; as well as the dimensions of scale: from individual, through community and national, to international scales.

While there are greater benefits that may accrue to an individual, issues pertaining to equity and extent of participation (voice) will limit this potential (Table 4.1). These issues are better addressed under the community level and it is not surprising that there have been calls for collective action and group marketing of products (see Box 4.3). At national level the potential is fully realized but will be subject to national policies (see section below). Surprisingly, the contributions made by NWFPs often go unnoticed and undervalued (Angelsen and Wunder, 2003). Dry forest and woodland dependent people, and especially the poor, extract a variety of NWFPs from these zones (Sunderlin et al, 2005; Vedeld et al, 2007; Kamanga et al, 2008). More systematic and comparative research on the linkages within and between poverty, the poor, NWFP use, commercialization, resource trends and ownership rights, is needed that can lead to a greater appreciation of NWFPs.

Key Challenges

With regards to the role of NWFPs role in livelihoods and their potential for poverty reduction, a number of challenges are recognized as described below.

Climatic conditions

The diversity and availability of NWFPs in the dry forests and woodlands of Africa are threatened by changing ecological conditions which will invariably affect the development, utilization, domestication, sustainability and commercialization of these products. It is worth noting that most dry forest and woodland countries have been hit by successive droughts from the 1970s, through 1990s. Consequently, natural resources have undergone serious deterioration and depletion and no respite is expected under climate change (Hulme et al, 2005).

Poor management of natural resources

The status of natural resource management in general, and dry forests and woodlands in particular, in the zones covered under this book is poor. The poor management of dry forests and woodlands which constitute the resources base of NWFPs could be attributed to the fact that policymakers are unaware of:

- the extent of the uses or the values of these resources;
- the economic importance and the contribution of NWFPs to the informal sector and national economy;
- the magnitude of dependence of the rural poor on the resource for food security and income.

If values were recognized, the sustainability of the dry forests and woodlands through state sponsored management plans might be better implemented (Alden-Wily and Mbaya, 2001).

Non-wood forest products and regulatory frameworks

Forest policies and regulations do affect the social and economic success of NWFPs management at the local and national levels. They define who has access to which kind of resources and in which kind of forests. They also further determine how benefits of forest management, collection and trade are shared among stakeholders. Forest policies and regulations are therefore essential in the determination of the social and economic attributes of forest management, including NWFPs collection and use. In most forest and woodland areas, NWFPs collectors are often people belonging to local forest and woodland-dependent communities. Subsistence products are collected by various social categories, including children, women and elders, whereas the collection of commercial products is usually dominated by young men.

Access regulations to NWFPs are generally more clearly embedded in the customary rights than in specific 'modern legal frameworks' and are thus often not regulated statutory by instruments and state control mechanisms (Michon, 2005). This may not be true for some of the important NWFPs e.g. mopane worms and the grapple plant in Botswana where the state has a keen interest. In most dry forest and woodland countries, competition for different land uses is leading to a poor consideration of NWFPs' production potential. Despite the safety net role of NWFPs, forests and woodlands are often perceived as a land reserve or available resource, even or especially for poor farmers.

Non-wood forest product policies should therefore pay particular attention to all kinds of local initiatives for NWFPs intensified management. They should particularly consider the relevance and legitimacy of management rules underlying cultivation, including:

- Access and property rules, including a blend of individual and collective rights and obligations, including rules concerning lands, but also on specific portions of space, or on trees or other resources.
- Customary management rules and practices.
- Local economic, but also social attributes of forests and woodlands: role in livelihood strategies, role in social cohesion and/or stratification of families and village communities. This implies the acknowledgement of local management as a specific domain of productive activity, independent of the legal domain in which it takes place.

Regulations at local level: Are they compatible with national policies?

The first level for non-wood forest products regulation concerns customary rights and institutions. Although these systems can, by essence, be quite variable from one place to another, they generally show the following characteristics:

- They do not emphasize uniform rights, but bundles of specific rights tailored according to resources, users, or uses.
- They involve rights and obligations.
- They are quite flexible and adaptive, and easily evolve as the context or the needs change.
- They are recognized and acknowledged by all people concerned in the community.
- They are usually neither understood nor recognized by national constitutions/legislations.

It is important to state that these local regulatory systems do not necessarily target or guarantee sustainability, unless there is a feeling of threat on the possibilities to continue the economic activity.

The main problems arise when national dry forest and woodland regulation systems ignore these customary systems. Conflicts between national laws and local systems entail, among others, unsustainable management or mining of forest and woodland resources, abuse of power and grabbing of economic benefits by local elites, and social disintegration. Therefore, policies targeting sustainable NWFP management have to assess the relevance of local management systems, and to find theoretical and practical ways to accommodate local rights in national systems.

Non-wood forest product harvesting and resource conservation

The analysis of the environmental benefits of promoting NWFPs has also advanced with regard to earlier optimistic assessments by Peters (1994). It is generally accepted that the harvesting of NWFPs tends to maintain forest cover, particularly when compared with other alternative land uses (Ruiz-Pérez et al, 2004). The effects on biodiversity are variable; NWFP based activities generally maintain a substantial amount of the species naturally occurring, although it certainly affects them, especially those most sensitive to human presence or those which are also collected in parallel with the commercial gathering of the main NWFPs (Peters, 1994; Freese, 1997; Bennett and Robinson, 2000; Ticktin, 2004). This extraction can also seriously affect the populations being exploited, particularly in the context of wild gathering and market expansion. The promotion of commercial uses of NWFPs can then be viewed as a double-edged sword, with potential and risks (Redford, 1992).

Marketing of non-wood forest products

It has been observed that the gathering and sale of NWFPs rarely generates enough income to sustain households. This has been presented as a conservation and development strategy but for such a strategy to work NWFP commercialization must be transparent, equitable and sustainable. There must be a positive impact on poverty reduction, gendered equality, resource access and tenure management. For this to work, local, regional and external markets must be accessed and to do this frameworks that enable closer collaboration within and among producers, processors and traders must be developed. New markets must be sought while the processing of locally gathered NWFPs should be encouraged for the purpose of adding value.

That said, the situation is not all bleak regarding the potential of NWFPs as a means to help people move out of poverty, and there are examples of successful NWFP commercialization. Moreover, a number of recent trends and new market opportunities hold fresh promise for the development and growth of the NWFP trade and for enhancing its livelihood benefits, particularly within an enabling policy environment and with government and other stakeholder recognition and support. Some of these trends include:

- A burgeoning demand for low-cost forest products such as wood-based fuels, foods (bushmeat, fruits, alcoholic beverages) and medicines, amongst the urban poor as African populations urbanize (Williams et al, 2000; UNEP, 2002; Scherr, 2004).
- A growing domestic market for products with traditional and cultural significance, particularly those that cannot be readily substituted (Cocks, 2006).
- The emergence of new niche markets for certified organic and fair-traded natural product-based health, food and cosmetic commodities. Enhanced consumer demand for such products is resulting in growing interest by private sector industrial companies in some of the more high value forest products and the possibility of working directly, or through intermediary partners, with local groups of producers (e.g. PhytoTrade Africa, 2006; Box 4.1).
- Widespread changes in forest governance that favour strengthened local rights over forest resources and more secure land tenure with positive impacts for access, sustainable resource use, and management and intensification of production (Scherr, 2004; Belcher, 2005).
- Greater interest by development agencies (NGOs, donors) in NWFPs and small-scale forest enterprises as an attractive option for reaching some of the poorest people with limited livelihood options (e.g. women, HIV/AIDS affected households), while many conservation agencies continue to view forest and woodland-based enterprise development as an important route to achieving both conservation and livelihood objectives (Box 4.2). Consequently, there is a significant degree of interest in facilitating forest product commercialization and trade.

Box 4.3 Shea (*Vitellaria paradoxa*) products in Burkina Faso

Ousseynou Ndoye, CIFOR

Shea products (nuts and butter) are very important NTFPs in Burkina Faso. The value of their exports in 2000 was estimated at US$7 million and they provide employment to 300,000–400,000 rural women (Harsch, 2001). A study of 102 women who collect, process and market shea showed that they receive, on average, a monthly income of US$5–40, which is higher than the estimates obtained from Benin (Schreckenberg, 2004) and other provinces of Burkina Faso (Crelerot, 1995, cited by Boffa, 1999). Women use about 36 per cent of their shea revenues to buy food and medicines, which is clearly indicative of the importance of shea to household economies. Thus food security is a major goal for households involved in shea production, processing and trade. Children's education is the third most important item on which the revenue received from shea sale is spent (13 per cent).

In Burkina Faso, farmers protect old and young shea trees when they clear the woodland for cultivation. The tree can live up to 300 years and produces fruits from year 15 (Kanguembega, 2003). In a few areas of Burkina Faso, shea trees are harvested for charcoal production, which is very destructive and requires urgent action from the government to stop this practice. This trend is a result of increasing population pressure to satisfy both food and energy needs. In addition, the productivity of shea is low due to the aging of trees which reduces the level of supply and the benefits for households involved in shea collection, processing and marketing.

There is an urgent need to accelerate the domestication and mass planting of shea trees to increase productivity and to enable Burkina Faso to play a more important role in the world market and further improve the livelihoods of rural dwellers. The availability of improved varieties of shea trees will enable farmers to replace aging ones and reduce the pressure on the wild population of trees in the woodlands. This will enable Burkina Faso to respond better to the decision of chocolate companies to use more shea butter rather than cocoa to make chocolate.

Market opportunities need to be enhanced and returns to primary producers increased by value addition, particularly through quality improvement and initiatives such as certification and fair trade labelling (Masters et al, 2004). Forest products, particularly NTFPs, are in general overlooked in national statistics. They also do not feature in poverty reduction strategies defined by many African governments. There is a need to engage a dialogue with policy-makers and to sensitize them on the need to recognize the contribution of NTFPs like shea to rural livelihoods and to include them in their poverty reduction programmes.

- Considerable advances in communication technology (e.g. cellular phones) are providing new opportunities for improved flow of information and better linkages between small-scale entrepreneurs and the markets – a major hurdle in the past. For example, in Africa, the mobile telecommunications sector has grown by an average of 78 per cent per annum over the last ten years. The positive benefits of this technology for small-scale entrepreneurs have been well demonstrated by a study from Ghana in which it was concluded that access to cellular phones had decreased informal traders' transaction and transport costs, created a higher profit margin for them, increased their efficiency, and enhanced trust building within trade networks (Overa, 2006).

Where some of these opportunities have been harnessed, there is evidence to indicate that livelihood and financial benefits can be both raised and extended to much larger numbers of people, although it is often only in combination with other sources of income that they can provide a pathway out of poverty. This is well illustrated by the shea butter trade in West Africa (Box 4.3), PhytoTrade's initiatives in southern Africa (Box 4.2), and honey in Zambia (Box 5.3). However, scaling-up commercial production can be fraught with difficulties.

Conclusions

The above sections have highlighted the importance of a number of categories of NWFPs for livelihood security, in particular for food security and alleviating dietary deficiencies, and for assisting households to cope with, if not escape, poverty. In particular, these products have been shown to be important for women and children, both extremely vulnerable groups. The use of NWFPs by urban, in addition to rural households, has also been pointed out, with this likely to grow with the increasing urbanization of Africa's population (UNEP, 2002). In fact, urban demand helps to create sustainable markets for NWFPs, contributing to their potential as a means to earn cash income. Furthermore, in all sections the role of NTFPs in mitigating some of the devastating impacts of HIV/AIDS has been stressed. Both plant and insect wild foods are highly nutritious and could assist in meeting some of the nutrition requirements of HIV/AIDS sufferers. The demand for traditional medicines has also risen as a result of the AIDS pandemic, with potentially negative feedbacks on the forest and medicinal plant stocks. The significant links of many NWFPs to culture and identity, and their role in building social capital, have also been indicated, with this sometimes forming an important dimension of the rural-urban link. For all NWFP groups, reference has been made to the importance of traditional knowledge, passed orally from parent to child, in the use, processing and management of a range of species. However, particularly with respect to the latter, it has been mentioned that this is under threat due to escalating pressures on resources and the drive to commoditization.

A key issue across the dry forest and woodland region is that most NWFPs, and the greatest quantities of these products, are collected from communal lands, which, due to the breakdown of traditional management systems, effectively function as open access regimes. For a number of products, there is an increasing fear that local controls and customary rules have decreased in effectiveness. However, in other instances these have formed the basis for the establishment of new resource management systems. Generally, though, there is little evidence, with the exception of honey, of any formal government policies, strategies, legislation or regulations specifically geared towards the NWFPs discussed in this book (Chapter 5), except perhaps where the products may be harvested from protected species. In general, many of these products are still invisible to policy-makers and implementers.

In terms of scaling-up the poverty alleviation potential of resources by increasing market demand and trade, a number of cautions were stated. Firstly, this should not be at the expense of the subsistence use of these products for basic household needs (Wynberg et al, 2003). Secondly, it should not impact negatively on local markets which often involve the poorest participants who may be prevented from participating in more high paying markets by local elites (Shackleton et al, 2007). Thirdly, it should not increase pressure on the resource base and result in unsustainable harvesting. The model provided by PhytoTrade provides good examples of how new opportunities may be exploited for both the benefit of poor people and the management of forests. It is being increasingly shown, including in the honey projects mentioned in Chapter 5, that commercialization models that involve the private sector, local communities and development NGOs usually result in greater equitability, sustainability and benefits for the poorest groups than other arrangements. However, government needs to be more actively involved in overseeing these partnerships and developing policy that is pro-poor, equitable and that recognizes local knowledge and rights (Wynberg, 2006).

In summary, the social contributions of NWFPs and their potential for the poor need to be better integrated in crucial policies at national level. The elements to consider as priority at the policy level are:

- Equitable access to resources and to benefits.
- Regulation modalities, local accountability and transparency of markets (no 'informal taxes'):
  - simplification of procedures (administrative, legal, financial);
  - differentiation of the existing forest and woodland types and managed landscapes and analysis of the specific resource/product potential according to the various expected functions to be ensured by forests and trees;
  - focus on adaptive local governance processes based on the usually flexible customary rules and on collaborative networks rather than on 'perfect' state or market regulations;
  - ensure capacity building, facilitate trade associations and links to known/transparent market chain.

REFERENCES

Alden-Wily, L. and Mbaya, S. (2001) 'Land, people and forests in eastern & southern Africa at the beginning of the 21st Century: The impact of land relations of the role of communities in forest future', IUCN-EARO (monograph)

Aliber, M. (2003) 'Chronic poverty in South Africa: Incidence, causes and policies', *World Development*, vol 31, pp473–490

Alexiades, M. N. and Shanley, P. (eds) (2004) *Productos Forestales, Medios de Subsistencia y Conservación. Estudios de Caso Solore Sisternas de Manejo de Productos Forestales no Maderables. Volumen 3 – America Latino*, CIFOR, Bogor

Ambrose-Oji, B. (2003) 'The contribution of NTFPs to the livelihoods of the "forest poor": Evidence from the tropical forest zone of south-west Cameroon', *International Forestry Review*, vol 5, pp106–117

Angelsen, A. and Wunder, S. (2003) 'Exploring the forest-poverty link: Key concepts, issues and research implications', *CIFOR Occasional Paper No. 40*, CIFOR, Bogor

Araia, M. G. (2005) 'Revealing the forest's hidden value: The case study of Eritrea', MSc thesis, University of Stellenbosch, Stellenbosch, South Africa

Arnold, J. E. M. (2002) 'Clarifying the links between forests and poverty reduction', *International Forestry Review*, vol 4, pp231–234

Arnold, J. E. M. and Ruiz-Pérez, M. (2001) 'Can non-timber forest products match tropical forest conservation and development objectives?', *Ecological Economics*, vol 39, pp437–447

Arnold, J. E. M. and Ruiz Pérez, M. (1998) 'The role of non-timber forest products in conservation and development', in E. Wollenberg and A. Ingles (eds) Incomes from the Forest; Methods for the Development and Conservation of Forest Products for Local Communities, Center for International Forestry Research, Bogor, Indonesia, pp17–42

Ashley, C. and Maxwell, S. (2001) 'Theme issue: Rethinking rural development', *Development Policy Review*, vol 19, pp395–425

Barrett, C. B., Lee, D. R. and McPeak, J. G. (2005) 'Institutional arrangements for rural poverty reduction and resource conservation', World Development, vol 33, no 2, pp193–197

Bass, S., Hughes, C. and Hawthorne, W. (2001) 'Forests, biodiversity and livelihoods: Linking policy and practice', in I. Koziell and J. Saunders (eds) *Living off Biodiversity. Exploring Livelihood and Biodiversity Issues in Natural Resources Management*, International Institute for Environment and Development (IIED), London

Beck, T. and Nesmith, C. (2001) 'Building on poor people's capacities: The case of common property resources in India and West Africa', *World Development*, vol 29, pp119–133

Belcher, B. (2003) 'What isn't an NTFP?', *International Forestry Review*, vol 5, pp161–168

Belcher, B. M. (2005) 'Forest product markets, forests and poverty reduction', *International Forestry Review*, vol 7, pp82–89

Belcher, B. and Kusters, K. (2004) 'Non-timber forest product commercialisation: Development and commercialisation lessons', in Kusters, K. and Belcher, B. (eds) *Forest Products, Livelihoods and Conservation. Case Studies of Non-Timber Forest Product Systems*, vol 1 Asia, CIFOR, Jakarta

Bennett, E. L. and Robinson, J. G. (2000) 'Hunting of wildlife in tropical forests: Implications for biodiversity and forest peoples', *World Bank Biodiversity Series*, no 76, World Bank, Washington, DC
Bird, N. and Dickson, C. (2005) *Poverty reduction strategy papers: Making the case for forestry*, Overseas Development Institute (ODI), London
Boffa, J. M. (1999) *Agroforestry Parklands in Sub-Saharan Africa*, FAO conservation Guide 34, FAO, Rome
Bryceson, D. and Fonseca, J. (2006) 'Risking death for survival: Peasant responses to hunger and HIV/AIDS in Malawi', *World Development*, vol 34, pp1654–1666
Byron, N. and Arnold, M. (1999) 'What futures for the people of tropical forests?', *World Development*, vol 27, pp789–805
Campbell, B. M. and Luckert, M. K. (2002) *Uncovering the Hidden Harvest. Valuation Methods for Woodland and Forest Resources*, Earthscan, London
Campbell, B., Frost, P. and Byron, N. (1996) 'Miombo woodlands and their use: Overview and key issues', in B. Campbell (ed), *The Miombo in Transition: Woodlands and Welfare in Africa*, Center for International Forestry Research, Bogor, Indonesia, pp1–10
Campbell, B. M., Luckert, M. and Scoones, I. (1997) 'Local-level valuation of savannah resources: A case study from Zimbabwe', *Economic Botany*, vol 51, no 1, pp59–77
Campbell, B., Jeffrey, S., Kozanayi, W., Luckert, M., Mutamba, M. and Zindi, C. (2002) *Household Livelihoods in Semi-arid Regions. Options and Constraints*, CIFOR, Bogor
Cavendish, W. (2000) 'Empirical regularities in the poverty-environment relationship of rural households: Evidence from Zimbabwe', *World Development*, vol 28, pp1979–2003
Chipeta, M. E. and Kowero, G. (2004) 'Valuation of indigenous forests and woodlands: An international perspective', in M. J. Lawes, H. A. C. Eeley, C. M. Shackleton and B. G. S. Geach (eds) *Indigenous Forests and Woodlands in South Africa: Policy, People and Practice*, University of KwaZulu-Natal Press, Pietermarizburg
Choge, S. K. (2004) 'The woodcarving industry in Kenya', in T. Sunderland, and O. Ndoye (eds) *Forest Products, Livelihoods and Conservation. Case Studies of Non-timber Forest Product Systems*, vol 2 Africa, CIFOR, Bogor
Clarke, J. and Grundy, I. M. (2004) 'The socio-economics of forest and woodland resource use: A hidden economy', in M. J. Lawes, H. A. C. Eeley, C. M. Shackleton and B. G. S. Geach (eds) *Indigenous Forests and Woodlands in South Africa: Policy, People and Practice*, University of KwaZulu-Natal Press, Pietermarizburg
Clarke, B., Cavendish, W. and Coote, M. (1996) 'Rural households and miombo woodlands', in B. M. Campbell (ed) *The Miombo in Transition: Woodlands and Welfare in Africa*, CIFOR, Bogor
Coad, L., Campbell, A., Miles, L., Humphries, K. (2008) 'The costs and benefits of protected areas for local livelihoods: A review of the current literature', Working Paper, UNEP World Conservation Monitoring Centre, Cambridge
Cocks, M. L. (2006) 'Wild plant resources and cultural practices in rural and urban households in South Africa: Implications for bio-cultural diversity conservation', PhD thesis, Wageningen University, Wageningen, Netherlands
Corcoran, M. (1995) 'Rags to riches: Poverty and mobility in the United States', *Annual Review of Sociology*, vol, 21, pp237–267
de Beer, J. H and McDermott, M. (1989) *The Economic Value of Non-timber Forest Products in Southeast Asia*, second revised edition, Netherlands Committee for the World Conservation Union (IUCN), Amsterdam

de Merode E, Homewood, K. and Cowlishaw, G. (2003) 'Wild resources and livelihoods of poor households in Democratic Republic of Congo', ODI Wildlife Policy Briefings, no 1, November

den Adel, S. (2002) 'Use of marula products for domestic and commercial purposes by households in North-Central Namibia', CRIAA SA-DC, Windhoek and Centre for Ecology and Hydrology, Wallingford, UK, www.ceh-wallingford.ac.uk/research/winners/

Devereux, S. (1999) 'Making less last longer: Informal safety nets in Malawi', *IDS Discussion Paper 373*, Institute of Development Studies (IDS), University of Sussex, Brighton, UK

Devereux, S., Baulch, B., Macauslan, I., Phiri, A. and Sabates-Wheeler, R. (2006) 'Vulnerability and social protection in Malawi', *IDS Discussion Paper 387*, IDS, University of Sussex, Brighton, UK

Dove, M. R. (1993) 'A revisionist view of tropical deforestation and development', *Environmental Conservation*, vol 20, pp17–25

Dovie, D. B. K. (2004) 'Economic value of secondary resources in the context of total livelihoods' in M. J. Lawes, H. A. C. Eeley, C. M. Shackleton and B. G. S. Geach (eds) *Indigenous Forests and Woodlands in South Africa: Policy, People and Practice*, University of KwaZulu-Natal Press, Pietermarizburg

Dubouis, O. (2002) 'Forest-based poverty reduction: A brief review of facts, figures, challenges and possible ways forward', International Workshop on Forests in Poverty Reduction Strategies: Capturing the Potential, 2001–2002, October, Tuusula, Finland

Falconer, J. (1996) 'Developing research frames for non-timber forest products: Experience from Ghana', in M. Ruiz-Pérez and J. E. M Arnold (eds) *Current Issues in Non-timber Forest Product Research*, CIFOR, Bogor

FAO (1999) *The State of World Fisheries and Aquaculture 1998*, FAO, Rome

FAO (2003) *State of the World's Forests 2003. Part II. Selected Current Issues in the Forest Sector*, FAO, Rome

Fereday, N., Gordon, A. and Oji, G. (1997) 'Domestic market potential for tree products from farms and rural communities: Experience from Cameroon', *NRI Socio-economic Series 13*, Natural Resources Institute, University of Greenwich, London

Fisher, M. (2004) 'Household welfare and forest dependence in Southern Malawi', *Environment and Development Economics*, vol 9, pp135–154

Freese C. (ed) (1997) *Harvesting Wild Species: Implications for Biodiversity Conservation*, John Hopkins University Press, Baltimore

Frost, P. and Mandondo, A. (1999) 'Improving rural livelihoods in semi-arid regions through management of micro-catchments', *IES Working Paper 12*, Institute of Environmental Studies, University of Zimbabwe, Harare

Fusari, A. and Carpaneto, G. M. (2006) 'Subsistence hunting and conservation issues in the game reserve of Gile, Mozambique', *Biodiversity and Conservation*, vol 15, pp2477–2495

Hamilton, A. C. (2004) 'Medicinal plants, conservation and livelihoods', *Biodiversity and Conservation*, vol 13, pp1477–1517

Harsch, E. (2001) 'Making trade work for poor women: Villagers in Burkina Faso discover an opening in the global market', *Africa Recovery*, vol 15, pp6–13

Hedge, R. (2007) *Contribution of Miombo Woodlands to Household Economy and Socio-economic Determinants of Woodland Use: The Case of Mozambique*, CIFOR, Bogor

Hulme, M., Doherty, R., Ngara, T. and New, M. (2005) 'Global warming and African climate change: A re-assessment', in Pak Sum Low (ed) *Climate Change: Science, Technology and Policy for Africa*. Cambridge University Press, Cambridge

ITC (2006) 'International Trade Centre statistics', http://langues.p-maps.org/pmaps/index.php?err=sess, accessed November 2006

Juste, J., Fa, J. E., Perez del Val, J. and Castroviejo, J. (1995) 'Market dynamics of bushmeat species in Equatorial Guinea', *Journal of Applied Ecology*, vol 32, pp454–467

Kaimowitz, D. (2003) 'Not by bread alone...forests and rural livelihoods in sub-Saharan Africa', in T. Oksanen, B. Pajari and T. Tuomasjukka (eds) *Forestry in Poverty Reduction Strategies: Capturing the Potential*, European Forest Institute, Joensuu

Kamanga, P., Vedeld, P. and Sjaastad, E. (2009) 'Forest incomes and rural livelihoods in Chiradzulu District, Malawi', *Ecological Economics*, vol 68, pp613–624

Kamanga, P., Vedeld, P. and Sjaastad, E. (2009) 'Forest incomes and rural livelihoods in Chiradzulu District, Malawi', *Ecological Economics*, vol 68, pp613—624

Kanguembega, R. (2003) 'Les déterminants de l'offre des amandes de karité par les producteurs ruraux au Burkina Faso', Mémoire pour l'obtention du diplôme d'Etudes Approfondie en Economie (DEA), Unité de Formation et de Recherche en Sciences Economiques et de Gestion, Université de Ouagadougou, Ouagadougou

Kepe, T. V. (2002) 'Grassland vegetation and rural livelihoods: A case study of resource value and social dynamics on the Wild Coast, South Africa', PhD thesis, Development Studies, University of the Western Cape, Cape Town

Kusters, K. and Belcher, B. (eds) (2004) *Forest Products, Livelihoods and Conservation: Case Studies of Non-timber Forest Product Systems, Vol 1: Asia*, Center for International Forestry Research, Bogor, Indonesia

Lowore, J. (2003) 'Miombo woodlands and rural livelihoods in Malawi: An in-depth analysis and critical review based on literature research', Department of Forestry, Mzuzu University, Mzuzu

Mander, M. (1998) *Marketing of Indigenous Medicinal Plants in South Africa. A Case Study in KwaZulu-Natal*, FAO, Rome

Mander, M. and Le Breton, G. L. (2006) 'Overview of the medicinal plant industry in southern Africa', in N. Diederichs (ed) *Commercialising Medicinal Plants: A Southern African Guide*, Sun Press, Stellenbosch

Marshall, E., Schreckenberg, K. and Newton, A. C. (eds) (2006) *Commercialisation of Non-timber Forest Products. Factors Influencing Success. Lessons Learned from Mexico and Bolivia and Policy Implications for Decision Makers*, UNEP World Conservation Monitoring Centre, Cambridge

Masters, E. T., Yidana, J. A. and Lovett, P. N. (2004) 'Reinforcing sound management through trade: Shea tree products in Africa', *Unasylva*, vol 219, pp46–52

McCall, M. (1985) 'The significance of distance constraints in peasant farming systems with special reference to sub-Saharan Africa', *Applied Geography*, vol 5, no 4, pp325–345

McSweeney, K. (2003) 'Tropical forests as safety nets? The relative importance of forest product sale as smallholder insurance, Eastern Honduras', International Conference on Rural Livelihoods, Forests and Biodiversity, 12–23 May, Bonn, Germany

McSweeny, K. (2004) 'Tropical forests product sale as natural insurance: The effects of household characteristics and the nature of shock in Eastern Honduras', *Society and Natural Resources*, vol 17, pp39–56

Michon, G. (2005) 'Domesticating forests: How farmers manage forest resources', IRD-CIFOR-ICRAF, Bogor, Indonesia

Monela, G. C., Kajembe, G. C., Kaoneka, A. R. S. and Kowera, G. (1999) 'Household livelihood strategies in the miombo woodlands of Tanzania: Emerging trends', *Tanzania Journal of Forestry and Nature Conservation*, vol 73, pp17–33

Mortimore, M. (1998) *Roots in the African dust*, Cambridge University Press, Cambridge

Mutamba, M. (2007) *Farming or Foraging? Aspects of Rural Livelihoods in the Mufulira and Kabompo Districts of Zambia*, CIFOR, Bogor

Neumann, R. P. and Hirsch, E. (2000) *Commercialisation of Non-timber Forest Products: Review and Analysis of Research*, CIFOR, Bogor

Ndoye, O., Ruiz-Pérez, M. and Eyebe, A. (1997) *The Markets of Non-timber Forest Products in the Humid Forest Zone of Cameroon*, ODI Rural Development Forestry Network Paper 22c, London

Overa, R. (2006) 'Networks, distance and trust: Telecommunications development and changing trading practices in Ghana', *World Development*, vol 34, pp1301–1315

Peters, C. M. (1994) *Sustainable Harvest of Non-timber Plant Resources in Tropical Moist Forest: An Ecological Primer*, Biodiversity Support Program, Washington, DC

Petheram, L., Campbell, B., Marunda, C., Tiveau, D. and Shackleton, S. (2006) 'The wealth of the dry forests. Can sound forest management contribute to the Millennium Development Goals in Sub-Saharan Africa?', *Forest Livelihood Briefs*, no 4, October, CIFOR, Bogor

PhytoTrade Africa (2006) 'PhytoTrade Africa', www.phytotradeafrica.com

Redford, K. H. (1992) 'The empty forest', *BioScience*, vol 42, pp412–422

Rogerson, C. M. and Sithole, P. M. (2001) 'Rural handicraft production in Mpumalanga, South Africa: Organisation, problems and support needs', *South African Geographic Journal*, vol 83, pp149–158

Ruiz-Pérez, M., Belcher, B., Achdiawan, R., Alexiades, M., Aubertin, C., Caballero, J., Campbell, B., Clement, C., Cunningham, A., Fantini, A., de Foresta, H., Garcıa Fernández, C., Gautam, K., Hersch Martınez, P., de Jong, W., Kusters, K., Kutty, M., López, C., Fu, M., Martınez Alfaro, M. A., Nair, T. K. R., Ndoye, O., Ocampo, R., Rai, N., Ricker, M., Schreckenberg, K., Shackleton, S., Shanley, P., Sunderland, T. and Youn, Y. (2004) 'Markets drive the specialization strategies of forest peoples', *Ecology and Society*, vol 9, no 2, www.ecologyandsociety.org/vol9/iss2/art4/

Scherr, S. J. (2004) 'Building opportunities for small-farm agroforestry to supply domestic wood markets in developing countries', *Agroforestry Systems*, vol 61, pp357–370

Schreckenberg, K. (2004) 'The contribution of shea butter (*Vitellaria paradoxa* C. F. Gaertner) to local livelihoods in Benin', in T. Sunderland, and O. Ndoye (eds) *Forest Products, Livelihoods and Conservation. Case Studies of Non-timber Forest Product Systems,*vol 2 Africa, CIFOR, Bogor

Sen, B. (2003) 'Drivers of escape and descent: Changing household fortunes in rural Bangladesh', *World Development*, vol 31, pp513–534

Shackleton, S. E. (2005) 'The significance of the local trade in natural resource products for livelihoods and poverty alleviation in South Africa', PhD thesis, Rhodes University, Grahamstown, South Africa

Shackleton, S. E. (2006) 'Forests as safety nets for mitigating the impacts of HIV/AIDS in southern Africa', CIFOR, Bogor

Shackleton, C. M. and Shackleton, S. E. (2004a) 'Use of woodland resources for direct household provisioning', in M. J., Lawes, H. A. C. Eeley, C. M. Shackleton and B. G. S. Geach (eds) *Indigenous Forests and Woodlands in South Africa: Policy, People and Practice*, University of KwaZulu-Natal Press, Pietermarizburg

Shackleton, C. M. and Shackleton, S. E. (2004b) 'The importance of non-timber forest products in rural livelihood security and as safety nets: A review of evidence from South Africa', *South African Journal of Science*, vol 100, pp658–664

Shackleton, S. E. and Shackleton, C. M. (2005) 'The contribution of marula (*Sclerocarya birrea*) fruit and fruit products to rural livelihoods in the Bushbuckridge district, South Africa: Balancing domestic needs and commercialisation', *Forests, Trees and Livelihoods*, vol 15, pp3–24

Shackleton, C. M. and Shackleton, S. E. (2006) 'Household wealth, status and natural resource use in the Kat River Valley, South Africa', *Ecological Economics*, vol 57, pp306–317

Shackleton, S. E., Shackleton, C. M. and Cousins, B. (2000) 'Re-valuing the communal lands of southern Africa: new understandings of rural livelihoods', *Natural Resource Perspectives*, no 62, Overseas Development Institute (ODI), London

Shackleton, C. M., Shackleton, S. E., Ntshudu, M. and Ntzebeza, J. (2002) 'The role and value of savanna non-timber forest products to rural households in the Kat River Valley, South Africa', *Journal of Tropical Forest Products*, vol 8, no 1, pp45–65

Shackleton, S. E., Shanley, P. and Ndoye, O. (2007) 'Viable but invisible: Recognising local markets for non-timber forest products', *International Forestry Review*, vol 9, pp697–712

Smith, D. L. (1992) 'The impact of the friction of distance on the early settlement and development of Eyre Peninsula, South Australia', *Australian Geographer*, vol 23, no 2, pp149–157

Stack, J., Dorward, A., Gondo, T., Frost, P., Taylor, F. and Kurebgaseka, N. (2003) 'Mopane worm utilisation and rural livelihoods in Southern Africa', *Proceedings (CD) of the International Conference on Rural Livelihoods, Forests and Biodiversity, 19–23 May 2003, Bonn, Germany*, CIFOR, Bogor

Styles, C. V. (1995) 'The elephant and the worm', *BBC Wildlife*, vol 13, pp22–24

Sunderland, T. and Ndoye, O. (eds) (2004) *Forest Products, Livelihoods and Conservation. Case Studies of Non-timber Forest Product Systems*, vol 2 Africa, CIFOR, Bogor

Sunderlin, W. D., Angelsen, A. and Wunder, S. (2003) 'Forests and poverty alleviation', in *State of the World's Forests*, FAO, Rome, pp61–73

Sunderlin, W. D., Angelsen, A., Belcher, B., Burgers, P., Nasi, R., Santoso, L. and Wunder, S. (2005) 'Livelihoods, forests, and conservation in developing countries: An Overview', *World Development*, vol 33, pp1383–1402

Takasaki, Y., Barham, B. L. and Coomes, O. T. (2004) 'Risk coping strategies in tropical forests: Floods, illnesses, and resource extraction', *Environment and Development Economics*, vol 9, pp203–224

Terry, E. M. (1999) 'The economic and social significance of the handicraft industry in Botswana', PhD thesis, School for Oriental and African Studies, University of London, London

Ticktin, T. (2004) 'The ecological implications of harvesting non-timber forest products', *Journal of Applied Ecology*, vol 41, no 1, pp11–21

Twine, W., Moshe, D., Netshiluvhi, T. and Siphugu, V. (2003) 'Consumption and direct-use values of savanna bio-resources used by rural households in Mametja, a semi-arid area of Limpopo province, South Africa', Research Letters, *South African Journal of Science*, vol 99, pp467–473

UNEP (2002) *Africa Environment Outlook: Past, Present and Future Perspectives*, UNEP, Nairobi

Vedeld, P., Angelsen, A., Sjaastad, E. and Kobugabe Berg, G. (2004) 'Counting on the environment: Forest incomes and the rural poor', *World Bank Environmental Economics Series*, Paper 98, Washington, DC

Vedeld, P., Angelsen, A., Bojö, J., Sjaastad, E. and Kobugabe Berg, G. (2007) 'Forest environmental incomes and the rural poor', *Forest Policy and Economics*, vol 9, no 7, pp869–879

Walter, S. (2001) *Non-wood forest products in Africa: a regional and national review*, FAO, Rome

Wickens, G. E. (1991) 'Management issues for development of non-timber forest products', *Unasylva*, vol 42, pp3–8

Wiegers, E., Curry, J., Garbero, A. and Hourihan, J. (2006) 'Patterns of vulnerability to AIDS impacts in Zambian households', *Development and Change*, vol 37, pp1073–1092

Wiersum, K. F. and Ros-Tonen, M. A. F. (2006) 'The role of forests in poverty alleviation: Dealing with multiple Millennium Development Goals. North-South Policy Brief', Wagningen University, www.north-south.nl

Williams, V. L., Balkwill, K. and Witkowski, E. T. F. (2000) 'Unravelling the commercial market for medicinal plants and plant parts on the Witwatersrand, South Africa', *Economic Botany*, vol 54, pp310–327

World Bank (2000) *World Development Report (WDR) 2000/2001: Attacking Poverty*, World Bank Group, Washington, DC, http://siteresources.worldbank.org/intpoverty/resources/WDR/

World Bank (2005) *World Development Indicators 2005*, World Bank, Washington, DC

Wunder, S. (2001) 'Poverty alleviation and tropical forests – what scope for synergies', *World Development*, vol 29, no 11, pp1833–1917

Wynberg, R. P. (2006) 'Identifying pro-poor best practice models for commercialisation of southern African non-timber forest products', PhD thesis, University of Strathclyde, Glasgow, UK

Wynberg, R. P. and Laird, S. A. (2007) 'Less is more: Governance of a non-timber forest product, marula (*Sclerocarya birrea* subsp. *caffra*) in southern Africa', *International Forestry Review*, vol 9, pp475–489

Wynberg, R., Laird, S., Shackleton, S., Mander, M., Shackleton, C., du Plessis, P., den Adel, S., Leakey, R. R. B., Botelle, A., Lombard, C., Sullivan, C., Cunningham, T. and O' Regan, D. (2003) 'Marula commercialisation for sustainable and equitable livelihoods', *Forests, Trees and Livelihoods*, vol 13, pp203–215

5

Non-wood Forest Products: Description, Use and Management

Sheona Shackleton, Michelle Cocks, Tony Dold, Sarah Kaschula, Keith Mbata, Guni Mickels-Kokwe and Graham von Maltitz

Introduction

The FAO defines non-wood forest products (NWFPs) as all biological materials other than wood that are extracted from forests for human use. Drawing on this, an NWFP has been defined for the purposes of this book as any raw or processed product, excluding wood, that is produced from an indigenous or wild biological resource found within the African dry forests and woodlands and that is harvested for either domestic consumption or trade. In some instances, the resource may be cultivated or sourced from modified or non-natural systems (as for some edible leafy plants), but cannot be regarded as a conventional agricultural crop. Chapter 4 discussed NWFPs in the context of livelihoods and economy; this chapter is devoted to the description of some of the more important NWFPs and their management.

Wild Plant and Fungus Foods

Dry forests and woodlands in sub-Saharan Africa are an important source of numerous wild plant and fungus foods that includes fruits, leafy vegetables or edible herbs, woody foliage, roots and tubers, wild cereals and grains, seeds, nuts and kernels, edible fungi, and a range of processed products such as traditional beers and palm wines and fats and oils extracted from fruit nuts and kernels (Grosskinsky, 2000). This section focuses on three food categories, wild fruits and fruit products, green leafy vegetables and mushrooms.

Description, processing and use

Fruits

A wide diversity of dry forest and woodland species produce edible fruits. In southern Africa, O'Brien (1988) reported that of approximately 1365 woody species, some 265 (19 per cent) are edible. In the miombo woodlands of Malawi, more than 75 species of edible fruits have been recognized, although not all are in common use (Akinnifesi et al, 2004). Among the popular fruit species from the miombo region are *Uapaca kirkiana, Parinari curatellifolia, Strychnos cocculoides* (and other *Strychnos* species), *Flacourtia indica, Diospyros mespiliformis* and *Azanza garkceana*. Popular fruit species in the semi-arid regions include *Hyphaene thebaica* (dom palm), *Balanites aegyptiaca, Grewia tenax, Borassus aethiopum, Tamarindus indica, Adansonia digitata, Ziziphus* spp and *Cordia africana*. In Burkina Faso, 28 species of edible fruits have been identified (Kristensen and Balslev, 2003), while Addis et al (2005) recorded 68 species in four districts in Ethiopia. From an exhaustive list of tree species bearing NWFPs compiled for the Sudan, 90 out of 150 species provided food sources (Badi, 1993). In the savannas of northeast South Africa, Shackleton et al (2000a) recorded a total of 54 wild edible fruit species across nine villages, although most households used only between 6 and 10 of these with any regularity; 30 per cent of respondents reported that these indigenous fruits could still be found during times of drought. South African and Namibian studies indicate that *Sclerocarya birrea* (marula) is amongst the most commonly used wild fruit species in the semi-arid woodlands of these countries, with 59–77 per cent of households reportedly consuming marula fruits four to five times per week in fruiting season and some 90 per cent making marula beer (Shackleton et al, 2000a; Shackleton and Shackleton, 2005).

Data on the actual quantities of wild fruits consumed by individuals and households are scant, partly because of the snack nature of consumption and the fact that these fruits are seldom carried home. Data from ten studies in South Africa, based on informants' recall rather than direct measurement, have revealed masses of between 19 and 165kg of fruit consumed per household per year, with a mean of 104 ± 16kg (Cunningham and Shackleton, 2004). These amounts are low when compared to the tropical forests of South America, where forest Indians have been recorded as eating up to 5 tonnes of fruits and seeds per year (Melnyk and Bell, 1996).

Traditional methods of processing, preserving and fermenting wild plant foods can help to overcome some of the seasonal limitations on nutritional intakes. Besides being a valuable method of preservation, fermentation raises the levels of bioavailability of proteins, vitamins and minerals, and enriches food through the synthesis of some B vitamins (Grosskinsky, 2000). Thus, despite being alcoholic, many fruit derived wines and beers have notable nutritional value. Palm wine is a source of nicotinic acid and vitamin C, and contains valuable quantities of protein, thiamine and riboflavin (Cunningham and Wehmeyer, 1988; Ezeronye, 2004). Marula fruit juice contains approximately four times the vitamin C content of orange juice, while marula beer provides a source of B vitamins and retains valuable amounts of the fruit's rich vitamin C, protein and iron content (Moll, 1972; Sullivan and O'Regan, 2003).

Vegetables

The varying understanding and different definitions of what constitutes a 'wild' vegetable has resulted in wide discrepancies in the literature regarding the importance of these NWFPs in the diet. Restricting the definition to those plants occurring only in uncultivated or wild areas can result in a seemingly low use by local communities. For example, McGregor (1995) found that only 2 of a reported 39 gathered wild vegetable species came from the forests in Shurugwi, Zimbabwe. In Nigeria, Hart et al (2005) found that when they restricted their definition of wild vegetables to uncultivated, traditional foods, 'true' wild species were infrequently consumed, confined to small inputs in minor dishes and isolated to more remote rural areas. In Bulamogi, Uganda, Tabuti et al (2004) found that 'true' wild vegetables formed only a third of all edible wild plants used, and their consumption was linked to casual encounters, periods of food shortages, and supplements to major crops.

Studies that detect a higher number of wild foods in the diet tend to support more broad definitions of wild plant foods, and may include species that are cultivated as crops in other regions. For example, in Burkina Faso, *Hibiscus surattensis* is largely referred to as a cultivated species (Mertz et al, 2001). On the other hand, Madge (1994) refers to *H. surattensis* as a collected food which is heavily relied upon in The Gambia in drought times. In South Africa, a total of 45 wild vegetables were recorded in nine villages in the Bushbuckridge region; naturalized weeds (*Amaranthus* spp; *Conchorus* spp) and nurtured herbs within homesteads or arable fields were included in the definition (Shackleton et al, 1998). Similarly, Mertz et al (2001) identified 14 wild vegetable species consumed by two communities in the southeastern Sudanian woodlands of Burkina Faso, including *Conchorus* spp, which occurs spontaneously in disturbed areas. They found these wild vegetables constituted 35 per cent and 59 per cent of the total vegetable consumption in the two communities respectively. In an earlier study by Smith et al (1996) in Burkina Faso, wild vegetables were defined in a similar way and also found to constitute some 36

Box 5.1 Gender, wild foods and food security: An example of the Hazda in Tanzania

In a study by Murray et al (2001) on the diets of the Hazda in Tanzania, the authors used a multifaceted, ethnobotanical approach to understand the role of culture and gender in influencing nutritional intake. The authors found that Hazda men eat more meat and honey than women, while women depend more on plant foods, in particular baobab seeds. From a cultural perspective, men have the 'better deal': meat and honey are high status foods. But, from a nutritional perspective, men and women's diets are of similar value: baobab seeds are a good source of protein, fat and energy, equivalent to that of honey. Furthermore, from an access and seasonality perspective, women's diets are favoured. Although baobab seeds and fruit are not the dominant wild foods eaten by Hazda women, they can be consistently collected throughout the year from a variety of locations. These range from the baobab trees themselves over the baobab's long productive season, to foraging for discarded seeds in baboon dung piles during non-productive periods. When these factors are weighed-up, the authors conclude that the baobab provides the most important and nutritionally reliable food in women's diets, even though other plants are eaten in significant quantities, and other sources of available foods may be more nutritionally dense.

per cent of total vegetable consumption. Wild vegetables are usually prepared as a relish or sauce, sometimes mixed with onions, tomatoes, chillies, ground nuts and other condiments, and served with the staple.

The traditional knowledge to identify, harvest and process/prepare wild vegetables lies mainly with women, and is transferred by mothers to their daughters. Men will eat wild vegetables but usually when there are no alternatives and they rarely collect or prepare these (Shackleton and Shackleton, 2004; Box 5.1). Women also dominate in the trade in fresh and dried leafy vegetables and herbs, often in urban markets, as well as the sale of prepared foods using wild species (Ngwerume and Mvere, 2000).

Mushrooms

The African dry forests and woodlands are host to a large diversity of mushrooms, with over 30 species of edible mushrooms, including the highly palatable and most commonly consumed genera of *Cantharellus* and *Russula*, found in the miombo woodlands. *Amanita* and *Termitomyces* species are also consumed and traded. The occurrence of mushrooms is highly seasonal and erratic. For example, in the miombo woodlands, almost 80 per cent of production occurs in the wet months of January to March. This, combined with their essentially low calorific value, means mushrooms are relatively unimportant for food security, although high quantities of protein (including the essential amino acid lysine), trace minerals (such as selenium), and vitamins (especially the B vitamins) do help contribute to dietary quality (Chang, 1980). In Ethiopia,

Addis et al (2005) found that recourse to wild fungi use did increase during periods of food shortage and famine, but that a few of the reportedly edible species cause health problems sometimes leading to fatality.

Resource management aspects

Fruits

Collecting the fruits of established woody species is unlikely to have the same negative consequences as uses that involve more destructive harvesting. For example, for relatively abundant, highly productive species such as marula, it is generally held that the nature of harvesting, which involves the collection of fruits from the ground following ripening, does not present a direct environmental risk (Emanual et al, 2005). However, for wild fruit species under communal tenure, reports of declines in availability are not uncommon. In South Africa, 36 per cent of respondents in a survey by Shackleton et al (2000a) believed wild fruit resources had declined over the previous five to ten years. In the Zimbabwean miombo, Tyynela and Niskanen (2000) noted perceived declines in wild fruit trees similar to those reported for other NWFPs.

Vegetables

Within the dry forest and woodland areas 'wild' leafy vegetables and edible herbs are harvested from a range of habitat types and land use mosaics. These represent a continuum from wild habitats to cultivated areas as illustrated in Figure 5.1. Levels of husbandry of these wild foods decreases from cultivated areas to no tending in wild habitats.

Because many wild vegetables are annual weedy species or fast growing perennials, there is generally less concern regarding the unsustainable use and declining abundance of these plants than for other NWFPs. Many traditional vegetables have been mainstreamed into the wider economy by means of commercialization and crop improvement schemes. In other parts of eastern and western Africa, traditional vegetables such as *Hibiscus* spp, *Solanum* spp,

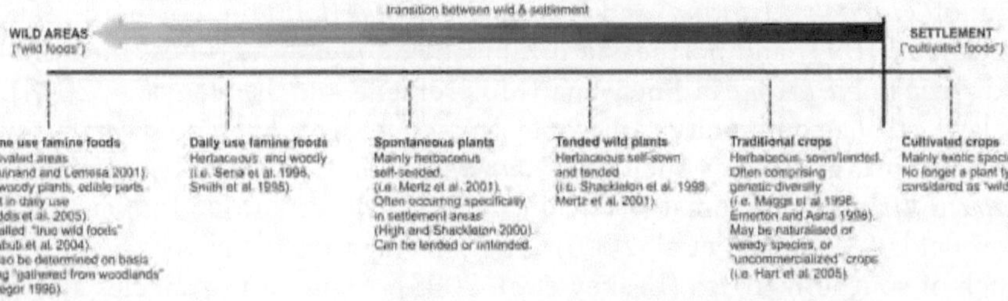

Figure 5.1 *A continuum of dry forest and woodland habitats and transformed landscape mosaics from which wild vegetables and edible herbs are harvested*

Source: Chapter first author

Vigna spp and *Amaranthus* spp have been successfully commercialized (Rubaihayo, 2004). Under such circumstances there is less reliance on natural production.

Mushrooms

Due to problems with supply consistency and the already optimal utilization of available wild mushrooms in informal markets, it is unlikely that wild gathered mushrooms will be a viable option for formal commercialization. The FAO has assisted numerous African countries in developing cultivated mushroom production, but with marginal success (Sene, 2000). Increasing the effectiveness of collection and processing of wild mushrooms would be an important step, both to expand the shelf-life, thus allowing the exploitation of wider markets and a longer trading season and because processed mushrooms have a higher value in local and external markets than fresh ones (Chirwa et al, 2006).

The effects of tree cutting and forest/woodland clearing on mushroom production have not been adequately studied. However, reduction in production can be expected for mushrooms that live in symbiosis with roots of host trees (mychorrhizae), especially if cutting and clearing result in death of the host trees. Under such circumstances, preserving host trees as standards, as is done for fruit trees, when clearing woodland might maintain mushroom production by symbiotic species, such as *Amanita zambiana* and *Cantharellus* spp in transformed landscapes.

Overview

The above findings need to be interpreted in a broader context of land degradation, overall population increases, poverty, land clearing and other drivers of environmental change such as HIV/AIDS. For small-scale harvesting of fruit for subsistence and local trade, there is evidence to suggest that community management practices can be effective. Seedling nurturing, deliberate planting and taboos regarding the harvesting of fruit species for timber are common in even the most heavily denuded areas, and may even persist when fruit trees are known to decrease arable crop yields. This has been observed with the protection of *Uapaca kirkiana* and *Parinari curatellifolia* in northern Zambia (Packham, 1993) and northeastern Zimbabwe (Tyynela and Niskanen, 2000), and *Azanza garkeana* in Botswana (Mojeremane and Tshwenyane, 2004). In Malawi, strong community management practices have likewise been observed, with important species such as *P. curatellifolia*, *Strychnos cocculoides* and *Uapaca kirkiana* customarily left uncut and scattered around homesteads or crop fields (Akinnifesi et al, 2006). The same applies to *Sclerocarya birrea* in much of southern Africa (Leakey et al, 2005). Many of the species also have good inherent regenerative potential. In common lands in Zimbabwe, Campbell (1987) found that wild fruit trees, as non-dominant woody species, can rapidly make up for high demands on their resources through opportunistic regeneration following land clearing, disturbance and selective harvesting by local users.

Box 5.2 Destructive harvesting of palm for wine

The most common way of tapping palm sap for wine production is by a method that generally results in the death of the stem, where a horizontal incision is made through the apical meristem (palm heart). Regular renewal of the incision (two to seven times a day) to stimulate sap flow eventually causes irreversible damage to the palm heart. For the single stemmed species that dominate the western African palm wine trade, such as *Raphia hookeri* and *Borassus aethiopum*, this can mean the entire plant dies following just a few days of tapping. Sambou et al (2002) note that in northwestern Guinea, full-time harvesters of *B. aethiopum* kill on average 56 trees a year. The authors note, however, that this 'lethal' practice is not widespread in regions where palm wine use is strongly engrained in the cultural history of the community. They cite ethnobotanical studies from Senegal, Ivory Coast and Burkina Faso, which have shown that tappers have avoided plant mortality by using less damaging techniques such as collecting sap from the inflorescences of female palms, tapping the palm heart by piercing the leaf sheaths or ceasing tapping activities before the apical meristem is completely destroyed. Amongst the Serer ethnic group in Senegal, where destructive harvesting is used, palm populations are maintained through the sowing of seeds.

In southern and eastern Africa, the much smaller palms *Hyphaene coriacea* and *Phoenix reclinata* are commonly exploited. Cunningham (1990) notes that these species are nearly always tapped through extreme apical meristem tapping, and that the tapped stem almost always dies. However, as *H. coriacea* and *P. reclinata* are generally multi-stemmed palms, the death of the tapped stem does not necessarily mean that the entire plant is destroyed. Moreover, these multi-stemmed species tend to coppice well. For these species, threats to sustainability may come from changes induced by tapping to plant size and age class distribution. Because tappers favour larger stems, subsequent declines in larger size classes can severely limit productivity and ultimately sustainability of tapping activities in the harvesting areas (Cunningham, 1990; Mollet et al, 2000).

However, increasing commercialization does raise concerns regarding the sustainability of resource harvesting as do competing uses for other products such as fuelwood, fibre, bark or sap.

Palm trees are particularly vulnerable because of the multiple use value of many of these species and the high market demand for palm wine, the production of which often involves destructive harvesting (Box 5.2).

Insect Foods

Description, processing and use

Insects are harvested from the dry forests and woodlands of sub-Saharan Africa primarily as food and medicine, although some produce important products, such as honey, beeswax and propolis (see Box 5.4). The use of insects as human food (entomophagy) dates back to time immemorial and is a custom that is deeply ingrained in the culture of many African ethnic groups (Mbata et al, 2002). Insects form a deliberate part of the diet of most rural, and to a lesser extent, urban dwellers, and are cherished as occasional snacks or delicacies, as well as dietary supplements served as relishes. The diversity of insect species consumed in sub-Saharan Africa is vast. Important commonly used edible insects include: termites (Isoptera); grasshoppers, locusts, crickets and katydids (Orthoptera); cicadas (Homoptera); bugs (Hemiptera); beetle and weevil grubs (Coleoptera); lake flies (Diptera); caterpillars (Lepidoptera); and bees and ants (Hymenoptera). Some insects are consumed whole, while others provide edible parts. About 80 per cent of all insect species consumed are in their immature stages, as eggs, nymphs/larvae or pupae (Ramos-Elorduy, 1996).

Insect foods are sourced from dry forests, woodlands and associated grasslands, as well as from cultivated and fallowed crop fields on private, communal and sometimes, even state lands. The abundance of edible insects and the quantities actually harvested vary greatly in relation to the insect species in question and the prevalence of its host plant(s) in a given area. Yields of as much as 3.3kg (dry weight) per host plant have been reported for the caterpillar *Imbrasia (Nadaurelia) oyemensis* on *Entandrophragma congolense* (sapelli) in western Africa (N'Gasse, 2003). In Malawi's miombo woodlands yields of 14.13kg per hectare of the caterpillars *Gynanisa maia* and *Imbrasia belina* have been recorded (Munthali and Mudgodgo, 1992).

Women, sometimes assisted by children, are mainly responsible for harvesting and processing edible insects for home consumption and/or sale, with a few exceptions. In many areas women harvest and process termites (Gelfand, 1971; FAO, 2004), grasshoppers (van der Waal, 2004) and caterpillars (Maviya and Gumbo, 2005; Fromme, 2005; Chileshe, 2005). Old women in Zimbabwe harvest soldier termites during the hot dry season for consumption (Wilson, 1990). Often men get involved in harvesting insects only when strenuous labour is required, such as when large branches or whole host trees need to be cut or felled to make caterpillars accessible for collection, or when traps have to be set to catch termites and grasshoppers (Wilson, 1990; Yagi, 1998; van der Waal, 2004). The exception is the harvesting by men of African palm weevil grubs (N'Gasse, 2003). In order to extract the grubs from the host plants, men cut and split the trunks of palm trees. Other insects such as the ground-dwelling tobacco cricket are hunted/collected by boys from their burrows in the soil (Chileshe, 2005).

Preparation of insects for consumption varies among different ethnic groups. Some insects, such as the thief ant are, in some areas like Zambia, eaten raw and/or fried in their own fat and served as snacks or relish (Mbata, 1995). Termites are generally salted in a little water in a pan/pot over a fire and then fried in their own fat, followed by brief sun drying and winnowing to remove any remaining wings and then served as snacks or side dishes (Mbata, 1995). The legs and wings and, for large specimens, also heads together with attached guts, are removed from grasshoppers, locusts, crickets and katydids before being roasted and eaten as snacks or cooked in water with salt for use as side dishes (van der Waal, 2004). The stinkbug, *Encosternum delegorguei* in South Africa, is eaten dry as a snack or fried in salt and a little water and consumed as a side dish (Toms et al, 2003).

In all countries, immediately after being gathered, caterpillars are eviscerated and then either taken home fresh for cooking in water or oil as relish, or they are sun- or smoke-dried for future consumption and/or sale (Maviya and Gumbo, 2005; Allotey and Mpuchane, 2003; Mbata et al, 2002). In Zambia, following evisceration, caterpillars are roasted in hot ash prepared from bonfires set in the woodlands, until hairs and spiny body adornments burn off and the caterpillars become hardened and then they are sun-dried. Caterpillar processing often takes place in the forest while collectors reside in temporary camps (Mbata et al, 2002). Sun- and smoke-dried caterpillars are handled the same way when being prepared for consumption; they are soaked in warm water for softening before being fried or boiled with various types of condiments (chilli, onion, pepper, tomato, pimento, etc.) to make a snack or relish. A groundnut meal may also be added to caterpillar relish in northern Zambia (Chileshe, 2005). Preparation of beetle and weevil grubs is similar to that of caterpillars.

The amounts of insect foods used vary depending on factors such as region, time of year, purpose for exploiting insects, economic standing and the food security situation of individual households. More insect foods are used during insect harvesting seasons than at other times of the year. In Bangui, Central African Republic, for example, an individual consumes approximately 131g of fresh caterpillars per day during the harvesting season, as opposed to 83g during other times of the year (N'Gasse, 2003). For edible grasshoppers, which are available throughout the year in the Limpopo Province of South Africa, in a good, wet year, a person will consume as much as 19.1 grasshoppers (about 14.1g) per day (van der Waal, 2004). Generally, poor individuals and households, as well as households that harvest the insect resource for sale, use more insect foods than the more well-off. Consumption of insect foods is also higher during the hunger period. In Zambia, the first two weeks of the caterpillar-harvesting season, which coincides with the start of the hunger period, are used to collect and stock caterpillars for home consumption.

Resource management aspects

The exploitation of edible insects involves complex knowledge-practice-belief systems among indigenous people, and often these are linked to the worship of ancestral spirits. The knowledge refers to the traditional knowledge that indigenous people possess on the edible insect resource, which is transmitted from parents to offspring through oral education. The practice part, refers to the practical skills in harvesting, processing and management of the resource, which is learnt hands-on by children. The belief component links to the role of insects in people's cultural practices and values. In the Mpika district of northern Zambia among the Bisa people (Mbata et al, 2002) and the Bemba people (Chileshe, 2005), for example, the harvesting of the two most popular commercial species of caterpillars in the country, *Gynanisa maja* ('*chipumi*') and *Gonimbrasia zambesiana* ('*mumpa*'), is linked to the worship of ancestral spirits. It involves specific ceremonies, rituals, rules and taboos. Similar systems occur elsewhere in countries with dry forests and woodlands in sub-Saharan Africa, such as among the Venda of South Africa (van der Waal, 2004) and the Yansi of the Democratic Republic of Congo (DRC) (Muyay-Tango, 1981).

Harvesting techniques for edible insects vary depending on the species being exploited and the substrates from which they are collected. Most are handpicked, although some may be gathered by cutting branches and/or felling host plants. The palm weevil is obtained by cutting down the host palms and splitting their trunks to extract the grubs (López and Shanley, 2004). These destructive harvesting techniques have negative impacts on forest biodiversity and can lead to the gradual disappearance of the insects in question, especially those with limited numbers of alternative host plants. Impacts of harvesting on the edible insect resource itself are variable depending on group and/or species. For many species, such as the palm weevil, harvesting of the grubs does not endanger weevil populations as grub production by females, which deposit as many as 800 eggs each, continues throughout the year. Regarding edible caterpillars, the situation is different. Overexploitation leads to a decline in abundances and even extinction of species in some habitats (Mbata et al, 2002; Chileshe, 2005). Bartlett (1996), for example, reported that due to overharvesting the mopane worm no longer occurs in some areas of Botswana.

Opportunities exist for increased production of insect foods through mass rearing. The mopane worm, for example, has already been successfully reared under laboratory conditions (Allotey et al, 1996; Greyling and Potgieter, 2004). What is now required is the adoption of the developed technologies to mass rearing situations. This could help solve the problems of fluctuating supply to the market (Greyling and Potgieter, 2004).

Many communities in sub-Saharan Africa have traditional practices for managing the insect food resources to ensure sustainability. These stem from customary regulations that determine:

- what developmental stages of edible insects are to be harvested;
- when to harvest;
- how long to harvest;
- how to ensure recruitment of the resource for the next harvest.

The presence of appropriate developmental stages to harvest, for example, is determined through continuous monitoring of the habitats. In Zambia, village scouts monitor the miombo woodlands and, when the correct stages appear, the traditional authorities are notified. The chief or other senior traditional leaders determine the start and stop dates for the harvest (Mbata et al, 2002; Chileshe, 2005; Maviya and Gumbo, 2005; Holden, 1991). The stop signal, in particular, is very important and when adhered to strictly by community members, ensures recruitment of caterpillars for the next harvesting season as enough caterpillars are left to pupate. Traditional beliefs and taboos associated with harvesting ensure discipline among harvesters and minimize damage to the forest resource and the environment. Other traditional practices for caterpillars include translocation of early stages of the caterpillars found on host plants far away from homesteads to host plants near homes (Silow, 1976; Malaisse, 1997; Maviya and Gumbo, 2005).

Honey and Beeswax

Description, processing and use

Honey bee products tend to be relatively more important in some parts of African dry forests and woodlands than others, notably in eastern and southern Africa (Walter, 2001). The miombo woodlands stretching across Angola, Zambia, southeastern DRC, western and southern Tanzania, northern Mozambique and northern Malawi provide ideal ecological conditions for the honey bee. The dominant miombo tree species of the genera *Julbernardia*, *Brachystegia* and *Isoberlinia* are good 'bee trees' providing abundant nectar, while species of *Parinari*, *Cryptosepalum*, *Guibourtia*, *Marquesia* and *Syzygium* ensure availability of bee fodder between the main seasons. In Zambia, the woodland indicator for good honey and beeswax production is given as 60 per cent main-flow species, 30 per cent second-flow species and 10 per cent gap-filling species (Holmes, 1964). Although agricultural crops (e.g. maize, sunflower, beans, pumpkin and sweet potato) and fruit trees (e.g. mango, guava) provide an important seasonal source of pollen and nectar, the bulk of bee fodder is derived from the woodlands.

Honey and beeswax are the main products harvested from the honey bee, *Apis mellifera*; other products include pollen and brood comb. The products are obtained either by honey hunting in the wild or beekeeping (apiculture). Historically, gathering of honey from feral colonies (honey hunting) was the more widespread practice and evidence from 16 dry forest and woodland countries shows this as a major human activity (Crane, 1999). In some cases honey hunting coexisted with apiculture which was a source of livelihood to specific ethnic groups (Fichtl and Adi, 1994; Makungwa, 1997; Deffar, 1998; Hussein, 2000). African honey hunters find wild hives by observing bees or following honey guide birds (*Indicator* spp). According to Crane (1999) the *Indicator* spp covered most of sub-Saharan Africa, south of the Sahara desert to northeast Namibia and Botswana. Practices developed of subduing wild bee colonies using narcotic and tranquilizer smoke derived from selected plants (Crane, 1999).

In some situations bees were suffocated, burnt and killed as honey hunters sought to pacify bees. In addition, tree trunks with nests were cut open as a means of accessing honey and this is a widespread approach to honey hunting which continues in many parts of sub-Saharan Africa even today. Honey hunting and beekeeping often coexist in adjacent communities. In Zambia's North-Western Province, the '*Lunda*' and '*Luvale*' people are accomplished beekeepers (Box 5.3), whilst the neighboring '*Kaonde*' prefer honey hunting.

Traditional beekeeping technology evolved in different ways depending on the environment. Hives are made from locally available materials such as logs, bark, twigs, bamboo, straw, reed, grass, calabashes and clay pots, mud and animal dung (Hussein, 2000). In woodland areas, bark and logs appear to be the preferred hive material. Nowadays, beekeeping is becoming increasingly widespread even in non-traditional areas. In the 1960s and 1970s, forestry and agriculture extension services promoted new plank hive technologies and intensive management practices. As government extension services weakened in the 1980s, the thrust of beekeeping promotion was taken over by various projects and non-governmental organizations (NGOs) with the aim of establishing income-generating opportunities amongst forest-dependent communities. Beekeeping is usually a subsidiary economic activity to farming. Most beekeepers in sub-Saharan Africa are small-scale producers producing a few buckets of honey per annum for home consumption and sale (Box 5.3).

Resource management aspects

Beekeeping is often perceived of as a benevolent, income-generating activity, intended to reduce rural poverty without harming the environment (Bees for Development, 2006). Whilst beekeeping often does have a positive impact on the environment, commercial beekeeping may also consume environmental resources as demonstrated by the Zambian example (Box 5.4).

Beekeepers across sub-Saharan Africa rely on access to the forests and woodlands for bee foraging. Depending on the forest type and condition, the optimal stocking rate of productive bee colonies ranges from 2 to 15 hives per km^2 (Hausser and Mpuya, 2004). Beekeepers rarely have control over other forest resource users and have few means of protecting their interests. Dry forests and woodlands are 'commons' under various forms of legal jurisdiction ranging from wholly state-owned land in Mozambique to various forms of traditional customary tenure in Zambia. With the exception of Tanzania, the

Box 5.3 Beekeeping in North-Western Province, Zambia

Guni Mickels-Kokwe

In 2004, in North-Western Province, the most prominent traditional beekeeping area in Zambia, the number of beekeepers was estimated at 18,000 (Mickels-Kokwe, 2004b). Nearly all of them men, the beekeeping households account for between 30 and 50 per cent of rural, forest-dependent households (Guveya, 2006) and traditional beekeepers consider beekeeping the second most important household source of income after farming. Beekeepers own on average between 20 and 75 bark hives, with an occupancy rate of 45–80 per cent. The hives are hung on branches in the forest, at a distance of up to 40km from the homestead. While beekeeping is fickle, and yields of honey vary depending on the rains and the flowering of the forest trees, it provides much desired cash income at very good returns on labour time. Almost all honey is sold: only 5–10 per cent is retained for home consumption (Mickels-Kokwe, 2004b). The average beekeeper produces about 100kg of liquid honey in a year, with the area having exports of about 400 tonnes in 2005. Honey beer, 'mbote', is made from the leftover honey rinsed from the combs in the process of making wax.

The labour requirements in traditional beekeeping are minimal. A beekeeper can construct several bark hives in a day. Most time is spent walking in the forest, going to hang, inspect or crop the hives. The heaviest work is transporting cropped honey in buckets, each weighing around 30kg, to the homestead by head-load. Transport is also required at the production stage and involves cash input, hiring helpers and bicycles to carry and transport the product.

The fact that hives are taken deep into the forests is perceived to restrict women from engaging in beekeeping. Women interviewed in 2004, expressed dislike of the idea of climbing trees, having to leave the homestead chores to go and camp in the forest, and were genuinely doubtful of the acceptance from their men folk of such behaviour. Instead, male beekeepers suggested that women be owners of hives and hire men to manage forest hives on their behalf.

legislation in most countries only marginally recognizes beekeeper access, usufruct and/or management rights to forests (Wily and Mbaya, 2001; Hausser and Mpuya, 2004).

The promotion of sustainable beekeeping has taken two main routes in eastern and southern Africa over the last few decades. In the first, producer and forest-based initiatives have aimed at strengthening traditional beekeepers forest rights and promoting the linkage between beekeeping and sustainable forest management. Examples include forest and product certification (Muzama Crafts Ltd, 1996; Kasuta et al, 2001; IUCN, 2002) and participatory forest management initiatives, such as joint forest management in Zambia and the bee reserves in Tanzania (Chiulukire, 2001; Hausser and Mpuya, 2004). Many of these initiatives have attempted to address the complex relationship between beekeeping and sustainable forest management (Box 5.5).

Box 5.4 Bee hives and impacts on the environment in Zambia

Despite modern box hive technology being introduced to Zambia in the 1970s to increase yield levels and to support forest conservation, beekeepers still prefer to use traditional bark hives. Box hives require a significant capital investment, more labour time in hive management and pose problems of pest management compared to bark hives. Cheap and easy to make from locally available materials, bark hives need minimal maintenance. Hung in the trees, the hives are attractive to wild bee colonies giving a high occupancy rate. Bark hives yield an average of 7.4kg honey per hive (Clauss, 1992). In North-Western Province, beekeepers make up for the limited yield by increasing the number of hives. The bark hive is one of the key elements in the supply chain contributing to the success of the organic honey export industry (Mickels-Kokwe, 2004a). At forest level, however, the use of the bark hive is controversial. Harvesting of hive material may have several negative consequences.

Five of the six tree species preferred as bark hive material are also important sources of nectar, among them *Julbernardia paniculata*. Usually only one, at the most two, hives are made from each tree. Therefore, Clauss (1992) estimated that an annual replacement rate of 30 hives per beekeeper resulted in the cutting down of about 351,000 nectar-yielding trees, on average 7.7 trees per km^2 per annum. Beekeepers look for trees in the age category 30–40 years to make bigger hives. Unfortunately, these are the very trees that produce the highest volume of nectar (Holmes, 1964). Heavy harvesting of mature trees may therefore lead to a gross loss of available bee forage. The selective harvesting of trees with a cross grain bark structure, an interwoven fibre pattern (Clauss, 1992), not only affects bee forage resources but also impacts on biodiversity. The miombo species with an interwoven fibre pattern constitute a significant genetic resource and heavy harvesting may lead to loss of genetic diversity. Furthermore in some areas, the shortage of hive material has resulted in bark harvesting from trees younger than the traditionally preferred age classes. This inevitably will increase the pressure on the regeneration of hive species. Strategies for sustainable harvesting of bark material therefore become critical elements in bark-hive beekeeping for sustainable forest management in Zambia.

Non-governmental and project initiatives, often in non-traditional beekeeping areas, have been directed at promoting beekeeping in communal lands, where bee forage rights are less controversial, usually tied to land rights determined through title, settlement or cultivation (Bradbear, 2004). Significant pro-poor benefits may be realized from beekeeping in this context, as it is accessible to vulnerable households, has low start-up costs, is gender-inclusive and has multiple market opportunities (Bees for Development, 2006). Poor forest-dependent communities, however, need support in establishing and maintaining bee forage rights outside the immediate vicinity of their homesteads.

The Beekeepers Association in Zambia has been instrumental in promoting beekeeper rights in Zambia's North-Western Province. The *modus operandi* of the North-Western beekeepers is vested in a mix of traditional authority and market mechanisms. The legal basis for the operation is the Beekeepers

Box 5.5 NWFP commercialization, conservation and livelihoods: An example from Malawi

Todd R. Johnson, Senior CBNRM Specialist,
Development Alternatives, Inc.

One of the foundation hypotheses of community-based natural resource management (CBNRM) is that tangible benefits accruing to rural communities from improved management will provide the incentive for further improving management. This implies that enterprises based on sustainable utilization of natural resources must be the driver of both rural economic growth and sustainable management of those natural resources. However, achieving sufficient scale of business in rural communities to be profitable is a recurring challenge in Africa.

The purpose of USAID/Malawi's Community Partnerships for Sustainable Resource Management (COMPASS II) project is to enhance household revenue from participation in CBNRM initiatives that generate income as well as provide incentives for sustainable resource use in Malawi. One of the ways that COMPASS II is achieving this purpose is through identifying and developing viable business models for small-scale commercialization of natural resource-based products that meet the selection criteria.

The criteria that COMPASS II uses in screening potential natural products or NWFPs are the following:

1. Have verified domestic and/or external markets sufficient to absorb increased production.
2. Have the possibility of significantly increasing the income of hundreds or thousands of households by an average of US$100 per year.
3. Form the basis for vibrant small-scale commercial production that provides incentives to revive rather than diminish the natural capital base.
4. To the extent possible, provide business opportunities for HIV-affected and other marginalized households (such as those headed by women or youth).

Once products are identified, the best available technical information is compiled into easily accessible formats (radio, print, video, drama) and disseminated as widely as possible, including to specifically targeted communities where opportunities for transformative utilization and reinvestment in natural resources are greatest. '*Chuma Chobisika*' (Hidden Treasure) radio and television programming has very successfully increased the public's awareness of opportunities from natural resource-based products. The print and video materials aim to expand the production base while also increasing product quality. The effectiveness of these informational materials is augmented by work at other leverage points along the value chains, such as input supply, product consolidation and distribution, value-added processing and market functionality, including access to credit from commercial banks and microfinance institutions. The most widely disseminated and well-known of the informational materials is a series of high-quality extension materials branded as Malawi Gold Standard production systems, consisting of print and DVD trainers' guides, illustrated practitioners' handbooks and producer group-level business management plans. The first title in the series, the Malawi Gold Standard Honey Production System was published in mid-2006 and is being sold at reprinting cost to beekeepers, private and public sector extension agents, and NGOs interested in promoting beekeeping.

COMPASS II is targeting 4000 beekeepers to adopt the Malawi Gold Standard Honey Production System by 2009. Each Gold Standard beekeeper may produce more than 200kg of honey per year, meaning a total of 800 tonnes worth US$600,000 (using a farm gate price of US$0.73 per kg), from a total apiary of just 400 hectares. This is a significant change from 2004 production of about 60 tonnes total from subsistence-level beekeepers earning less than US$65,000. Concurrent efforts at identifying the most productive bee pastures in the country are guiding extension agents toward the areas of highest return on investment.

Unfortunately for Malawi's beekeepers, the competition for their miombo woodlands is from charcoal producers. However, the latter earn a one-time income of only about US$150 per hectare versus the possibility of more than US$1000 per hectare every year from honey. This is one example of where conservation can indeed pay its own way.

Association which draws its' mandate from the beekeepers' biannual conferences, the legally registered North-Western Bee Products, the organic certification upheld by regular inspections from the UK Soil Association, and the close collaboration with local chiefs, one of whom chairs the association. Therefore, in 2002, the beekeepers warded off a threat of large-scale commercial farming from entering a forest zone certified organic, not by following official government procedures, but by presenting a plea to Chief Kanongesha, who refused to grant land rights to the investor (B. Malichi, personal communication).

In Tanzania, recent legal instruments such as the 1998 National Beekeeping Policy, the 2001 National Beekeeping Programme and the 2001 Beekeeping Act provide for the establishment of government and private 'bee reserves' to be managed jointly by stakeholders. The objective is to ensure not only species and habitat conservation, but also to empower beekeepers to address problems of reducing availability of bee fodder and disappearing wild bee colonies arising from land clearing, logging, fuelwood harvesting, bush fires, industrial expansion and uncontrolled use of pesticides. This is done in a framework of community-based natural resource management. Access to reserves is limited to a specified group of users who hold their right in common (Hausser and Mpuya, 2004). Initial results have been positive. As access, control and user rights are strengthened, management improves. Significant production increases have been recorded (Hausser and Mpuya, 2004). For example, honey production increased from 7 tonnes in 2001 to 200 tonnes in 2003 among beekeepers in 13 villages in Inyonga Division, mainly as a result of capacity-building support to institutionalized beekeeper interest groups following a village land demarcation process. Some of the challenges encountered include the need to coordinate natural resource management activities, the conflict over access to

Box 5.6 Honey Care Africa Ltd: A successful model for promoting an equitable and sustainable honey trade

Guni Mickels-Kokwe

Honey Care Africa Ltd is a Kenyan owned private company, established in 2000, that has helped to initiate a number of beekeeping projects in rural communities throughout Kenya. Four years later, the company had started expanding into Tanzania. By 2006, it employed close to 50 staff and had helped 9000 beekeepers earn a supplementary income of US$180–250 per annum (IFC, 2006). The company operates within a tripartite model between the private sector, development organizations and rural communities, drawing upon the core competencies of each. The private sector provides the sales and marketing channels for beekeepers and ensures that the project operates within realistic market expectations and conditions. The company guarantees to purchase all honey at a fair and fixed price. The development organization, usually an NGO, community-based organization or project, provides outreach into the rural areas and a conduit for the private sector operator to access communities and plays the role of primary arbitrator to ensure that an exploitative relationship does not develop. It may also provide start-up financing, training and assist in promoting sustainable natural resource management. The third partner comprises small-scale producers. This model creates a favourable environment for them to enter into beekeeping, through a combination of adequate training and extension, access to soft loans to acquire hives and equipment, access to a guaranteed market for the produce at a mutually acceptable price and cash-on-the spot mode of payment (Jiwa, 2002; Maurice, 2004). Honey Care Africa Ltd has obtained numerous international awards. As a socially responsible corporate player, the company has engaged in activities that are outside the traditional scope of private sector operators through its relationships with its partners, e.g. community mobilization, credit, training, production of training materials and the manufacture of beekeeping equipment. A major contributor to the success has been the company's ability to open new domestic markets. Many of the success elements of Honey Care Africa Ltd in Kenya correspond to findings from a study of best practices for a similar honey marketing chain in Zambia (Mickels-Kokwe, 2004a).

resources between beekeepers and trophy hunting companies, and the need to improve communication between government and communities and disseminate information on laws and legislation (Hausser and Savary, 2005).

The other type of initiative has aimed at forming alliances between beekeepers and socially responsible private sector companies, often mediated by local and international projects and NGOs. Examples are Honey Care Africa Ltd (Box 5.6) established in Kenya in 2000 (Maurice, 2004) and Forest Fruits Zambia Ltd established in 1998 (Mickels-Kokwe, 2005), which have successfully accessed third party organizational development support and financing to build long-lasting relationships with rural producers.

Medicinal Plants

Description, processing and use

Wild plants have been recognized as being an important component of health care throughout human history. In 1999, the FAO identified medicinal plants as being among the most valuable non-wood forest products (Osemeobo and Ujor, 1999). It is currently estimated that as many as 35,000 to 70,000 species of plants worldwide have been used at one time or another for medicinal purposes (Hamilton, 1992), resulting in thousands of kilogrammes of medicinal plants and/or their parts being collected and used everyday (World Bank, 1992; Lambert, 2001). Of the 40,000 or so flowering plants found on the African continent, an estimated 15–25 per cent are used in traditional medicines (Maundu et al, 2005).

Traditional medicines may be collected by users themselves, bought from traders or administered through consultation with a traditional healer. In many countries large markets for these NWFPs exist. For example, the informal trade in medicinal plants and products in southern Africa is dominated by between 400,000 and 500,000 traditional healers. The volume of plant material traded is estimated to be between 35,000 and 70,000 tonnes per annum, with a market value of US$75–150 million (Mander and Le Breton, 2006).

All plant parts are used for traditional medicines. These include roots, corms, bulbs, tubers, bark, wood, leaves, flowers, spores, fruits, seeds, seedlings and latex. Within the same plant the use of parts may vary depending on the treatment of the ailment (Osemeobo and Ujor, 1999).

Despite being such an important NWFP, information on the scale of use and the importance and value of medicinal plants to user households, traders and other actors is sparse. To date most research initiatives into the African medicinal plant trade have been carried out in southern Africa, including works by Cunningham (1991), Dauskardt (1990), Mander (1998), Williams et al (2000), Botha et al (2004), Dold and Cocks (2002), and Cocks (2004). Organized and documented information on the use and marketing of medicinal plants in other parts of dry forest and woodland regions in sub-Saharan Africa is largely fragmentary with the exception of some case material from Kenya, Nigeria (Osemeobo and Ujor, 1999) and Ethiopia (Deffar, 1998).

There are few quantitative data that record the amounts of medicines consumed by user households on an annual basis, and even fewer that attempt to place an economic value on this. In Eritrea it has been recorded that rural households consumed, on average, 6.8kg of medicinal plants each year (Araia, 2005). In South Africa, rural households utilized a mean amount of 3.9kg per annum and urban households 2.9kg. These quantities equate to approximately US$12.30 in monetary terms for urban households and US$8 for rural households. Only a slight decrease exists between the amount of material utilized between rural and urban areas, demonstrating that the use of medicinal plant material does not significantly decrease with increased urbanization (Cocks and

Dold, 2006). Mander (1998) has estimated that the average mass of medicinal plant material dispensed per visit to a traditional healer is 216g.

Resource management aspects

Throughout the African dry forest and woodland regions medicinal plant resources are sourced primarily from communal lands, but also private farms, protected areas, state forests and commercial forestry estates. For the latter, harvesting is often undertaken without the landowner's or state's consent and, being illegal, is done at the risk of being arrested by authorities (Mander, 1998; Cocks, 2004). Research results across the region attest to increasing harvesting pressures on traditional supply areas linked to a growing shortage in supply of popular medicinal plant species (Cunningham, 1991; Mander, 1998; Williams et al, 2000; Dold and Cocks, 2002; Maundu et al, 2005). In southern Africa this has led to the virtual extinction of species, such as wild ginger (*Siphonochilus aethiopicus*) and pepper-bark tree (*Warburgia salutaris*) outside of protected areas (Maundu et al, 2005). A TRAFFIC study in eastern and southern Africa has identified approximately 100 species of conservation concern in one or more countries (Mulliken, 2003). In Kenya and Uganda, for example, the roots of *Mondia whitei* have become rare as a result of demand for this plant in urban centres (Maundu et al, 2005). In western Africa, *Griffonia simplicifolia* has been adversely affected by overharvesting for the production of drugs in Europe (Maundu et al, 2005). *Cinchona* species and *Cassia acutifolia* are examples of the most widely traded medicinal plant products from Africa. *Cinchona* species contain 6–7 per cent total alkaloids, the most important ones being quinine and quinidine. The main dry forest and woodland regions producing it include Tanzania and Kenya (Iqbal, 1993). *Cassia acutifolia* grows wild in the dry forests and woodlands of Sudan. About 700–800 tonnes of C. *acutifolia* leaves and pods are exported annually from Sudan. The leaves and pods contain a glycoside, which is used as a laxative (Iqbal, 1993).

In the past, sustainable use of medicinal plants was facilitated by several inadvertent or indirect controls and some intentional management practices such as taboos, and seasonal and social restrictions on gathering of medicinal plants (Cunningham, 1993). It has been documented that many of these restrictions, issued by traditional leaders and enforced by local headmen and traditional community policemen, reduced commercial exploitation of local medicinal plant resources in many areas. However with cultural change, increased reliance on the cash economy and rising unemployment these controls are being eroded (Maundu et al, 2005). This change is accelerated by the commercialization of the medicinal plant trade to supply burgeoning urban and international markets.

Often commercial gatherers, whether for the national or international trade, are poor and their main aim is not resource management but to earn an income (Maundu et al, 2005). This, in turn, is creating more demand and destruction of local habitats (Mander, 1998). The break down of regulatory mechanisms, particularly with regard to communally owned resources, is further exacerbated by very little or no state investment to develop or support local management systems (von Maltitz and Shackleton, 2004). This has led to a situation whereby rural communities are most often unable to take responsibility for the state of their natural resources, allowing individuals in the community, or outsiders, to harvest medicinal plant populations without restriction (Mander, 1998).

It has been stated that a failure to stabilize this situation will not only affect the environment negatively (Cunningham, 1991; Mander, 1998; Deffar, 1998; Sunderland et al, 1999; Dold and Cocks, 2002), but also the health of millions of people in the region (Marshall, 1998) as well as the welfare it provides for the people engaged in the industry (Mander, 1998; Dold and Cocks, 2002). Cultivation initiatives and new management programmes are considered essential if the use of medicinal plant resources is to reach any level of sustainability (see Cunningham, 1993; De Beer and McDermott, 1996; Leakey et al, 1996; Ruiz-Pérez and Arnold, 1996; Sunderland et al, 1999).

At an organizational and capacity development level, it has been suggested that the long-term interests of different sectors involved in the medicinal plant trade might be better served through increased information sharing, dialogue and cooperation. The need for more effective cross-sector collaboration to address medicinal plant conservation issues was brought to world attention through the International Consultation on the Conservation of Medicinal Plants convened by the World Wide Fund for Nature (WWF), IUCN, the International Union for Conservation of Nature and the World Health Organization (WHO) in Chiang Mai, Thailand in 1988. More recently 20 participants at the symposium for Medicinal Utilization of Wild Species signed a Joint Declaration for the Health of People and Nature, which emphasizes the need to address conservation concerns in a multi-disciplinary and collaborative manner. Signatories included representatives of the phyto-pharmaceutical industry, practitioners' associations and conservation organizations (Mulliken, 2003).

Unfortunately in many sub-Saharan African states, governments lack appropriate and specific policies for the conservation of medicinal plants and therefore enforcing sustainable harvesting or monitoring the status of the resource becomes difficult. For example, in eastern Africa governments still do not have laws and mechanisms to prosecute illegal exporters. *Aloe* species have been placed under CITES protection, but as yet there are no mechanisms in the East African countries to incorporate these decisions into their legal frameworks (Maundu et al, 2005). It is therefore felt that alternative methods need to be sought to ensure the sustainable use of these resources.

The Declaration of Belem (Posey, 1988) recognized an 'inextricable link' between biological and cultural diversity (Posey, 1999), and therefore the medicinal plant trade threatens not only plant species but also traditional knowledge and cultural practices surrounding the use of medicinal plants (Tabuti et al, 2003). Biodiversity conservation programmes and cultural heritage agencies should develop awareness campaigns that illustrate the link between cultural and biodiversity conservation. Local communities and individuals, as well as conservationists, need to be made aware not only of the link between the loss

Box 5.7 Management of medicinal trees based on responses to bark harvesting in Benin, western Africa

Stephen Devereux (1999) studied the recovery after bark harvesting of 12 medicinal tree species in the dry forests of Benin, western Africa, over a two-year period. Two species (*Khaya sengalensis* and *Lannea kerstingii*) showed complete wound recovery through edge growth and three species (*Afzelia africana*, *Burkea africana* and *Maranthus polyandra*) showed very poor edge growth; the rest of the species had intermediate recovery rates. The author therefore concluded that only two species were suitable for sustainable bark harvesting due to their rapid bark regrowth.

Based on this study, the author pointed out that harvesting bark requires species-specific techniques to make harvesting sustainable and that harvesting technique be based on the regeneration capacity of the species that also determines the possibility of subsequent harvests. Devereux recommended that harvesting in the wild should include strip harvesting for species with very good regrowth, such as *Khaya sengalensis* and *Lannea kerstingii*; for species with no or little recovery, full-tree harvesting (all utilizable bark from trunk and branches) should be preferred for fallen trees cut for other purposes, such as timber.

of the natural habitat and cultural practices, but also of the options for incorporating cultural values into biodiversity conservation.

Sustainable harvesting of plant parts for medicinal purposes should be based on the capacity of the species to regenerate the harvested parts. However, few studies that have assessed recovery after harvesting have been done. A study in Benin, western Africa, has made species-specific recommendations concerning bark harvesting for medicinal purposes based on the ability of tree species to regenerate bark following harvesting (Devereux, 1999; Box 5.7).

Fibres

Description, processing and use

Historically, natural fibres gathered from dry forest and woodland plants provided the raw material for a wide range of utilitarian products, such as clothing, ropes, basketry, fishing nets, brooms, mats and construction materials. Despite clear evidence that fibre use is common to almost all areas of dry forests and woodlands, there are relatively limited data available on the importance of this product from a subsistence perspective. On the other hand, more extensive literature on the commercialization of fibre products exists. Woven mats, ropes and basketry products are often traded within the communities in which they are made, and also appeal to non-traditional and tourist markets for their decorative and novelty value (Pereira et al, 2006). Indeed, the commercialization of handicraft has been viewed as an important development option for poor rural communities in many African countries, where the sale of baskets and other craftwork to tourists and the export market has helped increase rural incomes (Cunningham and Milton, 1987; Cunningham and Terry, 1993). Dry forests and woodlands lack the bamboos and rattans found in moist forests. Instead, palm leaves, bark, grasses, and reeds and sedges provide the primary source of fibre for weaving and other uses.

Woven baskets are a key fibre product and serve a variety of functions especially in the storage and processing of crops. In Uganda, for example, flat, circular baskets are placed adjacent to grinding stones to catch the ground grain; deep, bowl-shaped baskets are used to store millet while larger, shallower baskets are used for other grain; and large diameter nearly flat baskets are used for winnowing. Baskets are also employed to carry head-loads of crops and special baskets are made for tealeaf collection. Other uses include as storage places for linen, clothes and dirty laundry, and as baby's cots (Naluswa, 1993). A similar diversity of basket uses is reported from many other countries in the dry forest and woodland regions of sub-Saharan Africa.

The leaves of the *Hyphaene* and other palms are the fibre of choice for basketry and other woven products, and basketry is found in most places where palms occur (Cunningham and Milton, 1987; Terry, 1986; Cunningham and Terry, 1993; de Vletter, 2001; El Tahir and Gebauer, 2004). *Hyphaene coriacea* ('*iLala*' palm) is found along the Zululand and Mozambique coasts, *Hyphaene petersiana* ('*mokolwane*' palm) forms a narrow band stretching from Mozambique to Angola and runs along the borders of Zimbabwe, the southern border of Zambia, the northern border of Botswana and the southern border of Angola. *Hyphaene thebaica*, dom palm ('*saaf*'), occurs in the Sudan, Eritrea and Ethiopia. Hyphaene has a relatively restricted distribution occurring on alkaline, saline, sodic, clay-rich sands and heavy clays in hot, dry river valleys. As such, it is not common to all areas of dry forest and woodland. Other palm species from the dry forest and woodland regions, such as *Phoenix reclinata* (Senegal date palm) and *Borassus aethiopum*, are also used in basket making (Sambou et al, 2002).

The bark of a number of tree species is used as weaving fibre, with the bark of the baobab (*Adansonia digitata*) being particularly favoured for basketry, including the '*kiondo*' baskets of Kenya. In both Senegal and Ethiopia, the fibres from baobab are woven into waterproof hats that may also serve as drinking vessels (Salim et al, 2002). Unlike *Hyphaene*, which gives a hard and stiff finish, the fibres of baobab allow for the weaving of softer and stronger baskets or carry-bags. Baskets and bags for agricultural produce are made from *Oxytenanthera abyssinica* and are used for transporting vegetables and fruits (Hines and Eckman, 1993).

Baskets made purely for household use seems to be a dying art. A study of the '*xirundzu*' baskets of the Inhambane region in Mozambique suggests that the skills needed to make these baskets are rapidly disappearing. The younger generation lack interest in learning this art since they derive few benefits from it (de Vletter, 2001). In the Okavango region of Botswana, basketry had almost disappeared in the late 1960s, but was revitalized once commercialized (Arntzen, 1998).

The use of woven mats is widespread. These come in all shapes and sizes from sleeping mats to sitting mats, floor coverings, food mats, prayer mats, large wall panels and decorative wall hangings (Shackleton, 2005; Araia, 2005). Their uses range from practical to cultural. The intricate hand-made mats are produced using a simple frame, where stones or old batteries are used as weights on the strings and moved backwards and forwards over the reed fibre to create patterns. Unmarried girls assisted by older women traditionally made these mats, but nowadays many households buy mats from weavers who specialize in this activity. In Uganda, however, female household members still generally only sit on mats they make themselves (Howard and Nabanoga, 2007). The mat-rush, *Juncus krausii*, which grows in brackish wetland areas, is favoured for mat making, though other species may be used where *Juncus* is not available (Zakes Mazibuko, personal communication). In the Bushbuckridge area of South Africa it is estimated that each rural household owns between two and five mats, all of which are produced locally, mostly from *Cyperus*, *Schoenoplectus* and *Typha* species. This represents an annual demand of 70,000 to 100,000 mats in the region. As with the '*Zulu*', in Bushbuckridge mats are important in wedding ceremonies and as gifts, but are also used during several other rituals including initiations and burials (Shackleton and Shackleton, 1997; Shackleton, 2005). Palm fibres may also be used to make mats – with dom palm leaves being important in Eritrea (Araia, 2005), where most households own about ten mats, and *Phoenix reclinata* leaves in Uganda (Howard and Nabanoga, 2007).

A number of bark fibres are used to make rope and cordage. In Zambia, rope from the fibres of *Sesamum angustifolium* are used for stitching reed mats and making fishing nets and animal traps. Baobab fibre is credited with making excellent cordage, ropes, harness straps, mats, snares and fishing lines, fibre cloth, musical instrument strings, tethers, bedsprings and bow strings (Salim et al, 2002). Natural fibres are often used for simple tasks, such as tying up bundles of leafy vegetables in markets (Campbell et al, 1991). Thread from tree roots is used for sewing items, such as bags and sacks, and is harvested from *Acacia nilotica*, *Tamarindus indica*, *Cordia africana* and *Lannea schweinfurthii* (Hines and Eckman, 1993). A tree which is reputedly much appreciated by rural South Africans for its long, durable and tough fibres is the violet tree (*Securidaca longepedunculata*). Its fibres are used to make string and rope for fishing nets and lines, and bird and animal snares, thread to sew bark cloth, and bead string for necklaces (Maliehe, 1993). In the Sudan, fibres from *Adansonia digitata*, *Acacia nubica*, *A. senegal*, *Cordia africana* and *C. rothii* are used for making robes (El Tahir and Gebauer, 2004). Natural fabrics consisting of tough interlacing fibre that can be extracted from bark in layers and used as a substitute for cloth are obtained from *Ficus* species and such fabrics are commonly used in rural Kagera areas of Tanzania (Chihongo, 1993). Fibrous bark (e.g. from *Brachystegia spiciformis*) is important for beehive production (see Box 5.4).

Unlike food and medicinal NWFPs, natural fibres from dry forests and woodlands are playing a declining role in rural subsistence. A number of alternative products have replaced functions typically provided by fibre products. For instance, bought fabrics for clothing have tended to replace home produced natural fabrics. Cheap or free plastic bags and plastic or metal storage bins can substitute for baskets that require hours to make (de Vletter, 2001; Kgathi et al, 2005). Nylon ropes can replace hand woven bark rope. However, natural fibres are still extensively used for simple functional tasks, such as tying up bundles of vegetables or for the production of brooms. Today, the use of natural fibre products often appears more important for cultural and traditional purposes than utilitarian purposes. Mats, brooms and baskets all have cultural significance. Fibre use also remains an option to those in the community too poor to afford alternatives. Though difficult to substantiate with the data available, it is also probable that natural fibre remains more important to those in deep rural areas with limited access to alternatives and the cash economy. For example, communities living close to riverine forests in western Eritrea use 21 different household items made from dom palm, with the direct-use value of these being the highest of all NTFPs surveyed at US$80 per household per year (Araia, 2005).

Resource management aspects

There is a paucity of information relating to the harvesting and management of natural fibre species. In most countries, at national level, there is little knowledge of, or control over, the specific use of fibre products (Geldenhuys et al, 2005). National policy and legislation may provide a legislative framework in which forest product resource harvesting and management takes place, but to a large extent fibre products are low priority from a legislative perspective. Consequently, fibre harvesting tends to be unregulated or left to local authorities to regulate. Most control thus takes place at a local level through informal mechanisms, which are often location specific. For example, in central Uganda a number of spiritual beliefs are important in regulating the harvesting of *Phoenix reclinata* palm leaves (Howard and Nabanoga, 2007). A particularly interesting example that prevents harvesting from immature plants is the belief that young palms habour wetland spirits that only leave when the plant matures. Anyone cutting an immature palm may be 'swallowed' by the wetland.

In many instances the *de facto* use of fibre resources is as an open access resource. However, as the value of the resource increases this has, in some instances, led to renewed tribal authority control or to the privatization of individual trees, as has been the experience with fibre producing trees in Sudan (El Tahir and Gebauer, 2004).

Hyphaene palms are amongst the most important weaving species. Young leaves are selectively harvested as older leaves become too brittle. In *H. petersiana* new leaves are produced at a rate of 2.8 to 3.8 leaves per year (Cunningham and Milton, 1987) or four leaves per year (Sola, 1998). Harvesting should not exceed this production rate. According to Bishop et al (1994), historically there were customary rules relating to the harvesting of palm products in the Okavango region of Botswana, but by the early 1990s such rules and sanctions were no longer practised. During the 1990s the village headman of the islands of Oxge, near Danega, reintroduced rules for harvesting *Hyphaene petersiana*, and basket makers were advised not buy the fibre harvested with hoes or axes, as these devices are not selective and are therefore more destructive to the plant. These rules and sanctions were considered effective in managing the palm resource (Bishop et al, 1994). In Zimbabwe, rights to harvest palm leaves is open to all. Each village has a dedicated harvesting area, and the chief has overall control of harvesting which is administered through local headmen. To date, the sale of unprocessed leaves has been prevented. In Zimbabwe, household-use of fibre resources is not restricted, but a permit is needed if the use is for commercial purposes (Sola, 1998, 2004). In west Caprivi (Namibia), community resource monitors are teaching community members how to harvest palm leaves in the least destructive manner. This involves using a sharp knife and ensuring that a few young leaves are left behind (Suich, 2003; Murphy and Suich, 2006). This system seems to be successful as there is no obvious drop in volume or quality of craft. This is in contrast to studies in Botswana where a marked reduction in raw material had been recorded compared to 20 years ago (Mbaiwa, 2005; Kgathi et al, 2005). In east Caprivi, resources are already depleted in some areas forcing the weavers to travel greater distances to collect leaves.

Bark and tree root harvesting is potentially the most destructive form of fibre harvesting, though many trees seem to be able to regenerate bark provided that the harvesting is done in the correct manner (Geldenhuys, 2004). Baobabs have the potential to grow new bark directly from the cambium following the harvesting of young bark. Despite this, Romero et al (2001, 2002), who conducted a four-year study on baobab bark harvesting in Zimbabwe, suggest that as it is currently practised harvesting is likely to lead to the death of the trees.

In general, fibre harvesting has the potential to be sustainable. At a continental level the sustainable harvest limits on fibre products from the dry forests and woodlands have not been reached, though this may have been exceeded in some specific locations. There is also evidence that local controls and knowledge sharing around harvesting techniques can increase the sustainability of harvesting. Since many fibres can be harvested in non-destructive ways, there is also the potential for increased access to fibre species in conservation areas.

Grasses

Description, processing and use

Though largely overlooked, Shackleton (2005) has shown that the use of grass brooms is extensive. Twig brooms are used for sweeping outside whilst grass brooms are used to clean inside the house. In her study area in northeast South Africa, all households and many schools and local businesses use these brooms.

Brooms are not just for sweeping; there is also customary and spiritual significance attached to them. In Dighe administrative sub-zone in Eritrea, the average household owns eight dom palm brooms (Araia, 2005). Since most brooms last less than six months there is a high turnover in this product.

Thatch grass as a construction material (from genera such as *Eragrostis*, *Loudetia*, *Hyparrhenia* and *Hyperthelia*) remains important in most areas, even though there is a trend toward alternative building and roofing materials (Hawkes, 1992). Corrugated iron roofs are used instead of thatch or palm leaves, and though hotter, are more durable. In some tribes cultural norms dictate that at least one dwelling in the homestead must be thatched (Timmermans, 2004). In very poor areas, such as the Inhambane area of Mozambique, thatch remains the main roofing material on almost all dwellings. Palm leaves also are important in house construction for both walls (plaited sheets) and roofing. In Eritrea, the average traditional house requires some ten 'camel loads' of palm leaves a year for routine maintenance (Araia, 2005).

Resource management aspects

Reed and grass, on the other hand, both represent potentially sustainable resources, and in both cases harvesting may have a net beneficial impact on the resource as it reduces fire intensity. In Namibia, for example there are no rules to regulate the harvesting of grass (Katjiua, 1998 in Jones and Mosimane, 2000).

Improving Governance and Management of NWFPs

There are a number of ways in which the governance and management of NWFPs can be improved in sub-Saharan Africa. Some of these are presented below:

- The safety net as well as the income-generation role of NWFPs are threatened in the dry forests and woodlands of Africa in that some of the important species e.g. *Warburgia* spp, are fast depleting. The vicious cycle of increased poverty, with a reduced bargaining capacity of collectors, is likely to lead to an increase in unsustainable harvesting intensities as collectors try to harvest more to maintain current income levels. In general, NWFP management remains a highly unpredictable occupation for local forest- and woodland-dependent communities.
- The sustainable harvesting of NWFPs can only be realized if some of the following aspects are addressed. Firstly, the capacity for any NWFP population to withstand harvesting depends on the plant part which is harvested (e.g. bark, root, tuber, leaves, sap, fruits, flower), the harvesting intensity, frequency and timing in relation to annual phenological development, and

therefore on the species' reproductive and/or regrowth capacity. Generally, local people have detailed knowledge of these aspects. Secondly, the actual harvesting intensity, frequency and timing must be according to the species' capacity to reproduce/regrow, and this is highly dependent on the interest and effective possibility of the harvesters to restrict harvesting intensity to levels which do not hamper growth and long-term sustainability. Thirdly, the ecological requirements of the NWFP species need to be optimally maintained, e.g. if shade is required and the surrounding forest or woodland is harvested for timber, then irrespective of the care which local harvesters may give to appropriate harvesting intensity of the NWFP, the deteriorating ecological conditions will not allow the species to develop or reproduce. For all these aspects, traditionally developed harvesting rules often exist. These have to be respected and integrated in any development of new regulatory regimes.

- There are increasing calls for developing NWFP resource specific inventories and monitoring schemes, but as these are being developed note should be taken of the fact that forest inventory techniques have been developed for timber and are largely irrelevant for most NWFPs. 'Scientific' solutions are to be found more in complex plant ecology methodologies which have only been applied to a few NWFP situations so far, because of their very high cost to meaningful result ratio. Further, local people need to be able to apply the inventory and monitoring techniques if these are to contribute to their decisions over harvesting intensity. Local traditional harvesters have often developed their own indicators to assess the sustainable harvesting potential of an NWFP population, and any new methodology should consider these carefully and probably combine traditional knowledge and more modern scientific methods (Baker, 2001; Wong, 2000). It is generally assumed that the extraction of NWFPs is carried out with adequate knowledge about the resource base, silvicultural requirements and the seasonal variation in their productivity or else their entire process will result in resource depletion and severe environmental damage. The current extraction methods of some NWFPs in the dry forest and woodlands are detrimental to the sustainability of the resource base and therefore pose a threat to the existence of the species.
- Management systems that promote and sustain the multiple use of dry forests and woodlands for timber, NWFPs, grazing, ecological services and other non-consumptive uses, such as cultural importance and tourism, are required. In most countries the bias is still towards timber.
- Forestry departments need to work in partnership with local communities and rural resource users to derive appropriate systems for the co-management of forest resources on both communal and state-owned land. There is a need to learn from and build on what has already been undertaken in a number of dry forest and woodland countries, such as Malawi, Zimbabwe and Zambia, particularly in the wildlife management sector. Associated with this, mechanisms for providing rural resource users with more secure tenure to resources and land are required.

- The potential impact of extra-sectoral policies, including urban, land, agricultural and economic development, health and welfare, and energy policies, on dry forest and woodland use and management need greater consideration. Often pressures on NWFPs come from external sources rather than the actual harvesting of these species for their products.
- Indigenous or traditional knowledge of customary laws, norms, beliefs and practices relating to the use and the management of specific dry forest and woodland species needs to be included in any efforts to improve forest and species management. Furthermore, the interaction, and potential synergies, between such informal controls and more formal mechanisms, such as government imposed restrictions, etc need to be understood. The links between culture and forest product conservation are particularly important in some contexts and should be more explicitly explored. There are indications that cultural value may be as, or more, important than economic value in providing an incentive for sustainable management (Sambou et al, 2002). Indeed, the latter is often controversial with a high market value sometimes resulting in precisely the opposite outcome to that desired, i.e. overexploitation rather than conservation.
- Rural extension services need to include greater attention to NWFPs, their management and potential for intensified production and domestication. In some areas NWFPs are as important for rural households as livestock and agricultural production (Shackleton et al, 2000b; Mutamba, 2007) but rarely receive the same level of investment and support.

The above approaches, if adequately developed and implemented are more likely to improve governance and management of NWFPs in dry forests and woodlands in sub-Saharan Africa to the benefit of households, local communities and national governments.

REFERENCES

Addis, G., Urga, K. and Dikasso, D. (2005) 'Ethnobotanical study of edible wild plants in some selected districts of Ethiopia', *Human Ecology*, vol 35, pp83–118

Akinnifesi, F. K., Kwesiga, F. R. C. T., Mhango, J., Mkonda, A., Chilanga, T. and Swai, R. (2004) 'Domesticating priority miombo indigenous fruit trees as a promising livelihood option for smallholder farmers in southern Africa', *Acta Horticulturae*, vol 632, pp15–30

Akinnifesi, F. K., Kwesiga, F., Mhango, J., Chilanga, T., Mkonda, A., Kadu, C. A. C., Kadzere, D., Daka, J. D., Ramadhani, T. and Dhliwayo, P. (2006) 'Towards the development of miombo fruit trees as commercial tree crops in southern Africa', *Forests, Trees and Livelihoods*, vol 16, pp103–121

Allotey, J. and Mpuchane, S. (2003) 'Utilization of useful insects as a food source', *African Journal of Food, Agriculture, Nutrition and Development*, vol 3, pp1–8

Allotey, J., Mpuchane, S. F., Siame, A. B., Gashe, B. A. and Teferra, G. (1996) 'Biological studies on the mopane worm, *Imbrasia belina* (Westwood) (Lepidoptera: Saturniidae) under laboratory conditions', in B. A. Gashe and S. Mpuchane (eds)

Proceedings of the First Multidisciplinary Symposium on Phane, 18 June 1996, pp71–75, Department of Biological Sciences, UB/KCS

Araia, M. G. (2005) 'Revealing the forest hidden value: The case study of Eritrea', MSc thesis, University of Stellenbosch, Stellenbosch, South Africa

Arntzen, J. (1998) *Economic Value of Communal Rangelands in Botswana: A Case Study*, CREED Working Paper Series no 17, International Institute for Environment and Development, London

Badi, K. H. (1993) 'Study on the consumption of forest products: An exhaustive list of forest species bearing non-wood forest products in the Sudan', GCP/SUD/049/NET, Khartoum, Sudan

Baker N. (ed) (2001) 'Developing Needs-based Inventory Methods for Non-Timber Forest Products', Workshop report organized by the European Tropical Forest Research Network, www.etfrn.org/etfrn/workshop/ntfp/ntfpproceedingsfinal.pdf

Bartlett, E. (1996) 'Hold the turkey', *New Scientist*, vol 152, pp58–59

Bees for Development (2006) 'The African honey trade: Unlocking the potential', Observer presentation at UNCTAD expert meeting Enabling Small Commodity Producers in Developing Countries reach Global Markets, UNCTAD Commodities Branch, http://r0.unctad.org/commodities

Bishop, J., Cunningham, T., Pimbert, M., Scoones, I. and Terry, B. (1994) *Beer and Baskets: The Economics of Women's Livelihoods in Ngamiland, Botswana*, International Institute for Environment and Development (IIED), London

Botha J., Witkowski, E. T. F. and Shackleton, C. M. (2004) 'Markets profiles and trade in medicinal plants in the lowveld, South Africa', *Environmental Conservation*, vol 31, pp38–46

Bradbear, N. (2004) 'Determination of the improvements required for the production and export of quality honey from Rwanda', Chemonics International Inc./ADAR project, Kigali

Campbell, B. M. (1987) 'The use of wild fruits in Zimbabwe', *Economic Botany*, vol 41, no 3, pp375–385

Campbell, B. M., Vermeulen, S. J. and Lynam, T. (1991) 'Value of Trees in the Small-scale Farming Sector of Zimbabwe', IDRC-MR302e, Ottawa, Canada

Chang, S. T. (1980) 'Food from microbes: Mushrooms as human food', *Bioscience*, vol 30, pp399–401

Chihongo, A. W. (1993) 'Pilot country study on NWFP for Tanzania', in Non-Wood Forest Products: A Regional Expert Consultation for English Speaking African Countries, 17–22 October, Arusha, Tanzania

Chileshe, A. (2005) 'Land tenure and livelihoods in Zambia: Case studies of Kamena and St Joseph', PhD dissertation, University of the Western Cape, Cape Town, South Africa

Chirwa, P. W., Mwabumba, L. and Ngulube, M. R. (2006) 'Socio-economic and cultural issues in community based management of the Miombo woodlands in Malawi', Paper presented at the SANPAD project workshop: Links between natural resources, livelihoods and poverty alleviation, Grahamstown, South Africa, 2–7 February

Chiulukire (2001) 'Chiulukire Local Forest: Joint Forest Management Plan', Department of Forestry, Joint Forest Management Steering Committee of Katete, Chiefdom of Undi, Villages around Chiulukire Local Forest represented by Chief Mbangome, Cooperative League of the US, USAID, Lusaka

Clauss, B. (1992) 'Bees and beekeeping in the North-Western Province of Zambia', Report on beekeeping survey, German Volunteer Service (DED), Integrated Rural Development Programme (IRDP) and Forest Department, Kabompo, Zambia

Cocks, M. (2004) 'A new broom', *Veld and Flora*, December, https://www.ru.ac.za/documents/ISER/A%20new%20Broom%20Veld%20&%20Flora.pdf
Cocks, M. L. and Dold A. P. (2006) 'Cultural significance of biodiversity: The role of medicinal plants in urban African cultural practices in the Eastern Cape', *South Africa Journal of Ethnobiology*, vol 26, no 1, pp60–80
Crane, E. (1999) *The World History of Beekeeping and Honey Hunting*, Routledge, New York
Cunningham, A. B. (1990) 'Income, sap yield and effects of sap tapping on palms in south-eastern Africa', *Southern African Journal of Botany*, vol 56(2), pp137–144
Cunningham, A. B. (1991) 'The herbal medicine trade: Resource depletion and environmental management for a hidden economy', in E. Preston-Whyte and C. Rogerson (eds) *South Africa's Informal Economy*, Oxford University Press, Cape Town, South Africa
Cunningham, A. B. (1993) 'African medicinal plants: Setting priorities at the interface between conservation and primary healthcare', People and Plants Working Paper 1, People and Plants Initiave, UNESCO, Paris
Cunningham, A. B. and Milton, S. J. (1987) 'Effects of basket-weaving industry on mokola palm and dye plants in northwestern Botswana', *Economic Botany*, vol 41, pp386–402
Cunningham, A. B. and Shackleton, C. M. (2004) 'The use of fruits and seeds from indigenous and naturalised plant species', in M. J. Lawes, H. A. C. Eeley, C. M. Shackleton and B. G. S. Geach (eds) *Indigenous Forests and Woodlands in South Africa: Policy, People and Practice*, University of KwaZulu-Natal Press, Pietermarizburg
Cunningham, A. B. and Terry, M. E. (1993) 'Basketry, people and resources management in southern Africa', in F. Ganry and B. Campbell (eds) *Sustainable Land Management in African Semi-arid and Subhumid Regions*, Proceedings of the SCOPE workshop 15–19 November, Dakar, Senegal
Cunningham, A. B. and Wehmeyer, A. S. (1988) 'Nutritional value of palm wine from *Hyphaene coriacea* and *Phoenix reticulata* (Arecaceae)', *Economic Botany*, vol 42, pp301–306
Dauskardt, R. (1990) 'The changing geography of traditional medicine: Urban herbalism on the Witwatersrand, South Africa', *GeoJournal*, vol 22, pp257–283
De Beer, J. H. and McDermott, M. (1996) *The Economic Value of Non-Timber Forest Products in Southeast Asia*, second revised edition, Netherlands Committee for the World Conservation Union (IUCN), Amsterdam, The Netherlands
Deffar, G. (1998) 'Non-wood forest products in Ethiopia', EC-FAO Partnership Programme (1998–2000). Tropical forestry Budget line B7-6201/97-15/VIII/FOR Project GCP/INT/679/EC, www.fao.org/docrep/003/X6690E/X6690E00.HTM
Devereux, S. (1999) *Making Less Last Longer: Informal Safety Nets in Malawi*, IDS Discussion Paper 373, Institute of Development Studies (IDS), University of Sussex, Brighton, UK
de Vletter, F. (2001) 'Coping with extreme poverty through traditional skills: The case of the xirundzu basket makers of Mozambique', ILO/SAMAT discussion paper no 18, International Labour Organisation, Southern African Multidisciplinary advisory team (ILO/SAMAT), Harare, Zimbabwe
Dold, A. P. and Cocks, M. L. (2002) 'The trade in medicinal plants in the Eastern Cape Province, South Africa', *South African Journal of Science*, vol 98, pp589–597
El Tahir, B. A. and Gebauer, J. (2004) 'Non-timber forest products: Opportunities and constraints for poverty reduction in the Nuba Mountains, South Kordofan, Sudan',

Conference on International Agricultural Research for Development, Berlin, 5–7 October

Emanual, P. L., Shackleton, C. M. and Baxter, J. S. (2005) 'Modelling the sustainable harvest of *Sclerocarya birrea* subspecies *caffra* fruits in the South African lowveld', *Forest Ecology and Management*, vol 214, pp91–103

Ezeronye, O. U. (2004) 'Nutrient utilization profile of *Saccharomyces cerevisiae* from palm wine in tropical fruit fermentation', *Antonie van Leeuwenhoek*, vol 86, pp235–240

FAO (2004) *Programme des products forestiers nonligneus: Contribution des insects de la forêt a la sécurité alimantaire – L`example des chenilles d`Afrique Centrale*, FAO, Rome

Fichtl, R. and Adi, A. (1994) *Honey bee flora of Ethiopia*, National Herbarium, Addis Ababa University and DED German Development Service, Addis Ababa and Berlin, Margraf Verlag, Weikersheim

Fromme, A. (2005) 'Edible insects', *Zoogoer*, vol 34, p6

Geldenhuys, C. J. (2004) 'Meeting the demand for *Ocotea bullata* bark', in M. J. Lawes, H. A. C. Ealey, C. M Shackleton and B. G. S. Geach (eds) *Indigenous Forests and Woodlands in South Africa: Policies, People and Practice*, University of Kwazulu-Natal Press, Scottsville, South Africa, pp 517–552

Geldenhuys, C. J., Silenga, A., Riehmer, H. B. and Falcão, M. P. P. (2005) 'Sustainable use of wood, bark and fibre products from woodland tree species in southern Africa: An overview of the situation in Namibia, Zambia and Mozambique', Report Number FW-02/05, FORESTWOOD, Pretoria, South Africa

Gelfand, M. (1971) *Diet and Tradition in an African Culture*, E & S Livingstone, Edinburgh

Greyling, M. and Potgieter, M. (2004) 'Mopane worms as a key woodland resource: The use, trade and conservation of *Imbrasia belina*', in M. J. Lawes, H. A. C. Eeley, C. M. Shackleton and B. G. S. Geach (eds) *Indigenous Forests and Woodlands in South Africa: Policy, People and Practice*, University of KwaZulu-Natal Press, South Africa, pp575–589

Grosskinsky, B. (2000) 'Nutritional contribution of IWFP', in B. Grosskinsky and C. Gullick (eds) *Exploring the Potential of Indigenous Wild Food Plants in Southern Sudan. Proceedings of a workshop held in Lokichoggio, Kenya*, USAID, Nairobi

Guveya, E. (2006) 'Forest Resource Management Project mid-term baseline survey', Ministry of Environment, Tourism and National Resources, Lusaka, Zambia

Hamilton, A. (1992) 'International trade in medicinal plants: Conservation issues and potential roles for botanic gardens', in *Proceedings of the Third International Botanic Gardens Conservation Congress*, Botanic Gardens, Conservation International, www.bgci.org/congress/congress_rio_1992/hamilton.html

Hart, A. D., Azubuike, C. U., Barimalaa, I. S. and Achinewhu, S. C. (2005) 'Vegetable consumption patterns of households in selected areas of the old river state in Nigeria', *African Journal of Food Agriculture and Nutritional Development*, vol 5, online publication

Hausser, Y. and Mpuya, P. (2004) 'Beekeping in Tanzania: When the bees get out of the woods – an innovative cross-sectoral approach to community based natural resource management', *Game and Wildlife Science*, vol 21, pp291–312

Hausser, Y. and Savary, J. (2005) 'A cross-sectoral approach to beekeeping support', *Bees for Development Journal*, vol 64, www.beesfordevelopment.org/info/info/development/a-cross-sectoral-approach.shtml

Hawkes, R. K. (1992) 'Thatching grass: Sources, adequacy and potential for local management in the Bulalimamangwe NRMP Area', CASS/MAT Working paper 1-92, Center for Applied Social Sciences, University of Zimbabwe, Harare, Zimbabwe

Hines, D. and Eckman, K. (1993) 'Indigenous multipurpose trees of Tanzania: Uses and economic benefits for people', FAO Working Paper Misc/93/9, FAO, Rome

Holden, S. (1991) 'Edible caterpillars – a potential agroforestry research? They are appreciated by local people, neglected by scientists', *The Food Insects Newsletter*, vol 4, pp3–4

Holmes, W. D. (1964) *Bark-hive beekeeping in Zambia*, Government Printer, Lusaka

Howard, P. L. and Nabanoga, G. (2007) 'Are there customary rights to plants? An inquiry among the Baganda (Uganda) with special attention to gender', *World Development*, vol 35, pp1542–1563

Hussein, M. H. (2000) 'Beekeeping in Africa. North, East, North-East and West African countries', *Apiacta*, vol 1, pp32–48

IFC (International Finance Corporation) (2006) 'Honey Care Africa Kenya and Tanzania. Grassroot Business Initiative Profile', www.ifc.org/gbi

Iqbal, M. (1993) 'International trade in non-wood forest products: An overview', Food and Agriculture Organization of the United Nations Working Paper Misc/93/11, FAO, Rome

IUCN (2002) 'Best practices in community-based natural resources management', www.iucnrosa.org.zw/news/best_practices.html

Jiwa, F. (2002) 'Honey Care Africa's tripartite modiel: An innovative approach to sustainable beekeeping in Kenya', *Apimondia*, www.apimondia.org/apiacta/slovenia/en/jiwa.pdf

Jones, B. and Mosimane, A. W. (2000) 'Namibia case study', in S. Shackleton and B. Campbell (eds) *Empowering Communities to Manage Natural Resources: Case Studies from Southern Africa*, SADC Natural Resources Management Programme and CSIR

Kasuta, E. T., Takawira, M. D., Njovu, F., Mulenga, O. and Kaumba, J. (2001) 'Testing and verification of the principles of sustainable CBNRM management for forest based CBNRM programmes. A case study of Uchi Mukula Trust Programme in the North-Western Province of Zambia', IUCN Regional Office for Southern Africa (IUCNROSA), Harare

Kgathi, D. L., Mmopelwa, G. and Mosepele, K. (2005) 'Natural resource assessment in the Okavango Delta, Botswana: Case studies of some key resources', *Natural Resource Forum*, vol 29, pp70–80

Kristensen, M. and Balslev, H. (2003) 'Perceptions, use and availability of woody plants among the Gourounsi in Burkina Faso', *Biodiversity and Conservation*, vol 12, pp1715–1739

Lambert, J. (2001) 'Ethiopia: Traditional medicine and the bridge to better health', *IK Notes*, vol 35, pp1–4

Leakey, R. R. B., Temu, A. B., Melnyk, M. and Vantomme, P. (1996) 'Domestication and commercialisation of non-timber forest products in agroforestry systems', *Non-wood Forest Products*, vol 9, pp1–297

Leakey, R., Shackleton, S. and du Plessis, P. (2005) 'Domestication potential of marula (*Sclerocarya birrea* subspecies *caffra*) in South Africa and Namibia: 1. Phenotypic variation in fruit traits', *Agroforestry Systems*, vol 64, pp25–35

López, C. and Shanley, P. (2004) *Riches of the Forest: For Health, Life and Spirit in Africa*, Center for International Forestry Research, SMK Desa Putera

Madge, C. (1994) 'Collected food and domestic knowledge in The Gambia, West

Africa', *The Geographical Journal*, vol 160, pp280–294
Makungwa, S. D. (1997) 'Assessment of grass, honey and edible caterpillars in some forest reserves in Malawi', in S. A. Crafter, J. Awimbo and A. J. Broekhoven (eds) *Non-timber Forest Products. Value, Use and Management Issues in Africa, Including Examples from Latin America*, IUCN, Nairobi
Malaisse, F. (1997) Se *nourrir en forêt claire Africaine. Approche écologique et nutritionelle*, CTA, Wageningen et Les Presses Agronomiques de Gembloux, Wageningen
Maliehe, T. M. (1993) 'Pilot country study – South Africa', a regional expert consultation for English speaking African countries, Arusha, Tanzania, 17–22 October
Mander, M. (1998) *Marketing of Indigenous Medicinal Plants in South Africa: A Case Study in KwaZulu-Natal*, Food and Agriculture Organization of the United Nations (FAO), Rome, Italy
Mander, M. and Le Breton, G. L. (2006) 'Overview of the medicinal plant industry in southern Africa', in N. Diederichs (ed) *Commercialising Medicinal Plants: A southern African Guide*, Sun Press, Stellenbosch
Marshall, N. T. (1998) *Searching for a cure: Conservation of medicinal wildlife resources in east and southern Africa*, TRAFFIC International, Cambridge, UK
Maundu, P., Kariuki, P. and Eyog-Matig, O. (2005) 'Threats to medicinal plant species – an African perspective', in *Proceedings of Biodiversity Loss and Species Extinctions: Managing Risk in a Changing World*, a Global Synthesis Workshop convened at the IUCN World Conservation Forum, 18–20 November, 2004, Bangkok, Thailand
Maurice, S. (2004) *Lessons from the Equator Initiative: Honey Care Africa's beekeeping in rural Kenya*, Centre for Community Based Resource Management, Natural Resources Institute, University of Manitoba, Winnipeg
Maviya, J. and Gumbo, D. (2005) 'Incorporating traditional natural resource management techniques in conventional natural resources management strategies: A case of mopane worms (*Amacimbi*) management and harvesting in the Buliliamamangwe district, Zimbabwe', www.jsd-africa.com/Jsda/Fall2005/ARC_opane%20Worms%20Magmt.pdf
Mbaiwa, J. E. (2005) 'Wildlife resource utilisation at Moremi Game Reserve and Khwai community area in the Okavango Delta, Botswana', *Journal of Environmental Management*, vol 77, pp144–156
Mbata, K. J. (1995) 'Traditional uses of Arthropods in Zambia: I. The food insects', *The Food Insects Newsletter*, vol 8, pp5–7
Mbata, K. J., Chidumayo, E. N. and Lwatula, C. M. (2002) 'Traditional regulation of edible caterpillar exploitation in the Kopa area of Mpika district in northern Zambia', *Journal of Insect Conservation*, vol 6, pp115–130
McGregor, J. (1995) 'Gathered produce in Zimbabwe's communal areas: Changing resource availability and use', *Ecology of Food and Nutrition*, vol 33, pp163–193
Melnyk, M. and Bell, N. (1996) 'The direct-use values of tropical moist forest foods', *Ambio* vol 25, pp468–472
Mertz, O., Lykke, A. M. and Reenberg, A. (2001) 'Importance and seasonality of vegetable consumption and marketing in Burkina Faso', *Economic Botany*, vol 52, pp276–289
Mickels-Kokwe, G. (2004a) 'Best marketing practice for natural resource derived products: Honey and beeswax', in M. Kokwe, G. Mickels-Kokwe and D. Mulolani (eds) *Survey on the State of Best Practices in the Agricultural and Natural Resources Management Sectors in Zambia*, PELUM, Lusaka, Zambia

Mickels-Kokwe, G. (2004b) 'Small-scale woodland-based enterprises with outstanding economic potential: The case of honey and beeswax in Zambia. Stimulating policy dialogue on sustainable management of Africa's dry forests – a CIFOR/SIDA initiative', ZAPE Research and Consultancy Report 3, Lusaka, Zambia

Mickels-Kokwe, G. (2005) 'The contribution of non-timber forest products to GDP', in E. Puustjärvi, G. Mickels-Kokwe and M. Chakanga, *The Contribution of the Forest Sector to the National Economy and Poverty Reduction in Zambia, Annex 5*, Forest Department, Government of the Republic of Zambia and Ministry for Foreign Affairs of Finland, Lusaka

Mojeremane, W. and Tshwenyane, S. O. (2004) '*Azanza garckeana*: A valuable edible indigenous fruit tree of Botswana', *Pakistan Journal of Nutrition*, vol 3, pp264–267

Moll, E. J. (1972) 'The distribution, abundance and utilisation of the Lala palm *Hyphaene oriacea* in Tongaland, Natal', *Bothalia*, vol 16, pp27–636

Mollet, M., Herzog, F., Behi, Y. E. N. and Farah, Z. (2000) 'Sustainable exploitation of *Borassus aethiopum*, *Elaeis guineensis* and *Raphia hookeri* for the extraction of palm wine in Cote d'Ivoire', *Environment, Development and Sustainability*, vol 2, pp43–57

Mulliken, T. (2003) 'Sustainable use of medicinal plants – a multi-sectoral challenge and opportunity', in *Harvesting of Non-Wood Forest Products*, FAO, Rome, www.fao.org/DOCREP/005/Y4496E/Y4496E42.htm

Munthali, S. M. and Mughogho, D. E. C. (1992) 'Economic incentives for conservation: Beekeeping and Saturniidae caterpillar utilisation by rural communities', *Biodiversity and Conservation*, vol 1, pp143–154

Murphy, C. and Suich, H. (2006) 'Basket cases: Individual returns from common property resources', http://dlc.dlib.indiana.edu/archive/00001448/00/Murphy_Basket_040527_Paper227.pdf#search=%22%22mud%20hut%20trading%22%22, accessed 16 September 2006

Murray, S. S., Schoeninger, M. J., Bunn, H. T., Pickering, T. R. and Marlett, J. A. (2001) 'Nutritional composition of some wild plant foods and honey used by Hazda foragers in Tanzania', *Journal of Food Composition and Analysis*, vol 14, pp3–13

Mutamba, M. (2007) *Farming or Foraging? Aspects of rural livelihoods in Mufulira and Kabompo districts of Zambia*, CIFOR Bogor, Indonesia

Muyay-Tango, F. A. (1981) *Les insects comme aliments de l'homme*, CEEBA Publications, Serie II, vol 69, Bandundu, Zaire

Muzama Crafts Ltd (1996) 'Environmental and site setting document', Muzama's Responsible Forestry Programme, Kabompo, Zambia

N'Gasse, G. (2003) 'Contribution des chenilles/Larves comsestibles à la réduction de l'insécurité alimentaire en République Centrafricaine (RCA)', Rapport de Consultation, FAO, Rome

Naluswa, J. T. (1993) 'Pilot country study, Uganda', in A Regional Expert Consultation for English Speaking African Countries, Arusha, Tanzania, 17–22 October

Ngwerume, F. C. and Mvere, B. (2000) 'Report on the findings of a socio-economic survey on the marketing, consumption and production of traditional vegetables in the urban and peri-urban areas of Harare, Zimbabwe', DR&SS-HRC/NRI Project AO892, Harare, Zimbabwe

O'Brien, E. M. (1988) 'Climate correlates of species richness for woody edible plants across southern Africa', *Monographs in Systematic Botany of the Missouri Botanic Garden*, vol 25, pp385–401

Osemeobo, G. J. and Ujor, G. (1999) 'Non-wood forest products in Nigeria. EC-FAO Partnership Programme (1998–2000) – Project GCP/INT/679/EC Data Collection and Analysis for Sustainable Forest Management in ACP Countries – Linking National and International Efforts', FAO, Rome

Packham, J. (1993) *The Value of Indigenous Fruit Bearing Trees in Miombo Woodland Areas of South-central Africa*, ODI, London

Pereira, T., Shackleton, C. and Shackleton, S. (2006) 'Opportunities and constraints to trade in reed-based craft products in rural villages in the Eastern Cape, South Africa', *Development Southern Africa*, vol 23, pp477–496

Posey, D. A. (1988) 'The Declaration of Belem', in D. A. Posey and W. Overal (eds) *Proceedings of the First International Congress of Ethnobiology*, Museu Paraense Goeldi, Belem

Posey, D. A. (1999) 'Cultural and spiritual values of biodiversity: A complementary contribution to the global biodiversity assessment', in D. A. Posey (ed) *Cultural and Spiritual Values of Biodiversity*, UNEP and Intermediate Technology Publications, London, pp1–19

Ramos-Elorduy, J. (1996) 'Rôle des insects dans l'alimentation en fôret tropicale', in C.M. Hladik, A. Hladik, H. Pagezy, O. Linares, G. J. A. Koppert and A. Froment (eds) *L'alimentation en fôret tropicale, interactions bioculturelles et perspectives de développement. Vol. 1: Les resources alimentaires: production et consummation*, UNESCO, Paris

Romero, C., Dovie, D., Luoga, E., Schmmidt, S., Gambiza, J. and Grundy, I. (2001) 'The effects of commercial bark harvesting on baobab *Adansonia digitata* in the Save Odzi valley, with considerations for management', in *Household Livelihoods, Marketing and Resource Impacts: A Case Study of Bark Products in Eastern Zimbabwe*, Institute of Environment Studies, Harare, Zimbabwe

Romero, C., Grundy, I., Campbell, B. M. and Cunningham, A. B. (2002) 'Baobab bark (*Adansonia digitata*)', in P. Shanley, A. R. Pierce, S. A. Laird and A. Guillen (eds) *Tapping the Green Market: Certification and Management of Non-timber Forest Products*, Earthscan Publications, London

Rubaihayo, E. B. (2004) 'The contribution of indigenous vegetables to household food security', *IK Notes*, May

Ruiz-Pérez, M. and Arnold, J. E. M. (eds) (1996) *Current Issues in Non-Timber Forest Products Research*, CIFOR, Bogor

Salim, A. S., Simons, A. J., Orwa, C., Chege, J. and Mutua, A. (2002) *Agro Forestry Database. A Tree Species Reference and Selection Guide*, World Agroforestry Centre, Nairobi

Sambou, B., Goudiaby, A., Ervik, F., Diallo, D. and Camara, M. C. (2002) 'Palm wine harvesting by the Bassari threatens *Borassus aethiopum* populations in north-west Guinea', *Biodiversity and Conservation*, vol 11, pp1149–1161

Sene, E. H. (2000) 'Forests and food security in Africa: The place of forestry in FAO's special programme for food security', *Unasylva*, vol 202, pp6–19

Shackleton, S. E. (2005) 'The significance of the local trade in natural resource products for livelihoods and poverty alleviation in South Africa', PhD thesis, Rhodes University, Grahamstown

Shackleton, C. M. and Shackleton, S. E. (1997) 'The use and potential for commercialisation of veld products in the Bushbuck ridge area', ARUDEC and the Department of Water Affairs and Forestry (DWAF), Nelspruit, South Africa

Shackleton, C. M. and Shackleton, S. E. (2004) 'Use of woodland resources for direct household provisioning', in M. J. Lawes, H. A. C. Eeley, C. M. Shackleton and B. G. S. Geach (eds) *Indigenous Forests and Woodlands in South Africa: Policy, People and Practice*, University of KwaZulu-Natal Press, Pietermarizburg, pp195–226

Shackleton, S. E. and Shackleton, C. M. (2005) 'The contribution of marula (*Sclerocarya birrea*) fruit and fruit products to rural livelihoods in the Bushbuckridge district, South Africa: Balancing domestic needs and commercialisation', *Forests, Trees and Livelihoods*, vol 15, pp3–24

Shackleton, S. E., Dzerefos, C. M., Shackleton, C. M. and Mathabela, F. R. (1998) 'Use and trading of wild edible herbs in the central lowveld savanna region, South Africa', *Economic Botany*, vol 52, pp251–259

Shackleton, C. M., Dzerefos, C. M., Shackleton, S. E. and Mathabela, F. R. (2000a) 'The use of and trade in indigenous edible fruits in the Bushbuckridge savanna region, South Africa', *Ecology of Food and Nutrition*, vol 39, pp225–245

Shackleton, S. E., Shackleton, C. M. and Cousins, B. (2000b) 'The economic value of land and natural resources to rural livelihoods. Case studies from South Africa', in B. Cousins (ed) *At the Crossroads: Land and Agrarian Reform into the 21st Century*, Papers from a conference held at the Alpha Training Centre, Broederstroom, Pretoria 26–28 July 1999

Silow, C. A. (1976) 'Edible and other insects of mid-western Zambia. Studies in ethno-entomology II', *Occasional Papers of the Intitutionen for och Janforonde Etnografivia Universitet*, vol 5, pp1–223.

Smith, G. C., Clegg, M. S., Keen, C. L. and Grivetti, L. E. (1996) 'Mineral values of selected plant foods common to southern Burkina Faso and to Niamey, Niger, West Africa', *International Journal of Food Sciences and Nutrition*, vol 47, pp41–53

Sola, P. (1998) 'The performance of *Hyphaene petersiana* under community-based utilisation in Sengwe communal lands, Zimbabwe', MSc thesis, University of Zimbabwe, Harare, Zimbabwe

Sola, P. (2004) 'Palm utilization for basketry in Xini Ward, Sengwe communal area Zimbabwe', in T. Sunderland, and O. Ndoye (eds) *Forest Products, Livelihoods and Conservation. Case Studies of Non-timber Forest Product Systems. Volume 2. Africa*, CIFOR, Bogor

Suich, H. (2003) 'Summary of partial results from the socio-economic household survey regarding community based natural resource management and livelihoods in Caprivi and Kunene', WILD Working Paper, MET, Windhoek

Sullivan, C. A. and O'Regan, D. P. (2003) 'Winners and losers in forest product commercialisation', Final Technical Report R7795, Centre for Ecology and Hydrology, Wallingford, UK

Sunderland, T., ngo-Mpeck, M., Tchoundjeu, Z. and Akoa, A. (1999) 'The ecology and sustainability of *Pausinystalia johimbe*: An over-exploitated medicinal plant from the forest of Central Africa', in T. Sunderland, L. Clarke and P. Vantomme (eds) *Current Research Issues and Prospects for Conservation and Development*, FAO, Rome

Tabuti, J. R. S., Lye, K. A. and Dillion, S. S. (2003) 'Traditional herbal drugs of Bulamogi, Uganda: Plants, use and administration', *Journal of Ethnopharmacology*, vol 88, pp19–44

Tabuti, J. R. S., Dhillon, S. S. and Lye, K. A. (2004) 'The status of wild food plants in Bulamogi County, Uganda', *International Journal of Food Sciences and Nutrition*, vol 55, pp485–495

Terry, M. E. (1986) *The Basket Industry of Gomare and Tubu*, Ministry of Commerce and Industry, Gaborone, Botswana

Timmermans, H. G. (2004) 'Rural livelihoods at Dwesa/Cwebe: Poverty, development and natural resource use on the Wild Coast, South Africa', MSc thesis, Rhodes University, Grahamstown, South Africa

Toms, R.B., Thagwana, M. P. and Lothole, K. D. (2003) 'The mopane worm – indigenous knowledge in the classroom', *Science in Africa: Africa's Online Science Magazine*, www.scienceinafrica.co.za/2003/june/mopane.htm

Tyynela, T. and Niskanen, A. (2000) 'Use and sustainability of miombo woodlands under community management in Zimbabwe', *Nordic Journal of African Studies*, vol 9, pp18–142

van der Waal, B. C. W. (2004) 'Woodland grasshoppers as a free source of protein to rural households', in M. J., Lawes, H. A. C. Eeley, C. M. Shackleton and B. G. S. Geach (eds) *Indigenous Forests and Woodlands in South Africa: Policy, People and Practice*, University of KwaZulu-Natal Press, Pietermarizburg

von Maltitz, G. P. and Shackleton, S. E. (2004) 'Use and management of forests and woodlands in South Africa: Stakeholders, institutions and processes from past to present', in M. J., Lawes, H. A. C. Eeley, C. M. Shackleton and B. G. S. Geach (eds) *Indigenous Forests and Woodlands in South Africa: Policy, People and Practice*, University of KwaZulu-Natal Press, Pietermarizburg

Walter, S. (2001) *Non-wood Forest Products in Africa: A Regional and National Overview*, FAO, Rome

Williams, V. L., Balkwill, K. and Witkowski, E. T. F. (2000) 'Unravelling the commercial market for medicinal plants and plant parts on the Witwatersrand, South Africa', *Economic Botany*, vol 54, pp310–327

Wilson, K. B. (1990) 'Biological dynamics and human welfare: A case study of population, health, and nutrition in southern Zimbabwe', PhD dissertation University College London, London

Wily, L. A. and Mbaya, S. (2001) 'Land, people and forest in eastern and southern Africa at the beginning of the 21st century. The impact of land relations on the role of communities in forest future', IUCN Eastern Africa Programme, Forests and Social Perspectives in Conservation, NRI – IUCN, Nairobi

Wong, J. L. G. (2000) *The Biometrics of Non-Timber Forest Product Resource Assessment – A Review of Current Methodology*, Project Report of ZF0077 Forest Research Program, Department for International Department, London

World Bank, (1992) *World Development Report 1992: Development and the Environment*, World Bank, Washington, DC

Yagi, S. (1998) 'Edible insects in East Africa', JIRCAS research highlights no 14, collaborative work with ICIPE, http://ss.jircas.affrc.go.jp/kanko/newsletter/nl1998/no.14/yagi.html

6

Timber and Wood Products

Almeida Sitoe, Emmanuel Chidumayo and Mário Alberto

INTRODUCTION

The use of wood from African dry forests and woodlands is as old as human history. Ancient civilizations used timber in construction and in carving household utensils and weapons. Cultural artifacts were also created using wood, for example, mythological gods with human faces but animal bodies. Wood constitutes a significant proportion of the products used in traditional buildings. It is used as poles, as components of the wall and in the roof. Construction wood is also used in animal enclosures, granaries and a variety of other structures. Household utensils, such as pestles used to grind grain foods, and various handles for utensils are also made from wood.

This chapter presents information on the use, trade and management of timber and wood obtained from dry forests and woodlands of sub-Saharan Africa. We distinguish 'industrial timber' from 'non-industrial wood'. Industrial timber can be sold directly as logs without further processing, or processed as sawn wood, plywood and veneer, therefore industrial extraction, processing and trade involve wood that enters the international and domestic markets. Non-industrial wood refers to wood products used locally, mainly for household purposes and this often makes up a large proportion of wood and timber produced and utilized that does not enter the market but circulates locally, contributing to local economies (construction, small-scale informal industry, woodcarving and household utensils). Wood that is harvested for firewood and charcoal is discussed in Chapter 7.

Non-industrial Wood and Wood Products

Although a large proportion of wood and timber produced and utilized in dry forest and woodland countries in sub-Saharan Africa is made up of non-industrial wood, comprehensive statistics on this wood use are generally lacking in almost all the countries. Often, available data are site-specific and based on limited household surveys, making extrapolation to national level difficult and inappropriate. However, we present some of these site-specific data only to highlight the significance of non-industrial wood to households and local communities. The focus is largely on the consumption of poles for construction.

Dry forests and woodlands in sub-Saharan Africa provide an important source of construction poles for rural people. However, the extent of use of construction poles varies with architectural style, which is influenced by cultural preferences and the variety of materials available (Cunningham, 1993). Although quantitative data on consumption of poles for construction are scarce, the relative use of poles from dry forests and woodlands is perceived to be generally high. For example, a study in rural Zimbabwe found that over 70 per cent of households used poles to construct buildings (walls, trusses and fowl runs) and fences (home fences and livestock pens) (Zharare and Mudavanhu, 2000). In Zambia a national census in 1969 revealed that 47 per cent of all houses in the country had walls made of poles (Central Statistics Office, 1973) and the use of poles for construction is still significant. A study conducted in the Central Province of Zambia in the late 1990s revealed that 17 per cent of all timber and wood use in households annually was in the form of construction poles (Kasumu and Ng'andwe, 1996). In Mozambique, rural construction, including housing, granary and animal enclosures is estimated to involve about 200,000m^3 of wood per year (Sitoe and Alberto, unpublished).

Another study in Zimbabwe found that wood consumption for five household structures (house, garden fence, homestead fence, livestock pens and granary) required 47.4m^3 (Gumbo, unpublished). Assuming an average pole size of 0.1m^3 (Werren et al, 1995), this represents a total of 474 poles per household per year. With an estimated density of poles in indigenous woodlands, such as miombo, ranging from 347 to 558 per hectare (Werren et al, 1995), the harvesting of poles would require 0.85 to 1.37 hectares of woodland per household per year. In Owambo, northern Namibia, a single palisade fence surrounding the main homestead was made of 7700 poles while 500 large poles and 9000 small sticks were used for fencing, and in total, construction and fencing for a single homestead required nearly 21,600 poles or removal of more than 100m^3 of construction wood (Cunningham, 1993). In this case most of the wood (43m^3) is used for palisade fencing with the preferred species being *Colophospermum mopane* and *Combretum* trees.

Local wood also plays a key role in local livelihoods. Timber products constitute the base for small-scale industries in many communities. For example, a study conducted in Mozambique in 2001 found that there were 147 wood-based industries employing an average of 60 workers each (Eureka, 2001).

Apart from the direct use of local wood, dry forests and woodlands can also contribute to livelihoods through benefit sharing mechanisms in the timber industry. For instance, in Mozambique the forest and wildlife regulation requires that 20 per cent of the revenues from resource exploitation be returned to the local communities living in the area where the resources have been extracted. In addition, the same regulation requires that 50 per cent of the value of the fines collected upon transgression of the legislation shall be given to forest patrolling agents and community members participating in law enforcement activities. In 2006 the 20 per cent revenue sharing mechanism resulted in more than US$400,000 being paid to local communities for local development projects (Sitoe and Tchaúque, 2007). The establishment of industries to produce wood products in the rural areas is often accompanied with infrastructure development in the form of hospitals, schools, roads, etc. Accessibility to remote rural areas is made easier as a result of the development of forest roads.

Industrial Timber

African timber entered international trade during the colonial period. In the late 19th and early 20th century, the timber industry played an important role in the development of transport infrastructure through the use of wooden sleepers for railways. As part of the grand plan for a Cape-to-Cairo transport link, the railway line from South Africa reached the border between Southern Rhodesia (now Zimbabwe) and Northern Rhodesia (now Zambia) at Victoria Falls in 1902 and was extended to the copper belt of Northern Rhodesia when the first copper mine began production in 1910. The potential of Zambezi teak, *Baikiaea plurijuga*, a hard wood found in dry deciduous forest (see Chapter 2), for use as surface and underground railway sleepers was soon realized but exploitation for this purpose had actually already started in Southern Rhodesia in 1908 (Judge, 1986). In Zambia the production of *B. plurijuga* timber has been documented for the period 1933–1982 (Figure 6.1a), which shows extraction of large volumes of this species. In addition the mining sector in Zambia relied heavily on timber from other indigenous trees in miombo woodland for the supply of sawn wood and mining poles to the copper mines (Figure 6.1b). At local community level commercial exploitation of indigenous timbers also played a significant role in generating revenue from concession fees (Figure 6.2). This reliance on indigenous timber was widespread in many sub-Saharan countries.

An article of the *New York Times* dated 29 June 1920 presents information on the availability of 1.22 million hectares of forests in British East Africa. The then government had the task of improving harbour facilities to assist export of large quantities of timber to be used for railway construction, wagon making, furniture and mine props. Buckle (1959) indicates that in western Africa, British enterprises started operations in 1890, shipping timber to Europe from the port of Sapele.

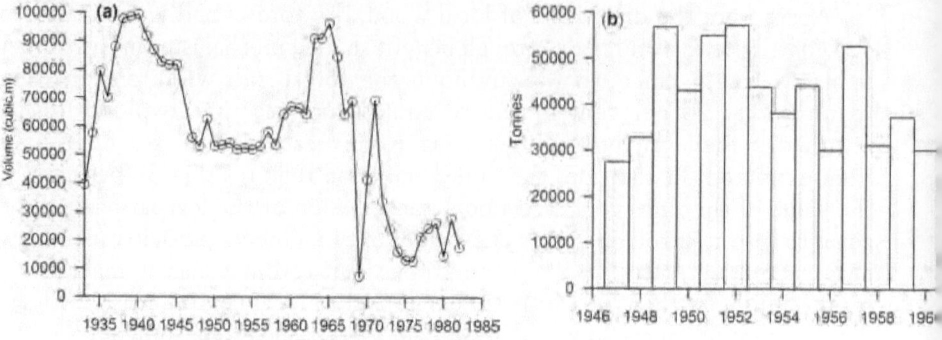

Figure 6.1 *(a) Quantities of* Baikiaea plurijuga *timber harvested from southwest Zambia and (b) multiple species sawn timber and poles harvested in miombo woodland for the copper mines in the Zambia*

Source: (a) based on Huckabay, 1986; (b) based on Lees, 1962

Although dry forests and woodlands contain valuable timber species, the contribution of commercial timber production to national economies is far less than that realized in tropical wet forests (NationMaster.com, 2004). Nevertheless, forest products play an important role in the international market. According to one estimate by the FAO (FAO, 2001), the global market for timber and timber products is estimated to be valued at over US$140 billion and is expected to grow steadily. Another study by FAO also showed that the total value of forest products (wood products) traded in the international market has increased from around US$60 billion to US$257 billion (FAO, 2006). Although there is no precise information on the forest area that is available to be managed for timber production (FAO, 2006), a large proportion of

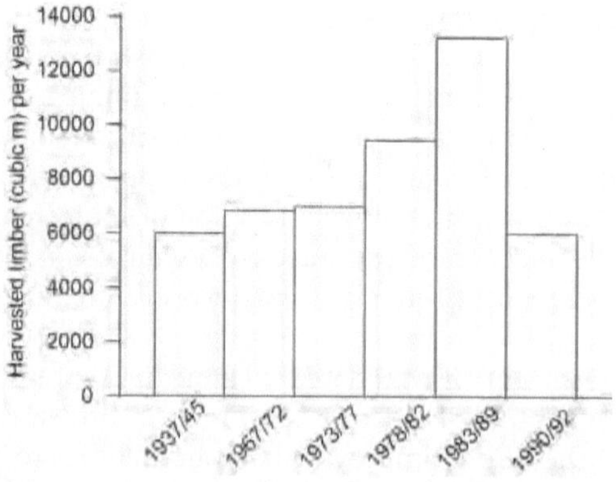

Figure 6.2 *Annual timber harvest from indigenous woodlands in Tsholotsho communal area in Zimbabwe*

Source: Based on Mushove, 1993

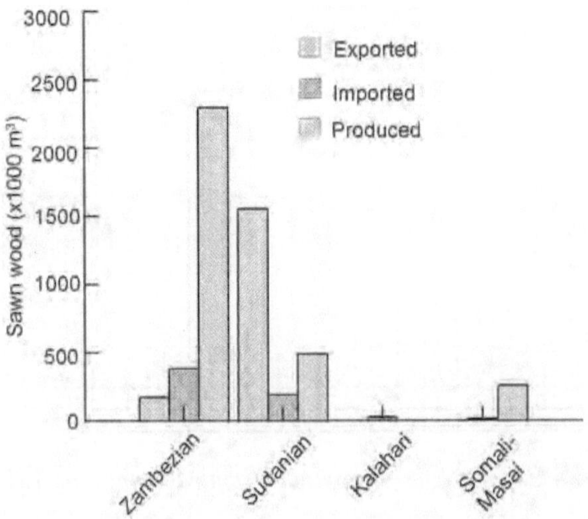

Figure 6.3 *Sawn wood production and trade in woodland phytoregions in sub-Saharan Africa in 2000*

Note: Figures for the Zambezian phytoregion exclude South Africa.
Source: Based on FAO (2005a)

the total natural forest area is classified (from the point of view of forest product industry), as non-productive, or as economically non-accessible. Another study carried out by ITTO (2006) showed that even though global supply of industrial wood is still dependent on productive forest plantations, the African forest product industry is still dependent on wood supply from natural forests, specially the plywood industry (100 per cent) and the sawn wood industry (85 per cent). It is difficult to disaggregate data on industrial round wood and sawn wood by source but there is no doubt that natural forests and woodlands have contributed to the overall trade in industrial timber, especially that of sawn wood which is believed, in the case of Africa, to come mainly from natural forests and woodlands. The Zambezian phytoregion has the highest sawn wood production per country among the dry forest and woodland regions in sub-Saharan Africa although the major exporters of sawn wood are countries in the Sudanian phytoregion (Figure 6.3). Both the Kalahari and Somali-Masai phytoregions have low sawn wood production and trade.

The timber trade in African countries with predominantly dry forests and woodlands remains low although the potential for more trade exists. For example, following the end of civil war, in the early 1990s, timber production in Mozambique increased steadily, reaching about US$65 million in wood products exports in 2005 (Figure 6.4). Much of the wood products in Mozambique come from miombo woodlands, mopane, acacia-combretum savannas and other dry forests.

The timber industry contributes to GDP in producer countries and the export of wood products has a great impact on balance of payments. For

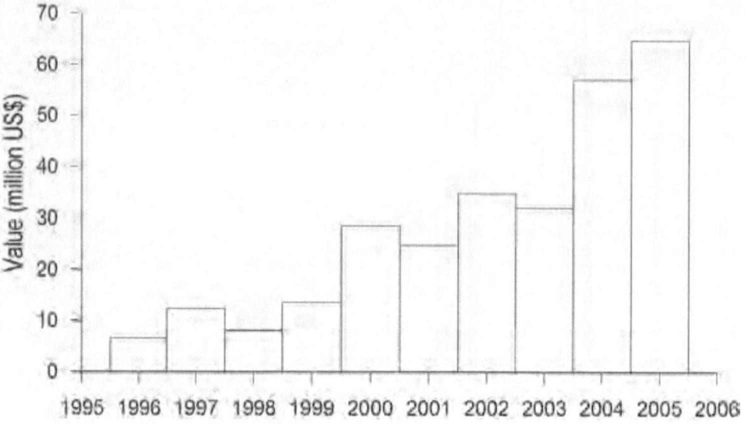

Figure 6.4 *Export of wood products from Mozambique 1996–2005*

Source: Based on data collected by authors

example, the timber industry contributed about 3 per cent to GDP in Mozambique in 2004 and made a similar contribution in Zimbabwe (Mabugu and Chitiga, 2002).

Timber Species, Stocks and Yield

A plant from which timber is extracted is referred as a timber species. There are about 62 timber tree species in dry forests and woodlands in sub-Saharan Africa (Table 6.1).

In spite of a large diversity of timber species in African dry forests and woodlands stocking rates are generally low. A recent country-wide inventory of forests in Zambia revealed that out of a total gross volume of nearly 2941million m^3 only 12.4 per cent was made up of 19 nationally classified commercial tree species with a commercial volume of 6.8m^3 per hectare (Zambia Forestry Department and FAO, 2005–2008); the distribution of these timber trees is dominated by larger trees (Figure 6.5). In Mozambique a national forest inventory also found that commercial timber species were only one to two mature trees or 5m^3 per hectare, representing 7 per cent of the standing volume (Marzoli, 2007). Stocking is only higher for gregarious species, such as *Millettia stuhlmannii* for which figures reach about 9m^3 per hectare (Sitoe, unpublished data). Similar observations have been made in western Africa where, for example, Orthmann (2005) found that the density of timber species in the Sudanian woodlands in central Benin ranged from one to ten per hectare.

The majority of African dry forest and woodland trees have diameter increments of only 0.03–2.6cm per year (Table 6.2). Zambezi teak valued for its hard heavy timber, is a slow-growing species with growth rates (mean annual diameter increment over bark) estimated at 0.13–0.20cm per year in western

Table 6.1 *Commercially valuable timber tree species in dry forest and woodland countries in sub-Saharan Africa*

Afzelia africana	*Juniperus procera*
Afzelia bipindensis	*Khaya anthotheca*
Afzelia pachyloba	*Khaya grandifoliola*
Afzelia quanzensis	*Khaya ivorensis*
Antrocaryon micraster	*Khaya senegalensis*
Aningeria altissima	*Lophira alata*
Anogeissus leiocarpus	*Lovoa trichilioides*
Aucoumea klaineana	*Lovoa swynnertonii*
Autranella congolensis	*Mansonia altissima*
Baikiaea plurijuga	*Microberlinia bisulcata*
Brachylaena huillensis	*Milicia excelsa*
Colophospermum mopane	*Milicia regia*
Combretum spp	*Millettia laurentii*
Cordia millenii	*Millettia stuhlmannii*
Dalbergia melanoxylon	*Nauclea diderrichii*
Daniellia oliveri	*Nesogordonia papaverifera*
Diospyros mespiliformis	*Pericopsis elata*
Entandrophragma angolense	*Populus ilicifolia*
Entandrophragma candollei	*Pouteria altissima*
Entandrophragma caudatum	*Prunus africana*
Entandrophragma delevoyi	*Pterocarpus angolensis*
Entandrophragma utile	*Pterocarpus erinaceus*
Eribroma oblonga	*Swartzia fistuloides*
Gossweilerodendron balsamiferum	*Terminalia ivorensis*
Guarea cedrata	*Terminalia superba*
Guarea thompsonii	*Testulea gabonensis*
Guibourtia coleosperma	*Triplochiton scleroxylon*
Guibourtia ehie	*Turraeanthus africanus*
Hallea stipulosa	*Vitellaria paradoxa*
Hallea ledermanni	*Vitex keniensis*
Haplormosia monophylla	*Warburgia salutaris*

Source: Compiled by Chapter authors from various sources

Zimbabwe (Crockford et al, unpublished). Sitoe (unpublished data) found average diameter growth rates of 0.3cm per year for the commercial timber species, *Millettia stuhlmannii*, and estimated that the fastest growing trees would mature in at least 60 years while trees with average growth rate would need about 105 years to mature in the miombo of Mozambique.

Of the average growth rates of 0.4–1.6m^3 per year per hectare for all species (including non-commercial species) in Mozambican forests, only about 4 per cent of this growth was due to timber species (Saket, 1994). This suggests very low harvesting rates must be applied to achieve sustainability. In miombo woodlands of Tanzania, Maliondo et al (2005) found that commercial trees species of over 50cm diameter (at breast height) accounted for 4 per cent of the total stock, 23 per cent of the basal area and 25 per cent of volume.

Cutting cycles have to be generally longer than 30 years to allow a sustainable yield. In the teak forests of Zimbabwe, Calvert (1986) calculated a basal area of 21m^2 per hectare with a growth rate of 0.17m^2 per hectare per year, 80

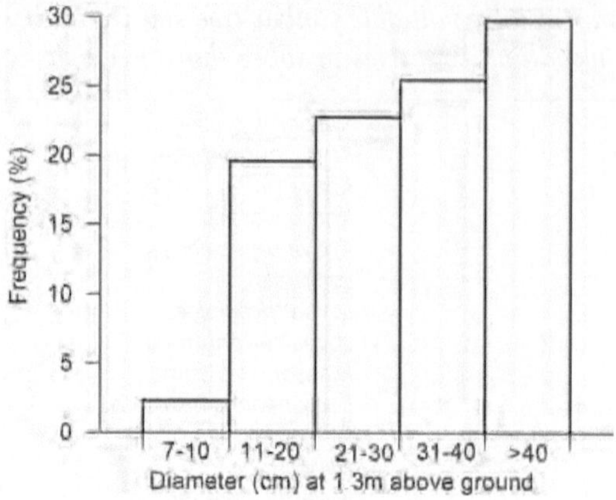

Figure 6.5 *Distribution of timber trees by diameter size in Zambia*

Source: Based on Zambia Forestry Department and FAO (2005–2008)

per cent of which was of exploitable timber species (*Baikiaea*, *Guibourtia* and *Pterocarpus*). For stands of such a dry forest and woodland, standing volume estimates were put at 72–86m^3 per hectare based on volume tables, with a mean increase ranging from 0.78m^3 per hectare per year (1952–1956) to 1.2m^3 per hectare per year (1935–1940); the temporal difference perhaps being due

Table 6.2 *Annual diameter increment at breast height of some African dry forest and woodland timber trees*

Species	*Diameter increment (cm)*	*Source of data*	*Study site*
Afzelia africana	0.392	Schöngart et al (2006)	Central Benin
Anogeissus leiocarpus	0.374	Schöngart et al (2006)	Central Benin
Anogeissus leiocarpus	0.454	Schöngart et al (2006)	Northeast Ivory Coast
Brachystegia spiciformis	0.24–0.33	Trouet et al (2006)	Western Zambia
Brachystegia spiciformis	0.03–0.27	Grundy (2006)	Central Zimbabwe
Brachystegia spiciformis	0.31	Holdo (2006)	Western Zimbabwe
Burkea africana	0.17	Holdo (2006)	Western Zimbabwe
Daniellia oliveri	0.312	Schöngart et al (2006)	Central Benin
Daniellia oliveri	0.318	Schöngart et al (2006)	Northeast Ivory Coast
Diospyros abyssinica	0.214	Schöngart et al (2006)	Northeast Ivory Coast
Erythrophleum africanum	0.16	Holdo (2006)	Western Zimbabwe
Isoberlinia doka	0.438	Schöngart et al (2006)	Central Benin
Millettia stuhlmannii	0.33	Sitoe (unpublished data)	Central Mozambique
Pterocarpus angolensis	0.03	Holdo (2006)	Western Zimbabwe
Pterocarpus angolensis	0.08–0.48	Boaler (1966)	Tanzania
Pterocarpus angolensis	0.30–0.41	Stahle et al (1999)	Western Zimbabwe
Pterocarpus erinaceus	0.374	Schöngart et al (2006)	Central Benin
Terminalia sericea	0.22	Holdo (2006)	Western Zimbabwe

Sources: As shown in the Table

to better rainfall or younger faster-growing stands. The Forestry Departments of Botswana, Zambia and Zimbabwe have tended to use a commercial cutting cycle of 40 years, and a minimum cutting size of 30cm diameter although these have since been reduced in a number of cases.

Timber and Wood Harvesting and its Impacts

Non-industrial wood

Axes and hand saws are commonly used to cut down small trees for poles; harvesting of construction poles tends to be selective and selection varies with vegetation type and occurs at three levels (Cunningham, 1993). First, selection is based on tree/shrub height, diameter and density; often favouring high density patches of tall straight poles of about 5–15cm dbh or laths of about 3–5cm dbh. Second, selection is based on access, e.g. adjacent to paths. Third, selection is based on resistance of species to termite attack and competing uses, such as non-edible fruit-bearing species. In the Kalahari-Highveld, *Colophospermum mopane, Combretum imberbe, Spirostachys africana* and *Terminalia sericea* are the major species used for building walls of homes while in the Zambezian miombo *Brachystegia boehmii, B. glaucescens, B. spiciformis, Combretum molle, Julbernardia globiflora, Terminalia sericea* and *Uapaca kirkiana* are used for building homes (Cunningham, 1993). At household level wood is also used for making implements, such as hoe handles, mortars, pestles, etc. and different tree/shrub species are preferred for the production of different household wood products (Table 6.3). This means that some highly preferred species may be depleted, thereby causing changes in the species composition of the forest or woodland. For example, Cunningham (1993) observed that the high wood consumption for construction resulted in changes in the structure and species composition of the mopane woodland in Owambo, northern Namibia.

Industrial timber

Logging in dry forests and woodlands is highly selective, based on species and size and a variety of harvesting techniques, from manual to mechanized, are used, but chainsaws are commonly used in commercial timber production. Fath (2002) found that commercial timber harvesting in Mozambique was based on low extraction intensity because of lack of harvesting preparation, low recovery rates, and poor working techniques for felling and crosscutting. Low recovery rates of about 10 per cent have also been recorded in western African dry forests and woodlands and in some cases a high number of trees of less than 45cm dbh were cut around the felled timber tree to use their trunks in sawing activities (Orthmann, 2005).

Table 6.3 *Tree species preferred for building and other wood products in miombo woodland by Bemba people of northern Zambia*

Purpose	*Tree species*
Fibre	*Brachystegia boehmii*
	Brachystegia longifolia
	Brachystegia taxifolia
	Julbernardia globiflora
Handles	*Burkea africana*
	Dalbergia nitidula
	Isoberlinia angolensis
	Julbernardia globiflora
	Lonchocarpus capassa
	Swartzia madagascariensis
Mortar	*Albizia antunesiana*
	Isoberlinia angolensis
Poles	*Erythrophleum africanum*
	Monotes sp
	Pericopsis angolensis
	Swartzia madagascariensis
Walling	*Brachystegia longifolia*
	Julbernardia globiflora
	Julbernardia paniculata
	Monotes sp
	Uapaca kirkiana
	Uapaca nitida
	Diplorhynchus condylocarpon
Roofing	*Marquesia macroura*
	Monotes sp
	Pericopsis angolensis
	Syzygium owariense
	Uapaca kirkiana
	Uapaca nitida

Source: Based on Holden (1988)

Primary transport distances varied from 2 to 49km, while secondary transport distances varied from 130 to 279km, with the former accounting for the largest share of the production costs of US$21–73 per m^3 in Mozambique.

From an ecological perspective, sustainable yield can be attained when the balance between forest growth rate and logging rate is positive or at least balanced. High logging rates pose a threat to most of the slow growing dry forest and woodland hardwood species. This suggests that under the current logging rates most dry forest and woodland timber species are not sustainably harvested. As a result, many of these species are considered threatened (see Table 3.3 for threatened dry forest and woodland timber trees). Slow growth rate, overharvesting by loggers and inadequate tree recruitment (see Figure 6.5) have been cited as possible reasons for the unsustainable timber industry based on *Pterocarpus angolensis* in Tanzania (Schwartz et al, 2002). Similar conclusions have been made about African blackwood, *Dalbergia melanoxylon* (Gregory et al, 1999). This species is a typical example of a highly valuable timber species whose populations have declined over time (Jenkins et al, 2002; Global Trees Campaign, 2007). In some dry forests and woodlands of western Africa logging in the 1950s utilized valuable species, such as *Ceiba pentandra, Khaya grandifolia, Antiaris africana* and *Chlorophora excelsa* but when suitable trees of these species were depleted logging shifted to the remaining smaller diameter trees and to new less valuable species, such as *Pterocarpus erinaceus, Diospyros mespiliformis* and *Pseudocarpus kotschyi* (Orthmann, 2005).

There is a need to promote lesser-known species. In Mozambique, for instance, Buster (1995) published a booklet on 52 commercial timbers, while the actual list of commercial timber species according to the forest regulation is 118. But in fact fewer than ten species are actually exploited for commercial purposes (Bila, 2004). Experiences from elsewhere show that after liquidating the most valuable timber, the lesser-known species are utilized.

The direct impacts result from the removal of trees. This leads to population reductions of preferred species. The indirect impacts result from the use of machinery, the opening of access roads to previously inaccessible areas, the destruction of habitats for other plant and animal species, and increased fire intensity (reduced canopy cover leading to increased grass biomass, e.g. Gambiza et al, 2000). Population reductions follow from the lack of silvicultural knowledge and practices (WWF, 2006). Species like *Dalbergia melanoxylon* and *Sterculia quinqueloba*, among others, were listed in southern Africa as vulnerable or endangered in the IUCN Red List due to extraction for timber production for commercial and local use (Izidine and Bandeira, 2002; WWF, 2006; see Table 3.3 in Chapter 3). Cutting cycles of 30 or more years are recommended for these species, which is not economically attractive, hence the unsustainable harvesting. Opening roads for timber extraction increases accessibility for other forms of land use. In Tanzania, for instance, the new bridge over the Rufiji River increased the flow of timber exports and the destruction of the richest remnants of miombo in the region (WWF, 2006).

Sustainable Forest Management for Timber and Wood Products

Sustainable forest management (SFM) is an old concept implying a continuous supply of forest goods and services through time. But in the late 1980s and early 1990s, SFM gained a new dimension before and after the Earth Summit in Rio in 1992. It was then that forests assumed an increasingly important role in sustainable development, with 'sustainable' having economic, environmental and social dimensions.

Sustainable forest management involves multi-scale initiatives. At the national level it involves clear definition of the national objectives of forest management, the establishment and support of national institutions for the implementation of SFM, the establishment of supportive policies, the monitoring of forests and forest activities, and the establishment of a favourable forest business environment (ATO/ITTO, 2003). The establishment of policies that favour SFM requires measures to facilitate access to forest resources, particularly the clear definition of property rights (Arnold, 1998). At the forest management unit level, three basic principles are defined: sustainable management with a view to supplying the required goods and services; maintenance of the main ecological functions; and contributions to the improvement of the economic and social well-being of workers and local populations.

Presently, illegal operations resulting from poor forestry governance and limited capacity of the forestry institutions, have characterized most African nations. The rise of demand from China and their apparent lack of interest in SFM have fuelled illegal forest operations. Inadequate political commitment and weak economies have limited the implementation of SFM (ITTO, 2004, 2005; Verbelen, 2002; Watts, 2005). Consequently, many timber species are threatened or near threatened (e.g. the African blackwood – *Dalbergia melanoxylon*) (Izidine and Bandeira, 2002; Jenkins et al, 2002; Pérez et al, 2005). There is little, if any, evidence that the current levels of timber exploitation are sustainable. Apparently, local use of timber does not seem to constitute a serious cause of forest degradation because of the large array of species used for local purposes (Table 6.3).

Logging can be carried out with little investment while returns are often immediate (Verbelen, 2002). This practice of mining forests is short-term and leads to the demise of the timber industry. Appropriate policy and its implementation are needed if a sustainable timber industry is going to be realized. Experience shows that although legal frameworks are in place in many countries, implementation has been limited. Such policy initiatives require justice, transparency and efficiency in the procedures for concession allocation, timber pricing, monitoring and community engagement. However, complex economic and legal instruments often hinder implementation (Sitoe et al, 2003).

Some of the specific factors limiting the success of the timber industry are as follows. First, the limited capacity for revenue collection, associated with weak technical capacity of forestry institutions and low royalties results in poor forest governance and low revenues. The current timber values of US\$0.26–5.00 per m^3 are considered extremely low by international standards (World Bank/WWF Alliance, 2002). Second, poor forest governance and monitoring stimulate illegal logging and illegal log exporting. This combination of factors discourages adherence to SFM.

Strategies have been put in place in several tropical countries to enable SFM. In Cameroon, for instance, the World Bank imposed a system to improve transparency in the forest sector with the aim of reducing illegal logging and fraud in allocation of forest concessions (Verbelen, 2002), although unlawful operations still take place. The World Bank/WWF Alliance (2002) in seven countries in western and southern Africa, including the dry forest and woodland countries of Tanzania and Mozambique, concluded that all countries are losing significant forest revenue (Table 6.4).

Table 6.4 *Estimates of revenue losses from poor regulation of timber production in selected African countries with dry forest and woodlands*

Country	*Estimated revenues lost per year (US$ million)*
Benin	1.0
Cameroon	5.3
Central African Republic	3.0
Ghana	37.5
Mozambique	2.5
Tanzania	4.3

Source: Compiled from various sources by Chapter authors

Legal and policy aspects

The forestry policy framework in African countries has been strongly influenced by colonial systems, with a focus on timber exploitation for international markets (WWF, 2006). There is limited revenue collection, weak institutional capacity to implement the regulations and little capacity to set acceptable prices. For example, Mozambique did not change its colonial forest regulation until 2002 (WWF/World Bank Alliance, 2002). Low and inappropriate taxation creates opportunities for corruption, does not encourage efficient utilization of the resource, and can cause inappropriate allocation of scarce investment funds.

Following the Rio Earth Summit, tropical forestry witnessed significant changes to accommodate the concepts of SFM. These changes stimulated many African countries to review their regulations and improve law enforcement (WWF/World Bank Alliance, 2002; Kowero et al, 2003). Policies and procedures related to forest concessions, logging licences and taxation systems were revised and new policy initiatives were introduced, including increased attention to conservation within production forests, community participation and log export bans. However, illegal logging is still rampant, and communities receive little or no benefit from the exploitation of forest products.

Kowero et al (2003) evaluated the linkages between forest policies and other major policies for the miombo woodland countries and observed that state-owned forests, typical of much of the region, are associated with weak forest governance and unsustainable timber exploitation. Agricultural policies, economic reforms and trade liberalization were identified as influencing the shape of forestry policies and their implementation.

Legal and policy strategies addressing timber harvesting in sub-Saharan African countries have focused on five main aspects:

1. Timber harvesting based on forest concessions. This strategy aims to increase access to forest resources and encourage investments through providing long-term exclusive access rights (Sitoe et al, 2003). This measure is expected to increase resource use efficiency and reduce the impact of illegal logging. It is also intended to increase community participation and local benefits.

2. Value addition. This aims to add value to forest products, create jobs and increase revenues. Forest operators, however, see this as a constraint, making it too expensive to invest in forest concessions. It is regarded as contradictory, when at the same time, annual logging permits are allowed.
3. Total or partial log export bans. These are implemented to promote domestic processing of wood products. Implementing quotas based on a percentage of production has proven difficult given that forest institutions do not have capacity to verify the quotas: as a result African exports of timber products are still dominated by unprocessed logs (FAO, 2005b; ITTO, 2005). The international market, and particularly China, favours low cost unprocessed logs.
4. Improving forest governance and monitoring. This focus is part of national capacity building to improve the management of forests. Countries have also implemented national stakeholder fora to improve public participation in forest governance. Low wages and poor working conditions in forest departments, as well as corruption and politically driven decisions, undermine these efforts (Kowero et al, 2003).
5. Community participation and benefit sharing. This has been adopted in most forest policies in dry forest and woodland countries, and is a strong part of the SFM concept. Sustainable forest management requires that forestry operations contribute to the improvement of the economic and social well-being of workers and the local population in forest concessions (ATO/ITTO, 2003). While colonial legislation limited local communities to subsistence use of forest resources, the new policy measures open up possibilities for communities to derive benefits from timber, either directly through being part of the business or through a variety of benefit sharing arrangements. Wily (2000) found that Tanzania had the most effective community access to forest resources in eastern and southern Africa. While it has proved difficult to manage forest concessions without taking into account the communities living within or near the forest, in general local communities have benefited little from forest operations (Sitoe and Tchaúque, 2007).

Forest concessions

Forest concessions and annual logging licences are two logging regimes commonly used in tropical forests, including dry forests and woodlands. The former have been recommended to promote SFM. Forest concessions are large areas of productive forest leased for a period of 5–50 years for the purposes of timber production. Sitoe et al (2003) compared the two regimes and observed that annual logging licences did not require detailed forest inventories and management plans; as a result this promoted forest mining and unsustainable use. In contrast, forest concessions promoted value-addition, community participation, resource use efficiency and long-term commitment. Although forest concession regimes are favoured, there are a number of elements that need to be put in place for them to be successfully implemented (Gray, 1999; Grut et al, 1991), as indicated below.

Forest zoning and allocation

Forest inventories are needed to define areas with productive forests that can sustain long-term timber production. The forestry institutions must have detailed information of these areas so as to be able to negotiate with concessionaires and establish an appropriate value of the forest. Forest concessions must be allocated in a transparent way and competition should be promoted to capture the appropriate revenue. In countries like Mozambique, the concessionaire conducts the inventory and submits a request for a concession to the forest department. The allocation of the concession is therefore not competitive and is based on who first submits the request (Sitoe et al, 2003).

Size of forest concession

The size is important as it contributes to the definition of 'allowable cut'. Dry forests and woodlands with low growth rates require large areas to maintain timber production on a sustainable basis.

Revenue and taxes

Grut et al (1991) found that African countries captured a very small proportion of the real value of the timber. For example, in 1987, in Cameroon, forest taxes amounted to only US$5.4 per m^3 (2–4 per cent of the log price), while in Ghana the taxes were US$0.38 per m^3 (0.5 per cent of the log price), although these figures may include timber mostly harvested from tropical forests in these countries. However, this has been changing in line with new regulations but the tax capture is still below the real potential (WWF, 2006).

Forest inventory and management plan

Forest management plans are key to SFM, as they represent the application of SFM principles to specific forest management units. They are also required for the issue of forest concessions. The plans should include an ecological evaluation of the resource, a socio-economic assessment of local communities and financial aspects of the forest business, among other aspects (Table 6.5). National level information of the forest area covered by formally approved management plans is scanty (FAO, 2001), but the area is probably increasing in many African countries following the review of forest policies and implementation of new regulations (Sitoe et al, 2003).

One of the most important components of a forest management plan is the definition of the allowable annual cut (AAC), or the sustainable yield. The estimation of this requires adequate forest inventory, information on tree growth rates and the definition of the cutting cycle. Such information is gener-

Table 6.5 *Content of forest management plans*

Forest information	Forest protection
Geographic location	Accessibility
Topography	Fire protection
Accessibility	Monitoring illegal logging
Socio-economy of the region	Social and environmental impact
Forest resource inventory	Impact mitigation
Forest types and species	Community rights and benefits
Timber volume	Agreements with local communities
Diameter distribution	Conflict resolution mechanism
Natural regeneration	Environmental management and monitoring
Non-timber forest products	Forest quality monitoring
Objectives of forest management	Forest industry environmental monitoring
Forest production	Research needs
Forest growth	Forest inventory
Cutting cycle	Forest growth and yield
Allowable cut	Impacts of logging
Forest zoning	Lesser-known species promotion
Forest operations	Administrative organization
Operational inventory	
Logging operations	
Post-logging treatments (coppicing, improvement planting, liberation treatment)	

Source: Adapted from Sitoe and Bila (2006)

ally lacking. Forest inventories by themselves are not sufficient for good decision-making and they are difficult to undertake because of cost, limited human capacity and the management plans derived from forest inventories being seen as a mere bureaucratic step towards getting a forest concession (Box 6.1). Forest inventories may cost about US$1 per hectare, requiring large sums of money for entire concessions. Technical capacity to conduct forest invento-

Box 6.1 Forest concessions and management plans in Mozambique

The situation in Mozambique has improved since 2002 when 24 concessions had been authorized but no management plans were in place (World Bank/WWF Alliance, 2002). But even where there are management plans, they may not be implemented because of limited capacity and negative attitudes towards the planning process. Technicians responsible for concession management of three approved forest concessions in Sofala in Mozambique stated that they knew of the existence of such forest management plans but had never used them to plan annual activities (Sitoe and Tchaúque, 2007). Lack of implementation of forest management plans usually leads to overexploitation (Pérez et al, 2005).

Source: Chapter authors

ries, and prepare and evaluate management plans, is lacking. In 2006, Mozambique had 111 authorized forest concessions, covering an area of about 4.5 million hectares, but only 60 had approved management plans (Sitoe and Tchaúque, 2007).

Information and monitoring

Monitoring is needed to promote good governance. It requires that appropriate statistics are kept and that reporting systems are in place within the concession system. Lack of transparency with information and procedures typifies forest concessions in many sub-Saharan countries. Independent monitoring systems have been introduced in some countries to improve the quality of information but they do not function in countries with high levels of corruption and political influence.

Timber certification

One mechanism that can contribute to SFM is forest certification, a voluntary, market-driven initiative that seeks to label products from forests that are managed according to internationally accepted stewardship principles. It is sometimes called timber certification, forest product labelling or forest management auditing. Certification can be a cornerstone of SFM. Certification involves two key activities. The first involves an independent assessment of the forest management operations, and typically includes an evaluation of the ecological health of the forest; the economic viability of the operation; and the social impact of the forest management activities. The second is the 'chain-of-custody' inspection and involves the verification of the flow of forest products from the forest, through milling and manufacturing processes, to the finished product. This process is currently carried-out by both non-profit and for-profit organizations, and is characterized as being an independent, objective and third-party process.

The objective of certification is to assure consumers that their purchases of forest products do not contribute to the destruction and degradation of the world's forests. Timber producers can benefit commercially from timber certification in two ways: first through the 'green premium' (consumers' willingness to pay a premium for certified timber) and second by averting losses of market share in the tropical timber market from not having timber certified. However, only a small number of schemes are operational at present and the volume of timber covered by them is minor (Baharudin, 1995). The organizations and initiatives active in forest certification include the following:

- Forest Stewardship Council (FSC);
- Pan European Forest Certification Council (PEFC Council);
- Canadian Standards Association (CSA);
- American Forest & Paper Association (AF&PA), United States;

- Malaysian Timber Certification Council (MTCC);
- Indonesian Ecolabeling Institute (LEI);
- Brazilian Association for Technical Standards (ABNT);
- Netherlands Timber Trade Association.

The Pan European Forest Certification Council and the Sustainable Forestry Initiative (SFI) of AF&PA are currently the leaders in forestry certification in terms of certified area. The largest concentrations of certified forests are in North America (49 per cent) and Europe (45 per cent). Only 7 per cent of the certified area is in tropical or subtropical countries (ITTO/ITC, 2005). The African Timber Organization (ATO), with assistance of the Center for International Forestry Research (CIFOR), has established the Pan African Certification scheme, which is based on principles, criteria and indicators that conform to those of the International Timber Trade Organization (ITTO). In the African Region, the number of areas of certified forests endorsed by FSC has reached 30, covering about 1.74 million hectares while a total of 136 chain-of-custody certificates have been endorsed by FSC, including in the dry forest and woodland countries of Mozambique, Namibia, South Africa, Zambia and Zimbabwe.

Lesser-known timber species

The growing concern regarding the sustainability of timber resources in African dry forests has motivated researchers to identify lesser-known species with potential to substitute the species that are under pressure. Among the identified species, bamboo was found to be of high potential not only because of its properties but also considering its wide occurrence over the dry forest and woodland countries of Africa, but this resource is underutilized. Taquidir and Cuco (2006) conducted an inventory within a 5000km^2 coastal area in Mozambique and found 2000km^2 of pure bamboo stands (*Oxytenanthera abyssinica*). Boko (2007) indicates that *Oxytenanthera abyssinica*, *Bambusa vulgaris* and *Andurinaria aplina* are the most common bamboos found in African woodlands. Kenya has large bamboo plantations, with 63,000ha of *Arundinaria alpina* (Boko, 2007).

In addition to bamboo, there are hundreds of other potentially valuable tree species whose physical and mechanical properties and end-uses are little known. These trees make up a larger portion of dry forests and woodlands than the currently utilized species (see above). Utilization of lesser-known species may help to alleviate the pressure on the diminishing primary or marketable species while contributing to the diversification of economic opportunities for communities involved in SFM (Vlosky and Aguire, 2001) (Box 6.2).

Box 6.2 Efforts to study and use lesser-known species in Mozambique

In Mozambique efforts have been made to market lesser-known species. For example, *Brachystegia spiciformis* was introduced to the European market as parquet strips. Although *Combretum imberbe* has been considered a secondary species with low commercial value, recently it has been exported to China as logs; this was followed by a log export ban because of unsustainable harvesting but it indicates the potential value of lesser-known species.

Product development research has also focused on the lesser-known species. For example, Alberto (2002) produced, on a laboratory scale, strong boards of *B. spiciformis* mixed with *Pinus patula*. Alberto et al (2001) also assessed the technical feasibility of some Mozambican secondary species for the manufacture of wood-cement composites. The study resulted in the identification of species compatible with cement without needing any treatment, such as *Amblygonocarpus adongensis*, *Brachystegia spicifomis* and *Brachystegia bohemii*, as well as the species becoming compatible after treatment of particles, such as *Erythropheum suaveleuns*, *Albizia adianthifolia* and *Sterculia appendiculata*.

Source: Chapter authors

Key Challenges

Local utilization of wood products is mainly limited to building material such as poles and bamboo. Although there are limited studies on the real impact and the characteristics of this sub-sector, its role in national poverty reduction strategies shows that it is highly significant and needs to be properly evaluated.

The introduction of environmental measures to protect African dry forests and woodlands is sometimes seen as an impediment to attracting investments in the timber sector. Promotion of lesser-known species represents a double-edged issue. On the one hand it is a measure to improve profitability while at the same time it assists in the protection of the depleted traditional valuable species. But on the other hand, since little is known about the utilization of these species, their use may also result in further reduction of forest cover, risking the degradation of forest ecosystems. The challenge is to undertake adequate studies to evaluate these species before bringing them under full exploitation.

The general perception about the timber industry in African dry forests and woodlands is that little is known about it. The fact that policies have been in place but their implementation is not effective suggests that there are information gaps that constitute impediments. Information availability can improve implementation and reformulation of existing policies. Research is therefore needed to provide information on:

- the potential and impacts of timber harvesting in these fragile ecosystems;
- the impact of promotion of lesser-known species;
- international timber trade and the paths of illegal timber;
- local use of timber products;
- impact of forest policies and the impediments to the implementation of these policies
- silviculture of timber species.

There is also need to assess requirements for training people and building capacity in sustainable forest management.

References

Alberto, M. M. (2002) 'Structural flackboards from *Brachystegia speciformis* with *Pinus patula*', *Discovery and Innovation* (Special edition), pp7–13

Alberto, M. M., Mougel, E. and Zoulalian, A. (2001) 'The suitability of some Mozambican secondary species for the manufacture of wood-cement composites', Department of Forestry, Eduardo Mondlane University, Maputo, Mozambique

Arnold, J. E. M. (1998) *Managing Forests as Common Property*, FAO Forestry Paper 136, FAO, Rome

ATO/ITTO (2003) 'ATO/ITTO principles, criteria, and indicators for sustainable management of African natural tropical forests', *ITTO Policy Development Series* no 14, ITTO, Japan

Baharudin, H. J. (1995) 'Timber Certification: An overview', *Unasylva*, vol 46, p183

Bila, A. (2004) 'Avaliação dos mercados de produtos florestais e faunísticos em Moçambique. Mimeo', MADER/Direção Nacional de Florestas e Fauna Bravia/PROAGRI, Maputo, Mozambique

Boaler S. B. (1966) *The Ecology of* Pterocarpus angolensis *in Tanzania*, Ministry of Overseas Development, London

Boko, G. (2007) 'Bamboo in Benin: Present state of knowledge', INBAR, www.inbar.int/documents/country%20report/Bamboo%20In%20Benin.htm, accessed 20 July 2009

Buckle, D. H. (1959) 'Timber operations in West Africa', *Unasylva*, vol 13, p1

Buster, J. (1995) *52 Madeiras de Moçambique: Catálogo Tecnológico*, Department of Forestry, Faculty of Agronomy, Eduardo Mondlane University, Maputo

Calvert, G. M. (1986) 'Growth summaries, sample plot 1, Gwaai Forest', in G. D. Pierce (ed) *The Zambezi Teak Forests*, Forest Department, Ndola

Central Statistics Office (1973) *Census of Population and Housing 1969: Final Report Volume 1 – Total Zambia*, Central Statistics Office, Lusaka

Cunningham, A. B. (1993) 'Low-cost housing needs, wood use and woodlands', in G. D. Pierce and D. J. Gumbo (eds) *The Ecology and Management of Indigenous Forests in Southern Africa*, Zimbabwe Forestry Commission and SAREC, Harare

Eureka (2001) *Inquérito á Industria Madereira*, DNFFB, Maputo

FAO (2001) *Proceedings of the International Conference in Timber Plantation Development*, FAO, Rome

FAO (2005a) *State of the World's Forests 2005*, FAO, Rome

FAO (2005b) *Trends in Wood Products 1961–2003*, FAO, Rome

FAO (2006) *FAOSTAT Database*,faostat.fao.org

Fath, H. (2002) 'Commercial timber harvesting in the natural forests of Mozambique', Forest Harvesting Case Study 18, FAO, Rome
Gambiza, J., Bond, W., Frost, P. G. H. and Higgins, S. (2000) 'A simulation model of miombo woodland dynamics under different management regimes', *Ecological Economics*, vol 33, pp353–368
Global Trees Campaign (2007) 'Tree profiles: African blackwood', www.globaltrees.org/tp_african_blackwood.htm, accessed 20 July 2009
Gray, J. (1999) 'Regime de propriedade florestal e valoração de florestas públicas no Brasil', Documento de trabalho 2, Ministério do Meio Ambiente, Programa Nacional de Florestas, Rio de Janero
Gregory, A-M., Ball, S. M. J. and Eziefula, U. E. (1999) *Tanzanian Mpingo*, Fauna & Flora International, Cambridge
Grundy, I. M. (2006) 'Age determination of miombo species *Brachystegia spiciformis* (Leguminosae-Ceasalpinadeae) in Zimbabwe using growth rings, *South African Forestry Journal*, vol 206, pp5–12
Grut, M., Gray, J. and Egli, N. (1991) 'Forest pricing and concession policies: Managing the high forests of West and Central Africa, *World Bank Technical Paper 143*, Washington, DC
Holden, S. T. (1988) 'Farming systems and household economy in New Chambeshi, Old Chambeshi and Yunge villages near Kasama, Northern Province, Zambia: An agroforestry baseline study', *Zambia SPRP Studies Occasional Paper Series A* (# 9), Noragric, Agricultural University of Norway, Ås
Holdo, R. M. (2006) 'Tree growth in an African woodland savanna affected by disturbance', *Journal of Vegetation Science*, vol 17, pp369–378
Huckabay, J. D. (1986) 'The exploitation of Zambezi teak in Zambia', in G. D. Pierce (ed) *The Zambezi Teak Forests*, Forest Department, Ndola
ITTO (2004) 'Reviving tropical plywood', *ITTO Technical Series* no 20, ITTO, Yokohama
ITTO (2005) *Annual Review and Assessment of the World Timber Situation*, ITTO, Yokohama
ITTO (2006) *Annual Review and Assessment of the World Timber Situation 2005*, ITTO, Yokohama
ITTO/ITC (2005) *International Wooden Furniture Markets. A review*, ITTO, Geneva
Izidine, S. and Bandeira, S. O. (2002) 'Espécies extintas e ameaçadas, Mozambique', in J. S. Golding (ed) *Southern African Plant Red Data Lists*, SABONET, Pretoria
Jenkins, M., Oldfield, S. and Aylett, T. (2002) *International Trade in African Blackwood*, Fauna & Flora International, Cambridge
Judge, J. G. (1986) 'The teak forests of Zimbabwe', in G. D. Pierce (ed) *The Zambezi Teak Forests*, Forest Department, Ndola
Kasumu, D. and Ng'andwe, P. (1996) 'Timber consumption and demand study of Central Province', Provincial Forest Action Programme, Forest Department, Ndola, Zambia
Kowero, G., Campbell, B.M. and Sumaila, U. R. (2003) *Policies and Governance Structures in Woodlands of Southern Africa*, CIFOR, Bogor
Lees, H. M. N. (1962) *Working Plan for the Forests Supplying the Copperbelt, Western Province*, Government Printer, Lusaka
Mabugu, R. and Chitiga, M. (2002) 'Accounting for forest resources in Zimbabwe', CEEPA Discussion Paper Series, Harare, Zimbabwe

Maliondo, S. M. S., Abeli, W. S., Meiludie, R. E. L. O., Migunga, G. A., Kimaro, A. A. and Applegate, G. B. (2005) 'Tree species composition and potential timber production of a communal miombo woodland in Handeni District, Tanzania', *Journal of Tropical Forest Science*, vol 17, pp104–120

Marzoli, A. (2007) 'Inventário florestal nacional. Relatório final. Direcção Nacional de Terras e Florestas', Ministério da Agricultura, Maputo, Mozambique

Mushove, P. T. (1993) 'Fifty-five years of timber commercialization in Tsholotsho communal land', in G. D. Pierce and D. J. Gumbo (eds) *The Ecology and Management of Indigenous Forests in Southern Africa*, Zimbabwe Forestry Commission and SAREC, Harare

NationMaster.com (2004) *Map & Graph: Africa*, www.nationmaster.com

Orthmann, B. (2005) 'Vegetation ecology of a woodland-savanna mosaic in central Benin (West Africa): Ecosystem analysis with focus on the impact of selective logging', PhD thesis, Rostock University, Germany

Pérez, M. R., de Blas, D. E., Nasi, R., Sayer, J. A., Sassen, M., Angoué, C., Gami, N., Ndoye, O., Ngono, G., Nguinguiri, J. C., Nzala, D., Toirambe, B. and Yalibanda, Y. (2005) 'Logging in the Congo Basin: A multicountry characterization of timber companies', *Forest Ecology and Management*, vol 214, pp221–236

Saket, M. (1994) 'Report on the updating of exploratory national forest inventory', National Directorate of Forests and Wildlife, Maputo, Mozambique

Schöngart, J., Ortrhmann, B., Hennenberg, K. J., Porembaski, S. and Worbes, M. (2006) 'Climate-growth relationships of tropical species in West Africa and their potential for climate reconstruction', *Global Change Biology*, vol 12, pp1139–1150

Schwartz, M. W., Caro, T. M. and Banda-Sakala, T. (2002) 'Assessing the sustainability of harvest of *Pterocarpus angolensis* in Rukwa Region, Tanzania', *Forest Ecology and Management*, vol 170, pp259–289

Sitoe, A. A. and Bila, A. (2006) 'Manual para a elaboração e implementação do plano de maneio da concessão florestal', MINAG, DNTF, Maputo, Mozambique

Sitoe, A. and Tchaúque, F. (2007) 'Trends in forest ownership, forest resources tenure and institutional arrangements in Mozambique: Are they contributing to better forest management and poverty reduction? A case study from Mozambique', FAO, Rome

Sitoe, A., Bila, A. and Macqueen, D. (2003) 'Operacionalização das Concessões Florestais em Mozambique', MADER/Direção Nacional de Florestas e Fauna Bravia/PROAGRI, Maputo, Mozambique

Stahle, D. W., Mushove, P. T., Cleaveland, M. K., Roig, F. and Haynes, G. A. (1999) 'Management implications of annual growth rings in *Pterocarpus angolensis* from Zimbabwe', *Forest Ecology and Management*, vol 124, pp217–229

Taquidir, M. and Cuco, A. (2006) 'Levantamento dos recursos do bambu usando técnicas avançadas de inventário com sensors remotos, para desenvolvimento de estratégia para conter a floração do bambu e conservação da biodiversidade', in IUCN (ed) *Fundo de Gestão dos Recursos Naturais e Ambiente - Resumos das iniciativas implementadas no período 2003–2005*, IUCN, Gland

Trouet, V., Coppin, P. and Beechman, H. (2006) 'Annual growth ring patterns in *Brachystegia spiciformis* reveal influence of precipitation on tree growth', *Biotropica*, vol 38, pp375–382

Verbelen F. (2002) *Role of the African Timber Trade in the Creation of Conflict and Poverty: A Call for Action, Illegal Logging Report*, Greenpeace, Gland

Vlosky, R. P. and Aguire, J. A. (2001) 'Increasing marketing opportunities of lesser known wood species and secondary wood products in tropical Central America and

Mexico', Louisiana Forest Products Development Center, Baton Rouge, US
Watts, J. (2005) 'China consumes forests of smuggled timber: Crime syndicates run huge rackets, say campaigners', *Guardian*, 22 April
Werren, M., Lowore, J., Abbot, P., Siddle, B. and Hardcastle, P. (1995) 'Management of miombo by local communities', Forestry Research Institute of Malawi and Department of Forestry, University of Aberdeen, Zomba, Malawi
Wily, L. A. (2000) 'Forest law in eastern and southern Africa: Moving towards a community-based forest future?', *Unasylva*, vol 203, pp19–26
WWF (2006) *Failing the Forests: Europe's Illegal Timber Trade*, WWF-UK, Godalming
World Bank/WWF Alliance (2002) 'Forest Law Assessment in Selected African Countries', SGS Global Trade Solutions (GTS), Washington, DC
Zambia Forestry Department and FAO (2005–2008) *Integrated Land Use Assessment (ILUA), Zambia, 2005–2008*, FAO, Rome
Zharare, G. E. and Mudavanhu, H. T. (2000) 'A case study of the use of wood from natural woodlands in two rural areas in Zimbabwe', *Silva Carelica*, vol 34, pp34–54

7

Woodfuel

Rogers Malimbwi, Emmanuel Chidumayo, Eliakim Zahabu, Stephano Kingazi, Salome Misana, Emmanuel Luoga and Jean Nduwamungu

Introduction

A high proportion of the population in sub-Saharan Africa is dependent on traditional energy sources (firewood, charcoal and organic wastes). The share of biomass fuels in national energy consumption in the majority of dry forest and woodland countries is large, with ranges of 35–75 per cent in many countries (e.g. Senegal, Togo, Ivory Coast and Angola) and over 75 per cent in many others (e.g. Sudan, Ethiopia, Kenya, Tanzania, Mozambique, Zambia, DRC and Nigeria). Access to modern or commercial energy sources by households is very low, at about 10 per cent, and the energy future does not appear bright for the majority of the African people. For example, the New Partnership for Africa's Development only proposes a modest 25 per cent increase in the number of households with access to reliable and affordable commercial energy supply in 20 years (from the current 10 per cent to a projected 35 per cent). Woodfuel (firewood and charcoal) will therefore play a significant role in the livelihoods of the majority of the people in the region.

In spite of the over-dependence on woodfuel, significant problems remain in estimating the amounts consumed. Methods of determining the amount of wood consumed vary considerably and the majority of statistics tend to be unreliable. As the Director of Forest Products and Economics Division of the Forest Department of FAO once commented 'determining the consumption of firewood or charcoal in a given country is always a trying experience because definitions are rarely consistent, the measurement units are different and discrepancies among reported values are so wide that one remains utterly confused' (Drigo, 2005).

This chapter attempts to describe the woodfuel situation and its dynamics in dry forest and woodland countries of sub-Saharan Africa. The chapter also considers options for sustaining the woodfuel system as the region struggles to expand household access to commercial energy sources.

Rural and Urban Use and External Trade

Dry forests and woodlands are a main source of woodfuel consumed in both rural and urban areas. In rural areas, the majority of households use firewood but much of this consumption is rarely recorded in national statistics; therefore the overall consumption of firewood in rural areas tends to be based on limited locality-specific household surveys. Except in wood deficit areas, firewood used in rural areas is obtained from dead wood or wood cut for other purposes and therefore represents a waste product from the forest. In contrast, firewood and charcoal consumed in urban areas are mainly produced from wood cut to supply urban markets. As a result there is greater official concern about the perceived negative environmental impacts of urban wood harvesting for the urban woodfuel markets; greater efforts have been made to quantify urban woodfuel trade, especially of charcoal, and in some cases attempts have been made to ban charcoal production for the urban market. For example, the government of Burkina Faso issued decrees against charcoal production in 2004 and 2005 in order to reorganize the charcoal industry but these have since been revoked. Nevertheless, due to its easy transportation and storage, charcoal is increasingly becoming the preferred urban household cooking fuel in many major cities. For example, of the 30 or so major cities in sub-Saharan Africa, charcoal is the main household cooking fuel in 70 per cent of these cities. Figure 7.1 presents the importance of firewood and charcoal as major urban energy sources in countries with dry forest. In over 50 per cent of the countries charcoal is the dominant urban household energy source; firewood is dominant in 20 per cent of the countries; in the remaining 28 per cent, both firewood and charcoal are important urban household energy sources. Much of what follows in this chapter will therefore be on charcoal with only occasional reference to firewood.

There is very little international trade in firewood but potential exists for charcoal export. However, international trade in charcoal has generally been discouraged in most countries because of its impact on forests and woodlands. For example, Kenya banned charcoal exports to countries in the Middle East in the 1970s because of the extensive deforestation that it caused in the coastal regions of the country. The Somali government also banned the export of charcoal to Saudi Arabia and Dubai for similar reasons but enforcement has been difficult from the 1990s due to the absence of a stable government (IRIN News, 2006; Oduori et al, 2006).

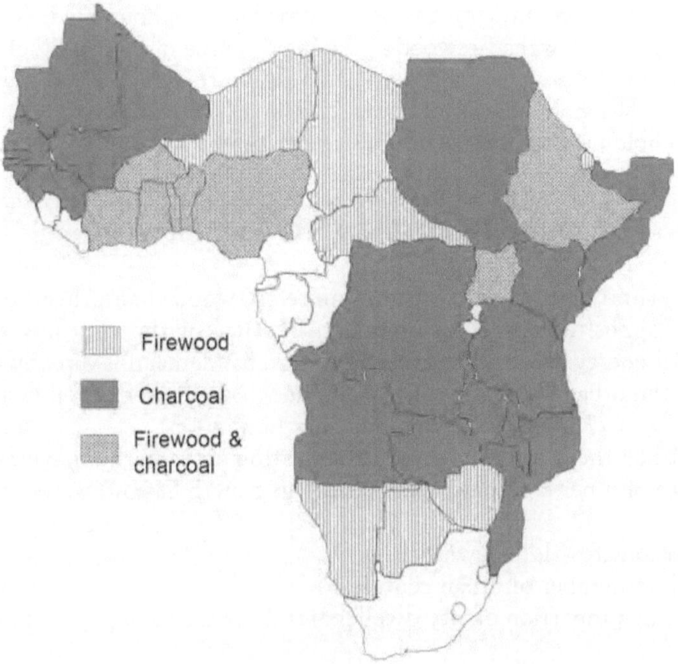

Figure 7.1 *Major types of household woodfuel used in urban areas in sub-Saharan Africa*

Source: Based on data and information from various sources

The per capita consumption of firewood varies between countries. In Tanzania, for example, firewood constitutes 97 per cent and 4 per cent of cooking and lighting fuel, respectively, in rural areas (Kaale, 2005). Subsistence firewood consumption in the miombo woodlands surrounding Kitulanghalo Forest Reserve in eastern Tanzania is 1100kg per capita per year (Luoga et al, 2000). In Nigerian urban areas approximately 97 per cent of households buy their firewood and only about 3 per cent collect it, while in the rural areas 45 per cent buy their firewood, and 55 per cent collect it (Alabe, 1994). In Kenya, firewood is the most common type of energy with close to 89 per cent of rural and 7 per cent of urban households reporting regular use of firewood, giving a national average of 68 per cent of all households. The average annual per capita consumption is approximately 741kg and 691kg for rural and urban households, respectively. In the Luapula Province of Zambia the total firewood consumption in 1996 was estimated at 353,100 tonnes (Kalumiana, 1996). Of this, 5539 tonnes are used for fish smoking annually, representing about 2 per cent of total consumption. Urban households consume 3 per cent of firewood, while rural households consume 95 per cent of firewood. The per capita annual consumption is estimated at 1025kg in rural areas and 240kg in urban areas (Kalumiana, 1996). In the city of Kano in Nigeria firewood consumption was estimated at 4000kg per household per annum with a wide variation between households (Cline-Cole et al, 1988). In Botswana firewood is the main source of biomass fuel used and total consumption was estimated at 1.42 million tonnes in 2000. On average each urban household consumes about 2230kg of firewood per annum while a rural household uses an average of 4820kg per annum (ProBEC, 2006).

Urbanization and Woodfuel

Generally, rural households consume more firewood than urban households while the opposite is true for charcoal. Because of the very low access to commercial energy sources the growth in charcoal demand is directly linked to growth in the urban population. For example, charcoal production increased by one-third from 1981 to 1992 (Kammen and Lew, 2005).

Africa has the world's highest urbanization rates with an average urban growth rate of 4 per cent per year. This urban growth has three aspects:

1. population growth of towns;
2. growth in number of urban centres;
3. increasing proportion of city dwellers *vis-à-vis* the total population.

The pattern of urbanization is generally similar throughout sub-Saharan Africa. The percentage of population residing in urban areas in Africa was 20–25 per cent in 1970, 35–40 per cent in 2000, and is projected to be 50–55 per cent in 2025 (UN Population Division, 1996). However, level of urbanization varies from region to region: 23 per cent in eastern Africa, 36 per cent in central Africa, 39 per cent in southern Africa and 40 per cent in western Africa. Of the total urban population in sub-Saharan Africa, almost 65 per cent is in countries in which dry forests and woodlands cover a significant percentage of the country.

In western Africa urban population grew sixfold from 1960 to 1990, from about 13 million to 78 million people, and the predicted future growth is that the urban population will reach 275 million in 2020 (Arnaud, 1993). Urbanization is occurring without industrialization and therefore is characterized by high levels of unemployment and dependence on traditional energy sources.

Africa's high urban growth rate is a result of rural-urban migration, population growth and in some areas, conflict. People leave rural areas because of declining agricultural productivity, lack of employment opportunities and lack of access to basic physical and social infrastructure. However, the expectation of higher incomes and standards of living in urban areas is rarely realized. This results in widespread urban poverty that continues to grow. For example, 45 per cent of urban households in southern Africa grow crops or raise livestock in urban environments in order to supplement their livelihoods (UNDP, 1996).

While firewood traditionally accounted for a major part of total wood energy consumption, the social and economic changes associated with urbaniza-

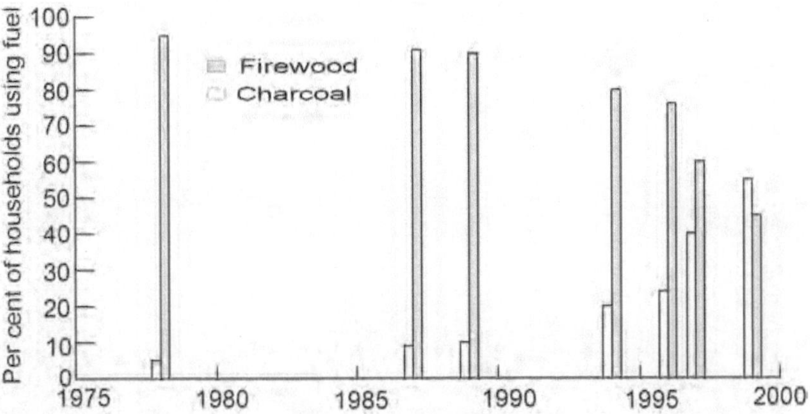

Figure 7.2 *Trends in the shift from household use of firewood to charcoal in the city of Bamako, Mali*

Source: Based on Girard (2002)

tion are leading to a significant shift from firewood to charcoal, as exemplified by the trends in Bamako, Mali (Figure 7.2). For urban dwellers charcoal is preferred to firewood because it is easy to store and more convenient to handle.

Urban Woodfuel Supply Issues

The growing and persistent urban poverty in sub-Saharan Africa can be characterized by an increasing dependence on cheap energy sources that are dominated by wood. This places heavy demands on forests and woodlands around urban centres. Since urbanization is affecting all towns, irrespective of size, each urbanizing area is depleting forest resources in its hinterland or catchment area. However, the size of the supply catchment is a function of city size. Often the source areas for woodfuel for larger towns may include catchments of one or more smaller towns. Many factors influence spatial and temporal dynamics in charcoal supply areas of major cities in sub-Saharan Africa, including forest depletion, regulation of charcoal production, land use and tenure, history etc.

The charcoal industry supplying large cities in sub-Saharan Africa operates over large areas and shifts when production areas are depleted, as in Senegal, Mozambique, Tanzania and Sudan. In Senegal, major areas of production supplying charcoal to Dakar shifted from within 200km in the 1950s to 200–300km in the 1960s and more than 300km in the 1970s and 1980s (Ribot, 1993) (Figure 7.3). Major supply areas of woodfuel to urban Maputo were within 50–60km in the 1980s but had shifted to 60–100 km in the 1990s and 150–200km in the 2000s, while charcoal came by railway from as far away as 600km (Pereira et al, 2001). The major sources of charcoal for

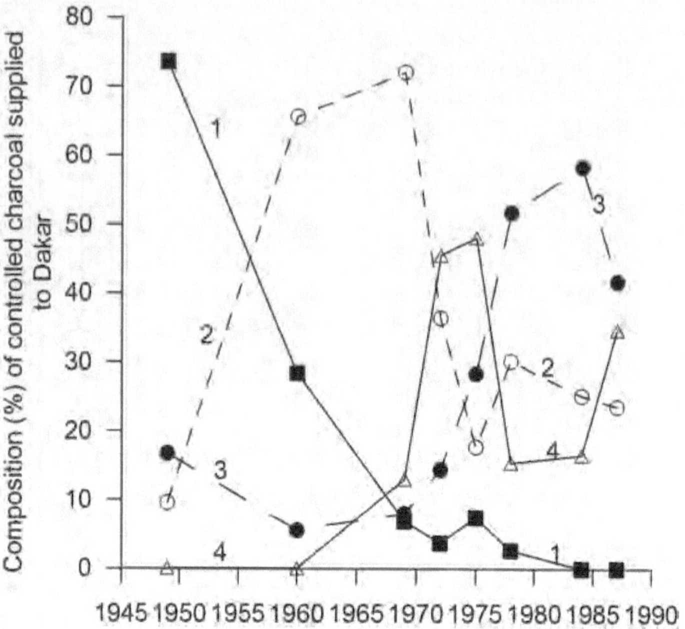

Figure 7.3 *Composition of charcoal supplied to Dakar from different regions in Senegal*

Note: Regions are based on distance from Dakar: (1) <200km, (2) 200–300km, (3) 300–400km and (4) >400km
Source: Based on Ribot (1993)

Dar es Salaam city in Tanzania used to be the neighbouring western and northern regions of Morogoro and Tanga (Luoga et al, 2002) but now the major supply source has shifted to the wood-rich southern districts of Rufiji and Kilwa following the opening of the Mkapa bridge. Similarly charcoal for Khartoum in Sudan comes from the more wooded southern regions (Khalifa, 1982). It is evident that the charcoal for large towns in dry forest and woodland countries can come from very distant source areas. The growth in the number of urban areas, that is part of the urbanization process in sub-Saharan Africa, also implies that there are now more islands of forest depletion than was the case 30–40 years ago.

However, what is most worrying about urbanization is that every 1 per cent increase in the level of urbanization is estimated to result in a 14 per cent increase in the consumption of charcoal (Hosier et al, 1993). The impact of urbanization has serious implications for the long-term well-being of rural communities. Sustainable forest management has a key role to play in maintaining charcoal supplies to urban areas while also supporting forest-based rural livelihoods.

Where does the wood come from?

Although it is generally recognized that dry forests and woodlands provide the bulk of resources for wood energy, the land use status of supply areas is rarely known. A study in Kenya found that whereas plantation forest/trees were a major source of wood for charcoal supplied to Kisumu, it was rangeland and natural forest clearings that were the main sources of charcoal supplied to Nairobi and Mombasa (Table 7.1). In contrast, firewood in Kenya is mainly from agroforestry or on-farm sources (84 per cent), trustlands (8 per cent) and gazetted forests (8 per cent) (Theuri, unpublished). Most of the charcoal in Tanzania and Mozambique is produced from general land under village control while in Burkina Faso, Mali and Niger communal and state forests are managed to supply woodfuel to major cities. However, in some other western African countries woodfuel supplies from farmed parklands are widely reported (Cline-Cole et al, 1988).

Table 7.1 *Origin of wood for charcoal production in Kenya in 1985*

Wood source	*Percentage of market share*			
	Nairobi	*Mombasa*	*Nakuru*	*Kisumu*
Plantation trees/forest	5	0	15	80
Rangeland clearing	38	40	40	0
Forest/range clearing	30	55	0	0
Forest clearing	20	0	35	0
Other	7	5	10	20

Source: Based on Kammen and Lew (2005)

The resource base and competing demands

Estimates of wood biomass in African dry forests and woodlands have been described in Chapter 2 but not all this biomass is utilized for energy purposes. A study in Kenya estimated that nearly 74 per cent of the above ground wood biomass in semi-arid dry forests was utilizable for firewood and charcoal (Western and Ssemakula, 1981). If this proportion is applied to all dry forests in sub-Saharan Africa, then the stocks potentially available for energy purposes in African dry forests and woodlands go from 11.7 tonnes per hectare in semi-arid dry forests in the Somali-Masai phytoregion to 136.3 tonnes per hectare in the sub-humid dry forests in the Congo-Zambezian phytoregion (Table 7.2).

However, dry forests and woodlands also need to be cleared for agricultural purposes (see Chapter 3). Cultivation has been a major source of woodland change and deforestation in southern Africa (Dewees, 1994) and the situation in eastern and western Africa is even worse (see Chapter 2). In Malawi, 24 per cent of the total forest area has been converted to arable land (Lele and Stone, 1989). In Zambia, cultivation was responsible for 95 per cent of the annual forest cover loss of 90,000ha in 1990 (Chidumayo, 1997). In Zimbabwe the

Table 7.2 *Utilizable wood for fuel in different dry forests in sub-Saharan Africa*

Phytoregion	*Dry forest type*	*Aboveground wood biomass utilizable for woodfuel (tonnes per hectare)*
Somali-Masai	Semi-arid dry forest	11.7 ± 1.81
Kalahari-Highveld	Semi-arid dry forest	20.6 ± 4.34
Sudanian	Warm dry forest	45.7 ± 4.29
Zambezian	Warm dry forest	68.4 ± 3.20
Congo-Zambezian	Sub-humid dry forest	136.3 ± 19.52

Note: Phytoregions and dry forests are described in Chapter 2
Source: Calculated by chapter authors from various literature sources

main decreases in woody vegetation between 1963 and 1973 were in areas of high to moderate population densities, particularly in communal areas where extensions of croplands and wood harvesting had resulted in annual cover losses of 3–10 per cent (Whitlow, 1980). In Tanzania charcoal production and cultivation were responsible for the degradation and deforestation of 25 per cent and 20 per cent of closed woodland, respectively, in the catchment area to the west and north of Dar es Salaam city (CHAPOSA, 2002). The agriculture sector is also a major user of wood energy. Often live trees are cut for fuel to cure tobacco and tea. Tobacco estates in Malawi account for 21 per cent of the national wood energy consumption (Moyo et al, 1993) and contribute nearly 47 per cent to forest cover loss caused by firewood harvesting. Another extensive use of wood biomass in agriculture is fencing of arable fields. In Botswana the fencing of fields to keep out livestock uses 1.5 times more wood than is used for cooking by farming households (Tietema et al, 1991). Other uses of wood energy include the curing of bricks, especially in rural areas. In Zimbabwe wood used for brick-making is said to equal that used for cooking in rural areas (Bradley and Dewees, 1993). Clearing natural forest for agriculture, grazing and harvesting of firewood from miombo woodlands for tobacco and tea curing, fish smoking, beer brewing, bread baking, salt-drying, and brick and lime making are considered to be the main agents contributing to deforestation. In Tanzania the magnitude of deforestation caused by unsustainable firewood consumption for tobacco curing and brick burning is estimated at around 20,000ha per annum (Kaale, 2005). To cure a hectare of tobacco requires approximately one hectare of woodland (Mnzava, 1981).

Charcoal Production

Charcoal is traditionally made in earth, brick or steel kilns. Although charcoal may be produced from plantations, it is often made from land cleared for agricultural purposes or from smaller areas specifically used for charcoal production. Where trees are abundant, of many species and variable in size, cutting

Figure 7.4 *A traditional surface earth-mound kiln for making charcoal in central Zambia*

Photo by Emmanuel Chidumayo

may be selective. The average cleared plot in central Zambia was 0.2ha and uncut trees represent about 7 per cent of the pre-cut biomass (Chidumayo, 1991). The majority (92 per cent) of uncut trees are small (dbh less than 10.0cm). The large uncut trees are either too hard to cut or make charcoal with poor burning qualities. In areas where forests are already much degraded and trees are in short supply there is no tree species selection for charcoaling. This is also true during the opening up of new agricultural land where trees of all species found in the area are cut and burnt into charcoal.

The bulk of the charcoal in sub-Saharan Africa is made in earth kilns of which there are two types: the pit kiln and the surface earth-mound kiln. The pit kiln is constructed by digging a pit or trench in the ground and then placing the wood in it after which the wood pile is covered with green leaves or metal sheets and then with earth to prevent complete burning of the wood. The earth-mound kiln is built by covering a pile of wood on the ground with earth (Figure 7.4). The mound is preferred over the pit kiln where the soil is rocky, hard or shallow, or the water table is close to the surface. In Somalia charcoal is made in both pit and surface earth-mound kilns with the former being more common in montane areas (Oduori et al, 2006).

The wood to charcoal conversion efficiency is dependent upon many factors, including kiln type, wood moisture content, tree species, wood arrangement and skill of the producers. Bailis (2003) reported that in the traditional

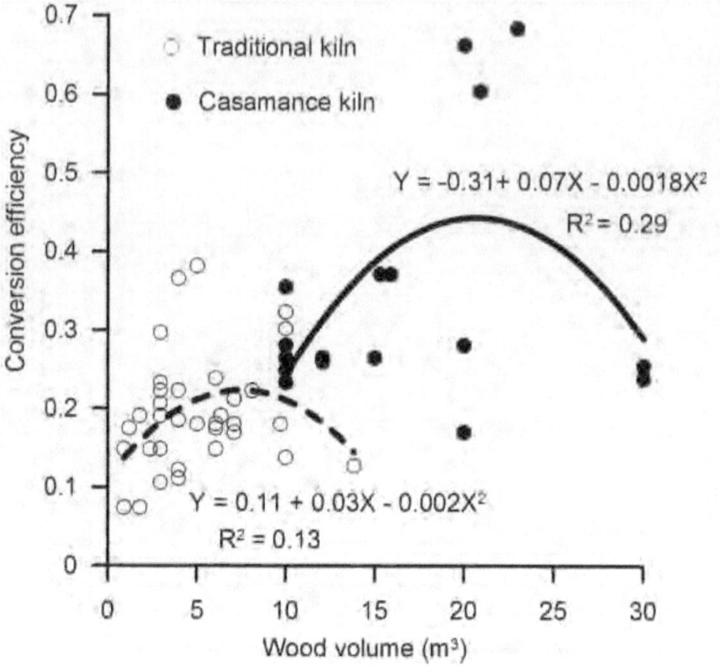

Figure 7.5 *Wood to charcoal conversion efficiency in the traditional earth kiln (○) and Casamance kiln (●)*

Note: The equations given describe the plots for each of the two kiln types
Source: Based on data in Kammen and Lew (2005)

earth-mound kiln 5–10 tonnes of wood are needed to make 1 tonne of charcoal (at a mass-based conversion efficiency of 10–20 per cent). Thus, using such traditional mound kilns, 60–80 per cent of the wood's energy is lost in the charcoal production process. Traditional mound kiln efficiencies in Zambia were in the range 20–28 per cent while in Mozambique the range was 14–20 per cent and in Tanzania it averaged some 19 per cent (CHAPOSA, 2002). These efficiency values are similar to those reported by Chidumayo (1991), Sawe and Meena (1994), Hofstad (1995) and Kaale (1998).

Efficiency tests conducted on a Casamance kiln, a modified surface earth-mound kiln with a chimney that improves air flow out of the kiln, and the traditional mound kilns in Senegal, showed that although the Casamance kiln is more efficient than the traditional kiln, there was a large amount of variation within each kiln (Figure 7.5), suggesting that other factors may be more important in determining efficiency. For example, lack of proper control during the carbonization process has been reported to reduce efficiency due to some complete combustion of wood (Ishengoma and Nagoda, 1991) and in Zambia experienced charcoal producers achieved higher wood-charcoal production efficiency than less experienced producers.

Attempts have been made to improve kiln efficiencies in many countries. In Tanzania, some of the proposed technologies and techniques include: portable steel kilns, improved traditional earth kilns, and 13 half-orange brick kilns. These methods have been under investigation for over 30 years (FORINDECO, 1992). Studies by the Tanzania Forestry Research Institute also enabled development of improved earth mound charcoal kiln reinforced with burned bricks but uptake of the results has not been successful due to the high cost involved (Sawe and Meena, 1994). The average carbonization efficiency of these improved technologies was estimated to be 27–35 per cent. Portable steel kilns achieve considerably higher conversion efficiencies but have not been adopted because of the higher initial cost, manpower and skill required for viable management. A problem with marketing of softwood charcoal (low calorific value), the production of which is technically most feasible using improved kilns, is also a limiting factor in the adoption of these technologies (FORINDECO, 1992).

Biomass Regeneration after Charcoal Production

Forest degradation refers to less obvious changes in the woody canopy cover while deforestation is the more or less complete loss of forest cover that is often associated with forest clearance (Grainger, 1999). Degradation therefore represents the temporary or permanent reduction in the density, structure, species composition or productivity of vegetation cover. Annual deforestation rates differ among countries: Tanzania (0.3 per cent), Zimbabwe (0.4 per cent), Zambia (0.2 per cent), Botswana (0.1 per cent), Mozambique (0.8 per cent), Malawi (3.3 per cent) and Angola (0.2 per cent) (Chambwera, 2004).

The degree of forest clearing for charcoal production varies considerably among countries and sites within each country. Generally the impact of producing a specified amount of charcoal depends on the stocking rate that varies among the different dry forests and woodlands (see Chapter 2 and Table 5.3). In Mozambique, charcoal production is characterized by a 'clear-felling system' since almost all species are used (Pereira et al, 2001). In the Morogoro region of Tanzania, there is little species and size selection in tree felling for charcoal, such that there is virtual clear-felling of the woodland around a kiln site (Luoga et al, 2002). However, harvesting intensity tends to decline with increasing distance from settlements and transportation routes. Chidumayo (1991) observed that 97 per cent of the standing wood biomass was harvested for charcoal production in central Zambia. The extent to which tree harvesting for charcoal production causes deforestation remains debatable. On one hand, Luoga et al (2002) reported that the removal of 4.64 tonnes per hectare for charcoal production in public lands in the Morogoro region of Tanzania exceeded the mean annual increment of 3.15 tonnes per hectare and therefore concluded that harvesting was changing the structure and composition of the vegetation and was unsustainable. On the other, Ribot (1999) argued that deforestation in the western African dry forests and woodlands due to wood fuel production has not been demonstrated. Under severe wood resource degradation, even stumps left over from previous charcoal production may be dug up and used to make charcoal (Oduori et al, 2006); this reduces the forest capacity to recover.

Clear cutting removes the entire canopy and for dry forests and woodlands that regenerate vegetatively from stump sprouts and resprout saplings, such as miombo woodland, this is the best way of encouraging regeneration (Chidumayo, 1997, 2004). Such forest regrowth usually has higher tree species diversity than old-growth forest that it replaces and the danger of species loss is minimized (Chidumayo, 1997). The main disadvantage of clear cutting is that it tends to increase run-off, and possibly soil erosion, from cleared catchments (Mumeka, 1986). In order to minimize this negative effect, cleared strips or coupes should alternate with shelterbelts to buffer cleared strips against increased run-off and soil erosion. Other environmental effects that are associated with sites that are clear cut and burnt include severe visible crusting and soil erosion (Hosier, 1993). Furthermore, clear cut sites are easily converted to cropland which discourages regeneration hence contributing to deforestation.

Africa's dry forests and woodlands have the potential to regenerate after clearing for charcoal (see Chapter 2). Observations made in eastern Senegal indicate that woodland areas were cut for charcoal in 1940, 1969 and again in the mid-1980s (Ribot, 1999). That charcoal makers return to harvest the same area after 9–12 years has also been reported in Mali, Niger and Burkina Faso (Nygård et al, 2004). In Zambia, return periods to previous clear cut areas for charcoal production range from 20 to 30 years (Chidumayo et al, 2001). In Kenya a 14-year rotational harvesting period has been proposed for regrowth acacia woodland (Okello et al, 2001). Woodlands in Tanzania appear to recover relatively well following harvesting for charcoal production (Hosier, 1993) but recovery periods of degraded woodland after charcoal production range from 8 to 23 years (Malimbwi et al, 2001). Even so, large areas for charcoal production are needed if production is to be sustainable because long periods are needed between harvest activities for regrowth to develop adequately. Apparently, the regeneration rates are a function of woodland type, rainfall, fire management and grazing intensity (Hosier, 1993; Ribot, 1993) and often the recovery period is prolonged through heavy grazing and uncontrolled burning.

Urban Woodfuel Trade

Charcoal use has socio-economic benefits for numerous actors along the chain from the producers in rural areas to the consumers in urban areas. In many African countries, the charcoal industry has the characteristics of a competitive market with large numbers of producers, transporters and sellers (both wholesalers and retailers) with little government regulation in practice (CHAPOSA, 2002). In Kenya the charcoal industry was estimated to include 30,000 full-time

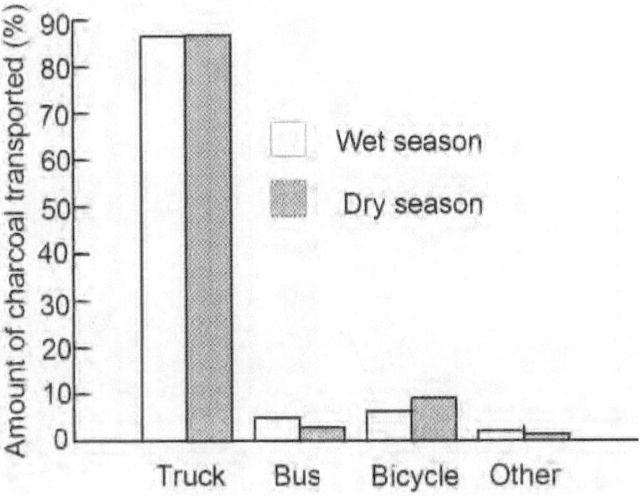

Figure 7.6 *Charcoal transportation into Dar es Salaam in 2000*

Source: Based on Malimbwi et al (2004)

producers, 4000 transporters and 800 retailers (Kammen and Lew, 2005) and represented an annual business of about KSh17 billion (US$213 million) in 2002, which was about 53 per cent of the 1998 bill for imported oil (Theuri, unpublished). Charcoal production in central Zambia in 1990 was estimated to involve 10,400 producers and about 3500 households in retail trade and the whole charcoal industry was worth ZK4.7 billion (US$15 million) or 2.3 per cent of GDP in 1991 (Hibajene et al, 1993) while the World Bank/UNDP (1990) projected that foreign exchange savings due to wood energy use in Zambia would be US$4 million annually for the period 1992 to 2000. In the Lusaka province of Zambia a study revealed that although the rural per capita income from forestry and crop agriculture declined from US$37.07 in 1990 to US$17.33 in 2000, the contribution of charcoal production had increased from 65 per cent in 1990 to 83 per cent in 2000 (Chidumayo et al, 2001). A similar study in Mozambique found that 65 per cent of the household income in the Licuati region that supplied charcoal to Maputo city was generated from charcoal production (Pereira et al, 2001). In some cases, income from charcoal sales is used to buy agricultural inputs and in this way, dry forests subsidize agricultural production and therefore contribute to household food security. According to CHAPOSA (2002), in areas with reasonable accessibility, charcoal is the main cash income source for rural households. In eastern Tanzania, in 1992 charcoal contributed substantially to the annual household income of communities adjacent to the Morogoro–Dar es Salaam highway who earned about US$176 per household annually (Monela et al, 1993).

Charcoal is transported to urban markets by a variety of means, including trucks, buses, bicycles and wheelbarrows. Trucks usually account for the largest charcoal volumes as has been observed in Tanzania (Figure 7.6) while in Zambia

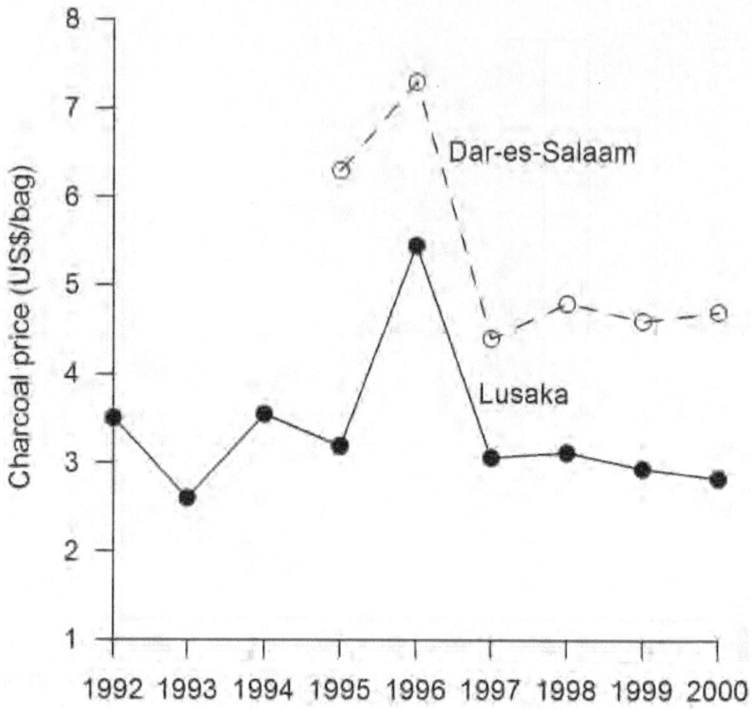

Figure 7.7 *Trend in charcoal price in Dar-es-Salaam in Tanzania and Lusaka in Zambia during the 1990s*

Source: Based on CHAPOSA (2002)

trucks were estimated to carry 97 per cent of charcoal to urban markets in 1990.

Charcoal retailing in many African cities is well structured so as to make charcoal accessible to different consumers. Charcoal prices have not generally increased in real terms. In most African cities for at least the last two decades, urban consumers have been paying slightly less than US$0.10 per kg of charcoal (CHAPOSA, 2002). Charcoal is retailed in a variety of different quantities. At the one extreme of the retailing system there are outlets for bulk purchase that are often along main city roads. At the other extreme near consumer households (often less than one or two minutes walk), there are small shops that sell charcoal in small tins or by small heaps (Mangue, 2000).

It is also surprising that the increasing cost of charcoal transportation due to increasing distances does not always appear to influence the price of charcoal in urban markets (Figure 7.7). One reason for this is that different modes of transportation are used for different distances. Often small trucks are used over short distances while larger trucks, and even trains, are used for longer distances which buffers the charcoal price against increasing distances to production areas due to economies of scale.

Woodfuel Policies and Legislation

Land tenure and woodfuel production

In the realm of land law, the term 'land' embraces not only the physical surface but also buildings and other infrastructure, as well as growing trees (Laltaika, 2005). The term tenure on the other hand refers to the bundle of "rights" (and obligations) that a person, community or private entity holds (Bruce, 1989). Some tenure systems confer property rights to standing trees but not the land on which they stand. Some specific concepts relevant to land tenure are:

- 'Freehold', a traditionally western concept implying the absolute right to control, manage, use and dispose of pieces of property.
- 'Leasehold', in which land belonging to one entity is by contractual agreement leased to another entity for a fixed period of time.
- 'Statutory allocations', a particular form of state land where such land by virtue of some statutory provision, is allocated for the use by some legally constituted body.
- 'Customary systems', in which tenure rights are presumably controlled and allocated according to traditional practice.

Generally speaking, most of the legislation relating to forestry in most of the sub-Saharan countries was inherited from the colonial governments. During the colonial period, indigenous peoples' rights to harvest and dispose of trees were significantly restricted. Similarly, after independence, forest policies in many developing countries have been characterized by the strong concentration of power over forest resources in the central state apparatus, and the corresponding lack of local participation in forest and tree management. Failure to recognize indigenous systems of forest management and indigenous rights to resources at policy level has led to:

- loss of incentives by the local communities to protect trees, encouraging indiscriminate tree felling;
- discouragement of local people to engage in tree planting and reforestation projects;
- excessive reliance by the state on punitive measures to enforce the law.

Lawry (1990) argues that where forests have little economic value to local people because of restrictive access rules, sustainable local management institutions are unlikely to emerge. Incentives for conservation by local people can be improved by increasing the value of the resource to local people by, for example, granting more access rights or by granting local communities a percentage of forest concession revenues. A few countries in the region, such as Tanzania, have implemented such measures. Insecurity of tenure, among other factors, has promoted open access to dry forests.

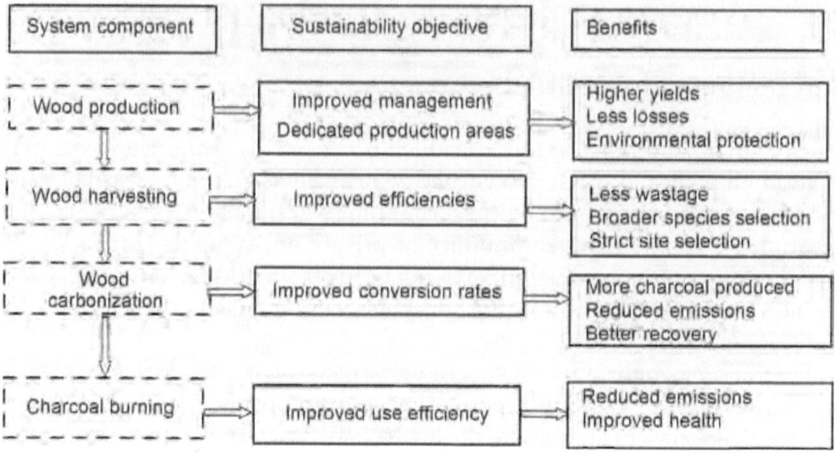

Figure 7.8 *Components of the charcoal energy system and how they contribute to the sustainability of charcoal production and use*

Source: Designed by Chapter authors

In many countries any person that makes charcoal for commercial purposes from wood taken from the forest should obtain a licence. A licence gives a buyer (holder) an exclusive right to utilize the forest produce for which the licence has been issued. However, in many countries illegal or unlicensed charcoal production predominates. For example, almost all charcoal that enters Maputo city is illegal and the licensing system is estimated to cover about 1 per cent of the biomass consumed in Maputo city (Pereira, 2000). In Tanzania only 25 per cent of the estimated 24,500 bags of charcoal consumed daily in Dar es Salaam city is accounted for at the road checkpoints (CHAPOSA, 2002). This situation occurs in many countries because law enforcement is weak and the procedure of securing a licence or permit is cumbersome.

Making woodfuel sustainable

The charcoal system consists of four components: wood production, wood harvesting, wood carbonization and charcoal burning, as shown in Figure 7.8. Each of these components has specific sustainability objectives as well as associated benefits as outlined in Figure 7.8. Achieving these objectives requires appropriate measures that cumulatively contribute to the sustainability of the urban wood energy system.

Wood production

At the wood production level, charcoal production may stimulate tree planting. In areas with high woodfuel demand tree planting in either woodlots or agroforestry systems is inevitable. In these areas woodfuel scarcity is the driving factor to motivate people to plant trees. Agroforestry is recommended where the land area is limited. The choice of tree species to plant is very important in order to avoid competition with agricultural crops. In areas where shortage of woodfuel has not reached a critical level, people are likely to be more motivated to plant trees that provide multiple utilities, such as timber, fruits and fodder, than those that provide only woodfuel.

Degraded areas have the potential to regenerate naturally if they are properly managed, especially by protecting them from disturbances such as destructive fires and overgrazing. Most African countries are currently introducing new forestry strategies that cater for greater participation by local communities and other players in forestry development. A fair benefit sharing arrangement between the government and local communities should be set in place as an incentive for the latter to invest in forestry. Clarification of land tenure and rights are also needed. This includes allocation of forestland to village communities for charcoal production according to approved management plans. In Tanzania local communities are now managing most general lands. This has shifted the situation from an open access regime to a common property system.

Wood harvesting

In terms of wood harvesting, two options exist. The first is selective harvesting of trees large enough for charcoal production in order to allow the small trees to grow. A necessary condition is that the small trees must not be severely damaged by the cut trees. The second is clear-felling in which all trees are cut for charcoal production. Both regimes require demarcation of the forest or woodland into blocks that can be allocated to different felling cycles. This latter system is already being implemented in Burkina Faso (Sawadogo et al, 2002; Nygård et al, 2004). Such systems would be best implemented as part of the move to involve local communities in forest management. To ensure successful regeneration in clear-felled areas, the areas should be protected against disturbance, such as cultivation, grazing and fires. There is usually a tendency for local people to cultivate a clear-felled area as the land is perceived to be more fertile and require less labour to clear. This should be discouraged if the objective is to dedicate such areas to charcoal production.

Wood carbonization

Improved kilns could contribute significantly to production efficiency. However, in spite of their efficiency, the use of improved kilns has failed due to lack of capital for kiln construction. The need to process the billets into specific sizes and transport them to kiln sites is an added cost that is limiting adoption. However, there is evidence that experienced producers who use traditional kilns achieve more efficiency than less experienced ones. This calls for studying the techniques used by the experienced producers and disseminating them to less experienced producers. For example, a manual for best practice for charcoal making based on the experience of charcoal producers has been produced in Zambia (CHAPOSA, 2002). In the current system there are no incentives for charcoal makers to adopt efficient production technologies, for a combination of reasons, including: unrealistic fees and royalties; open access systems; weak government monitoring of forests and reserves; haphazard issuance of permits (legal and otherwise); ignorance; and long term problems associated with land and tree tenure. The current situation does not provide the incentives for efficient charcoal production. In most cases the wood for charcoal making is regarded as a free commodity. A fresh set of initiatives is thus required to encourage the use of improved kilns to address the inadequacies and constraints cited above. Such initiatives need to encompass economic, legal and technological aspects.

Charcoal burning

Low-income households normally spend a higher fraction of their income on energy compared to other groups. In the case of charcoal, they pay more for the same level of energy service because their appliances are generally less efficient than those used by wealthier households, and also because low-income households usually purchase fuel in smaller amounts.

Direct energy subsidies are often provided to the poor but they are often ineffective since in most instances, the poor receive only a small fraction of the total subsidy. For example, in the case of kerosene use in Tanzania, subsidies largely benefited better off households who were able to purchase kerosene. Most electricity supply companies have a tariff structure that is meant to encourage low-income households to use electricity for cooking but such schemes often leave out the appliances to be used. Since efficient devices, such as electric or gas cookers or improved stoves, tend to have higher initial costs, poor households typically do not buy them. There may be room for subsidising cooking devices.

In order to reduce charcoal consumption it is important to encourage innovation, dissemination and adoption of efficient charcoal stoves. Reasons for the slow adoption rates include the higher initial costs of shifting from standard (less efficient) stoves to improved types. Furthermore, the current charcoal price does not exert sufficient pressure to cause the shift.

At the household level adoption of improved kitchen management skills is also essential in reducing woodfuel consumption (Kaale, 2005). Such skills include:

- use of dry firewood to increase burning efficiency;
- extinguishing woodfuel after cooking;
- pre-treatment of some food stuffs through soaking to reduce cooking time;
- construction of cooking shelters instead of cooking in the open to increase woodfuel utilization efficiency.

Possible alternative sources of energy for most African countries are hydroelectricity, kerosene, solar, coal, wind and biogas. Johnsen (1999) made a good review of these alternative sources of energy for the case of Tanzania. In his review he acknowledged that kerosene was the most common fuel for cooking next to woodfuel. However, this was not an interesting policy option because kerosene has to be imported, and subsidized by the government. This means that a switch to kerosene would add to the stress on both the state's finances and the country's scarce foreign currency. He concluded that none of these alternatives would become widespread enough to have any significant impact on the consumption of woodfuel in Tanzania. The reason for the limited success in implementation of alternative energy sources is that a large number of the renewable energy technology projects are 'beyond the financial reach of their target groups: the rural and urban poor'. Most countries that had much potential for hydroelectricity are now realizing that this is not a reliable source of energy. For example in 1990s Tanzania was said to have large hydropower potential estimated at 48,000MW, while only 300MW were utilized by the existed plants (Kjellstrøm et al, 1992 in Johnsen, 1999). Currently only 25 per cent of all the electricity in the country is generated from hydroelectricity and the major portion is from diesel and natural gas-powered engines. This is mainly caused by prolonged drought as has occurred in many sub-Saharan countries.

Key Issues and Challenges

Urbanization is an ongoing process in sub-Saharan Africa and does not appear to be associated with real industrialization. As a result, high levels of unemployment and over-dependence on traditional biomass fuels characterize urbanization. As cities grow larger, so does the demand for woodfuel and production areas progressively move further away into distant rural areas. Paradoxically, increasing distances to charcoal production areas does not necessarily translate into higher real fuel prices, possibly because of dynamic changes in the charcoal transportation system; larger vehicles replacing smaller vehicles as distances to forest areas increase. From a welfare perspective, this is a good thing as it maintains affordable charcoal prices among the urban poor. However, without a corresponding investment in proper dry forest and woodland management, the environmental costs of forest clearing are likely to be much higher than is currently perceived by policy-makers.

Dry forest and woodland clearing for woodfuel production does not necessarily result in permanent deforestation, and forest recovery is possible with good management. Good management requires dedicated production areas, development of harvesting plans and enforcement, efficient revenue collection, improvement in charcoal production efficiency, control of forest conversion to other land uses, protection of harvested areas from uncontrolled bush fires and overgrazing and, in some cases, tree planting, especially of fast growing indigenous species, such as those of the genus *Acacia*. Unfortunately, few countries are making adequate investment, both human and financial, in dry forest and woodland management. There is also a general lack of enabling legislation, policies and political commitment to encourage the participation of the private sector and local communities in sustainable forest management.

Woodfuel production contributes significantly, in some cases 60–80 per cent, to rural household income and therefore to poverty reduction in production areas. Sustainable dry forest and woodland management is therefore key to the maintenance of forest-based income generation in rural areas supplying charcoal to urban markets.

The following recommendations are made to support good dry forest and woodland management for the sustainable supply of charcoal to urban markets:

- Given that the environmental and social (loss of forest has negative impacts on welfare given that forests are important safety nets for the poor) costs of woodfuel production potentially outweigh the benefits, governments should review extra-sectoral policies, including urban, land and energy policies, and provide targeted subsidies that can shift energy use to alternative or more efficient energy systems.
- Governments should, through a process of participatory engagement with local communities and traditional leaders, set aside areas that can be dedicated and managed for sustainable woodfuel production. In addition, previous charcoal production areas that are not converted to other land uses should be surveyed, demarcated and brought under proper forest management to promote forest regeneration for future woodfuel production. Different management regimes, ranging from state to community and private sector, should be explored and implemented depending on the prevailing circumstances. For communities to actively and successfully participate in forest management, legal and administrative provisions that ensure that communities benefit from forest utilization and management should be implemented.
- Local communities are sometimes better placed to collect stumpage fees at production sites than forest departments that often wait for producers to come and pay the stumpage fee at forest department offices. Local communities should, therefore, be legally and administratively empowered, and local capacity developed, to collect and account for the revenue from stumpage for woodfuel production. A portion of the revenue should be invested in sustainable forest management.
- Governments should continue, and improve, the collection of woodfuel conveyance fees but should ensure that a portion of the revenue is spent on sustainable forest management.

References

Alabe, M. (1994) 'Rural and urban energy assessment in developing countries', Published Seminar Paper presented to the Faculty of Technology, Buyero University, Kano, Nigeria

Arnaud, M. (1993) 'Urbanization in West Africa: Local ways and local thinking', *OECD Working Document* no 8

Bailis, R. (2003) 'Environmental and Socio-economic Impacts of Charcoal Production in Kenya', Paper presented at the International Seminar On Bioenergy and Sustainable Rural Development, Morelia, Mexico, 26–28 June, Mexico City

Bradley, P. and Dewees, P. (1993) 'Indigenous woodlands, agricultural production and household economy in the communal areas', in P. N. Bradley and K. McNamara (eds) *Living with Trees: Policies for Forestry Management in Zimbabwe*, World Bank, Washington, DC

Bruce, J. W. (1989) 'Rapid appraisal of tree and land tenure', *Community Forestry Note* 5, FAO, Rome

Chambwera, M. (2004) 'Economic analysis of urban fuelwood demand: The case of Harare in Zimbabwe', PhD thesis, Wageningen University, Wageningen, The Netherlands

CHAPOSA (2002) *Charcoal Potential in Southern Africa (CHAPOSA)*, Stockholm Environment Institute, Stockholm

Chidumayo, E. N. (1991) 'Woody biomass structure and utilization for charcoal production in a Zambian miombo woodland', *Bioresources Technology*, vol 37, pp43–52

Chidumayo, E. N. (1997) 'Woodfuel and deforestation in Southern Africa – A misconceived association', *SEI Renewable Energy for Development*, vol 10, pp3–4

Chidumayo, E. N. (2004) 'Development of *Brachystegia-Julbernardia* woodland after clear-felling in central Zambia: Evidence for high resilience', *Applied Vegetation Science*, vol 7, pp237–242

Chidumayo, E. N., Masaileti, I., Ntalasha, H. and Kalumiana, O. S. (2001) 'Charcoal potential in southern Africa (CHAPOSA) – report for Zambia', Stockholm Environment Institute, Stockholm, Sweden

Cline-Cole, R. A., Falola, J. A., Main, H. A. C., Mortimore, M. J., Nichol, J. E. and O'Reilly, F. D. (1988) 'Wood fuel in Kano, Nigeria: The urban-rural conflict', *Social Forestry Network Paper* 7b

Dewees, P. A. (1994) 'Social and economic aspects of miombo woodland management in southern Africa. Options and opportunities for research', *CIFOR Occasional Paper* 2, Bogor, Indonesia

Drigo, R. (2005) *Interactive Wood Energy Statistics: i-WESTAT Update 2004*, FAO, Rome

FORINDECO (1992) 'Project proposal for the development of national woodfuel strategies and plans for Tanzania', Norwegian Agricultural University, Ås, Norway

Girard, P. (2002) 'Charcoal production and use in Africa: What future?', *Unasylva*, vol 53, pp30–35

Grainger, A. (1999) 'Constraints on modeling the deforestation and degradation of tropical open woodlands'. *Global Ecology and Biogeography*, vol 8, pp179–190

Hibajene, S. H., Chidumayo, E. N. and Ellegard, W. (1993) *Summary of the Zambia Charcoal Industry Workshop: Policy and Management Challenges for the Future, 10–14 May 1993, Siavonga, Zambia*, Stockholm Environment Institute, Stockholm

Hofstad, O. (1995) 'Wood and deforestation by charcoal supply to Dar es Salaam', *Journal of Environmental Economics and Management*, vol 33, pp17–37

Hosier, H. H. (1993) 'Charcoal production and environmental degradation', *Energy Policy*, vol 21, pp491–509

IRIN News (2006) *Somalia: Charcoal Trade on the Rise*, IRIN News.org/UN Office for the Coordination of Humanitarian Affairs, Mogadishu

Ishengoma, R. C. and Nagoda, L., (1991) *Solid Wood Physical and Mechanical Properties, Defects, Grading and Utilization as Fuel. A Compendium*, NORAD, Ås, Norway

Johnsen, F. H. (1999) 'Burning with enthusiasm: Fuelwood scarcity in Tanzania in terms of severity, impacts and remedies', *Forum for Development Studies*, vol 1, pp107–131

Kaale, B. K. (1998) 'An overview of the energy sector in Tanzania and the involvement of women as compared to men', Tanzania national consultation workshop on women and sustainable energy, TATEDO, Dar es Salaam

Kaale, B. K. (2005) 'Baseline study on biomass energy conservation in Tanzania', SADC Programme for Biomass Energy Conservation (ProBEC), Gaborone, Botswana

Kalumiana, O. S. (1996) 'Study of the demand and supply of firewood and charcoal, Copperbelt Province', Provincial Forestry Action Programme, Ndola, Zambia

Kammen, D. M. and Lew, D. J. (2005) 'Review of technologies for the production and use of charcoal', National Renewable Energy Laboratory, University of California, Berkeley, CA

Khalifa, K. O. (1982) 'Energy from the Sudan forests: Wood for the people', in A. Aworry (ed) *Report of the Seminar on Afforestation in Rural Development in Eastern Africa*, Environment Liaison Centre, Nairobi

Laltaika, E. (2005) 'Community based natural resources management and the law in Tanzania: A plea for action', A presentation made to the 5th Annual Scientific Conference of TAWRI held at Hotel Impala, Arusha, December

Lawry, S. W. (1990) 'Tenure policy towards common property natural resources in sub-Saharan Africa', *Natural Resources Journal*, vol 30, pp403–404

Lele, U. and Stone, S. W. (1989) *Population Pressure, the Environment and Agricultural Intensification: Variations on the Boserup Hypothesis*, World Bank, Washington, DC

Luoga, E. J., Witkowski, E. T. F. and Balkwill, K. (2000) 'Economics of charcoal production in miombo woodland of eastern Tanzania: Some hidden costs associated with commercialization of the resources', *Ecological Economics*, vol 35, pp243–257

Luoga, E. J., Witkowski, E. T. F. and Balkwill, K. (2002) 'Harvested and standing wood stocks in protected and communal miombo woodlands of eastern Tanzania', *Forest Ecology and Management*, vol 164, pp15–30

Malimbwi, R. E., Misana, S., Monela, G., Jambiya, G. and Nduwanungu, J. (2001) 'Charcoal potential in southern Africa (CHAPOSA), Final report for Tanzania', Stockholm Environment Institute, Stockholm, Sweden

Malimbwi, R. E., Nduwamungu, J., Misana, S., Jambiya, G. C. and Monela, G. C. (2004) 'Charcoal supply in Dar es Salaam city, Tanzania', *Tanzania Journal of Forestry and Nature Conservation*, vol 75, pp108–118

Mangue, P. D. (2000) 'Review of the existing studies related to fuelwood and/or charcoal in Mozambique', EC-FAO Partnership Programme (1998–2002), Project GCP/INT/679/EC

Mnzava, E. N (1981) 'Village industries vs savanna forests', *Renewable Energy for Development*, vol 7

Monela, G. C., Oktingáti, A. and Kiwele, P. M. (1993) 'Socio-economic aspects of charcoal consumption and environmental consequences along Dar es Salaam–Morogoro highway, Tanzania', *Journal of Forest Ecology and Management*, vol 58, pp249–258

Moyo, S., O'Keefe, P. and Sill, M. (1993) *The Southern African Environment: Profiles of the SADC Countries*, ETC Foundation/Earthscan Publishers, London

Mumeka, A. (1986) 'Effect of deforestation and subsistence agriculture on runoff of the Kafue River headwaters, Zambia', *Hydrological Sciences Journal*, vol 31, pp543–554

Nygård, R., Sawadogo, L. and Elfving, B. (2004) 'Wood-fuel yields in short-rotation coppice growth in the north Sudan savanna in Burkina Faso', *Forest Ecology and Management*, vol 189, pp77–85

Oduori, S. M., Alim, M. S. and Gomes, N. (2006) 'Environmental study of degradation in the Sool Plateau and Gebi Valley: Sanaag Region of Northern Somalia', Horn Relief and Oxfam, Netherlands

Okello, B. D., O'Connor, T. G. and Young, T. P. (2001) 'Growth, biomass estimates, and charcoal production of *Acacia drepanolobium* in Laikipia, Kenya', *Forest Ecology and Management*, vol 142, pp143–153

Pereira, C. (2000) 'Licensing and forest harvesting law enforcement for woodfuel production in Maputo province', CHAPOSA/FAEF/DEF, Maputo, Mozambique

Pereira, C., Brouwer, R., Monjane, M. and Falcão, M. (2001) 'Charcoal potential in southern Africa, Final report for Mozambique', Stockholm Environment Institute, Stockholm, Sweden

ProBEC (2006) 'Programme for Biomass Energy Conservation in Southern Africa', www.probec.org, accessed October 2006

Ribot, J. C. (1993) 'Forestry policy and charcoal production in Senegal', *Energy Policy*, vol 21, pp543–585

Ribot, J. (1999) 'A history of fear: Imagining deforestation in the west African dryland forests', *Global Ecology and Biogeography*, vol 8, pp291–300

Sawadogo, L., Nygard, R. and Pallo, F. (2002) 'Effects of livestock and prescribed fire on coppice growth after selective cutting of Sudanian savannah in Burkina Faso', *Annuals of Forest Science*, vol 59, pp185–195

Sawe, E. N. and Meena, J. S. (1994) 'Utilization of proven R & O results a case of TATEDO improved charcoal stoves and ovens', TATEDO, Dar es Salaam, Tanzania

Theuri, D. (unpublished) 'Rural energy, stoves and indoor air quality: The Kenyan experience', Intermediate Technology Development Group in Eastern Africa, Nairobi, Kenya

Tietema, T., Tolsma, D. J., Veenendaal, E. M. and Schroten, J. (1991) 'Plant responses to human activities in the tropical savanna ecosystem of Botswana', *Vegetation*, vol 115, pp157–167

UN Population Division (1996) *World Urbanization Prospects: The 1996 Revision*, United Nations Population Division, New York

UNDP (1996) *Urban Agriculture: Food, Jobs and Sustainable Cities*, UNDP, New York

Western, D. and Ssemakula, J. (1981) *A Survey of Natural Wood Supplies in Kenya and an Assessment of the Ecological Impact of its Usage*, Beijer Institute, Stockholm

Whitlow, J. R. (1980) *Deforestation in Zimbabwe. Supplement to Zambezia*, University of Zimbabwe, Harare, Zimbabwe

World Bank/UNDP (1990) *Zambia Urban Household Energy Strategy*, World Bank/ESMAP, Washington, DC

8

Livestock and Wildlife

James Gambiza, Emmanuel Chidumayo, Herbert Prins, Hervé Fritz and Petros Nyathi

Introduction

Dry forests and woodlands as defined in Chapter 2, make up most of the sub-Saharan grazing lands that support large populations and biomass of both wild (native) and domesticated (exotic) mammalian herbivores. In most African dry forests and woodlands the biomass of domestic livestock equals or exceeds that of wild herbivores (Cumming, 1982; Mott et al, 1985). For example, average biomass of wild and domestic herbivores in nine southern African countries (Angola, Botswana, Malawi, Mozambique, Namibia, South Africa, Tanzania, Zambia and Zimbabwe) is estimated at 370 ± 80 and 2620 ± 630kg per km^2, respectively (du Toit and Cumming, 1999). Similarly the biomass of 1543kg per km^2 of wild animals in Lolldaiga Hills ranch in Laikipia district in Kenya was lower than that of 6512kg per km^2 of cattle and sheep (Mizutani, 1999). However, the wild herbivore fauna is the most diverse, especially in eastern and southern Africa. In contrast, western African dry forests and woodlands currently support lower numbers and smaller biomass of wild herbivores due to a number of factors, including high hunting pressure, impoverished soils and low quality vegetation with a low carrying capacity (Frost et al, 1986). Carrying capacity determines the maximum livestock and/or wildlife population that a habitat or ecosystem can support on a sustainable basis. The carrying capacity concept has particular relevance in the management of wildlife and livestock in African dry forests and woodlands where animal production is largely based on free-ranging regimes. It is for this reason that this chapter concentrates on free ranging wildlife and livestock.

The chapter describes the diversity of wildlife and livestock, rangelands and their management and the contribution of livestock and wildlife to livelihoods and national economies. The emphasis is on livestock and wildlife management systems based on natural rangelands with no or minimal external inputs and those based on mixed crop-livestock production in which the livestock is free ranging. The last section highlights the key issues concerning wildlife and livestock in African dry forest and woodland ecosystems.

LIVESTOCK AND WILDLIFE SPECIES

Species diversity

The total number of wild ungulates in African dry forests and woodlands is estimated at 98 species. Dry forests and woodlands with the richest assemblages contain more than 30 large herbivore species (Prins and Olff, 1998). These have the most species-rich ungulate fauna on earth (Olff et al, 2002). According to Olff et al (2002), because larger herbivore species tolerate lower plant-nutrient content but require greater plant abundance, the highest potential herbivore diversity should occur in locations with intermediate moisture and high soil nutrients. Another explanation for wild ungulate species richness and high biomass may be due the high spatial heterogeneity of African dry forest and woodlands and savanna ecosystems.

The main livestock species in dry forests and woodlands of Africa are cattle, goats, sheep and donkeys. In Somalia, northern Kenya and southern Ethiopia, Arabian camels (or dromedaries) are also important to semi-nomadic herders. The only ungulate domesticated solely in Africa, is the donkey, which originated in northeast Africa. None of the other domesticated ungulates are indigenous to Africa. Domestic livestock have had a dramatic effect on land use, people's livelihoods and the ecology of dry forests, however. The increase in the population of cattle, goats and sheep from 1996 to 2005 in 12 African dry forest and woodland countries is shown in Figure 8.1. During this period the cattle population grew at about 2 per cent per annum while that of goats and sheep grew at about 2.5 per cent and 4.1 per cent per annum, respectively. These figures indicate that there has been a steady increase in these livestock populations in African dry forest and woodland countries. The livestock of equines includes horses, donkeys and mules. Large numbers of equines and camels are found in arid and semi-arid regions of Africa.

Food and feeding habits

Many authors have observed the dietary overlap between livestock and wildlife herbivores (Dekker, 1997; du Toit et al, 1995; Rees, 1974). Figure 8.2 shows the dietary separation between livestock and wildlife species. The species included in Figure 8.2 can be divided into three feeding groups: predominantly

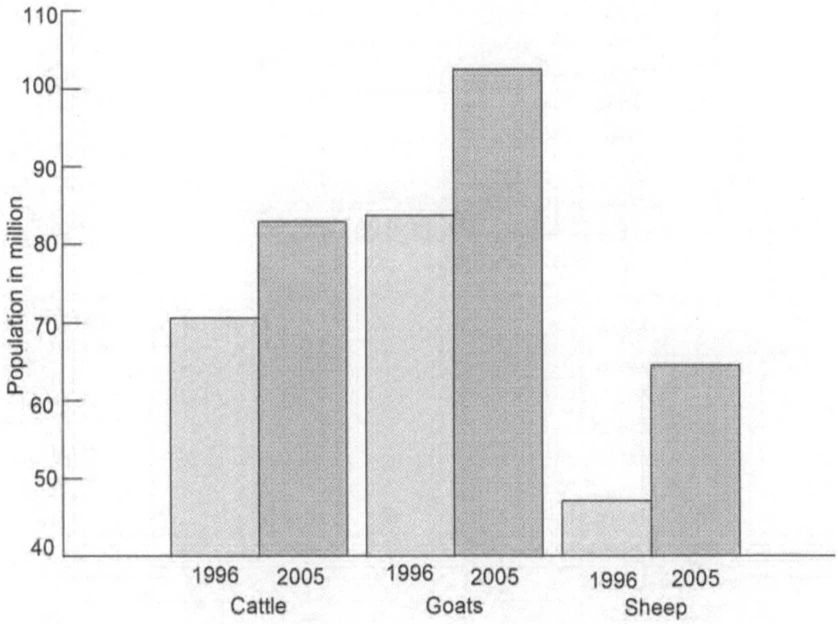

Figure 8.1 *Changes in cattle, goat and sheep populations between 1996 and 2005 in 12 African dry forest and woodland countries*

Note: The 12 countries are Botswana, Burkina Faso, Ghana, Kenya, Malawi, Mali, Namibia, Niger, Nigeria, Tanzania, Uganda and Zimbabwe.
Source: Based on FAO (2006)

grazers, mixed feeders and predominantly browsers. Cattle and sheep are grazers and mixed feeders, respectively, while goats are browsers. The diversity of feeding groups is larger among wild herbivores. However, beyond these three feeding groups, some species exhibit seasonal variations in their diet. For example, Field (1972) observed significant seasonal differences in the diet of most wild herbivores in Uganda. Consequently, diet-overlaps may vary seasonally and in some cases spatially. The existence of diet overlap may not always imply competition as different species may utilize the same food source at different heights and thereby reduce the degree of competition. For example, Arsenault and Owen-Smith (2008) observed that the grass height grazed by impala, wildebeest, zebra and white rhino, did not simply follow the pattern expected from body size differences because white rhino consistently utilized shorter grass than the three smaller grazers. Moreover, the smallest species, impala, tended to use grass heights intermediate between those grazed by wildebeest and zebra. Nevertheless, wildebeest generally used shorter grass than zebra. Hence the observed pattern of grass height use corresponded more to the relative bite width of these animals than body size.

In free-ranging systems, browse species constitute an effective insurance against seasonal feed shortage in the dry season, supplementing the quantity

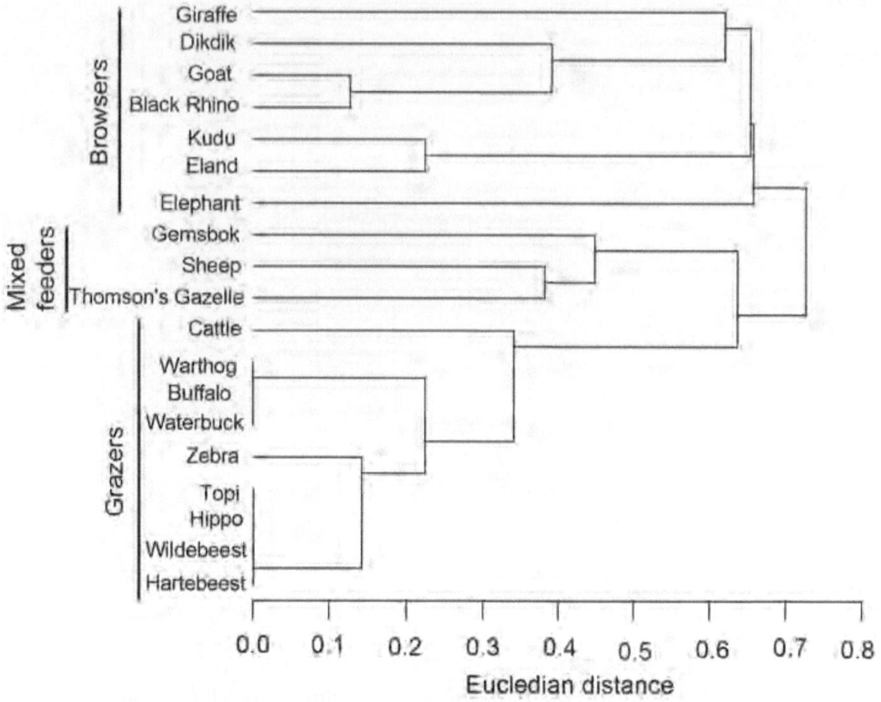

Figure 8.2 *Separation of livestock and wildlife species on the basis of diet composition*

Source: Based on data from various sources

and quality of pasture. However, the utilization of browse is limited by the high lignin content and the presence of anti-nutritional factors, including plant chemical defences produced after a period of browsing, which may be toxic to ruminants (Sanon, 2007). The most important factor is tannin, which decreases the digestibility in browse fodders. A level of tannin below 5 per cent seems to be tolerable for ruminant animals.

The accessible biomass varies according to the animal species, the plant species and the height of the plants. For example, goats browsing at a higher height had more edible biomass than sheep. The amount of accessible edible biomass as a percentage of total edible biomass decreased with increasing plant height for all the tree species studied. Cattle, sheep and goats although using the same rangelands showed different grazing behaviour (Figure 8.3), which also varied seasonally, depending on forage availability. The preference of cattle and sheep for grazing, especially in the rainy season, suggests that they could exert great pressure on the herbaceous cover when grazing together in the same area, resulting in degradation of this resource while cattle and goats grazing together could be advantageous (Sanon, 2007), although this also depends on stocking density, grazing practices and ecosystem health.

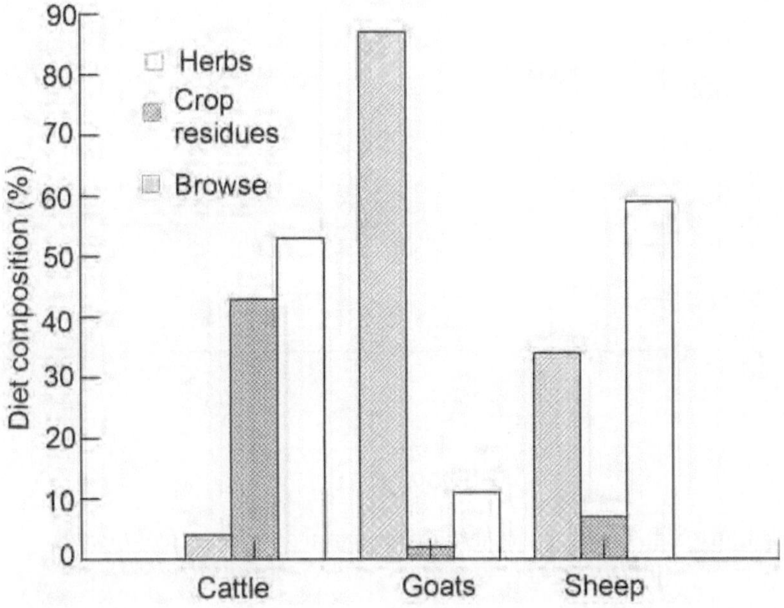

Figure 8.3 *Annual diet composition of livestock in semi-arid Mali*

Source: Based on Bellefontaine et al (2000)

Multi-species herbivore communities tend to utilize rangelands more uniformly than cattle alone (Walker, 1976, 1979). This occurs either through inter-specific niche separation (Lamprey, 1963) or overlapping and flexible habitat use (Walker, 1976). Consequently, it has been argued that wildlife production is ecologically the most rational form of land use in such areas (Child and Child, 1986), although this varies with amount of rainfall. In areas with rainfall of over 650mm per annum, a system of either livestock alone or a mixed wildlife-livestock would be more suitable than in areas with less than 650mm rainfall. Here wildlife would be the preferred form of land use under current global environmental constraints of keeping large numbers of livestock that contribute to the production and emissions of greenhouse gases, such as methane in livestock manure.

In terms of above ground biomass of herbaceous plants and woody leaf, there appear to be noticeable differences between the Sudanian and Zambezian phytoregions with the former having more herbaceous biomass than woody plant leaf biomass and the opposite pattern in the Zambezian region (Figure 8.4). The Kalahari has the highest herbaceous biomass and the Somali-Masai the least. If these data (from various sources) are a good representation of the situation on the ground, then one would expect a higher biomass of browsers in the Zambezian region than in the Sudanian region.

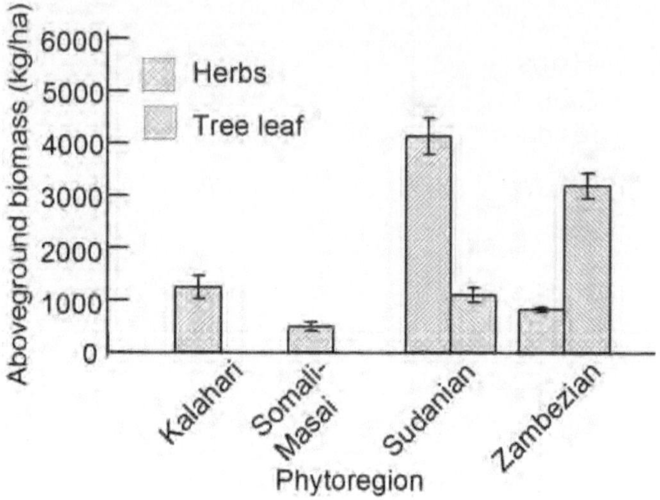

Figure 8.4 *Aboveground herbaceous and woody plant leaf biomass in different phytoregions in sub-Saharan Africa*

Source: Based on data from various sources

Complementarity or displacement

Surveys suggest that over 65 per cent of the original wildlife habitat in Africa's dry forests and woodlands has been lost (Kiss, 1990) as a result of agricultural expansion, deforestation, and overgrazing, which have been fuelled by rapid human population growth and poverty. As a result, protected areas are becoming increasingly ecologically isolated while wildlife on adjacent lands is actively eliminated (Newmark and Hough, 2000). In some cases protected areas have been encroached upon with people living inside protected areas and practising their livelihoods. This is placing significant pressures on resources and biodiversity. Of all livestock production systems, grazing systems that are based almost exclusively on livestock kept on rangelands (i.e. unimproved grasslands, shrublands, savannas) with no or only limited integration with crops and relying little on imported inputs (Steinfeld et al, 1997) are the most likely to coincide geographically with areas of high value for wildlife and other forms of biodiversity (Mearns, 1996).

Outside areas exclusively set aside for wildlife conservation, livestock and native herbivores share land, water, forage and diseases, and the fate of wildlife in such areas largely depends on both the interactions between wildlife and livestock (Grootenhuis and Prins, 2000) and on the people who live close to conservation areas. For example, in East Africa the majority of the populations of native herbivore species occur outside protected areas (Rannestad et al, 2006). This is in sharp contrast to western and southern Africa where most of the native herbivore species are largely confined to protected conservation areas (see Chapter 3). Furthermore, the majority of African protected areas are small which makes adjacent rangelands potentially important in the local survival of wild herbivores.

Historically, efforts to exterminate populations of wild species have taken place in many areas of Africa perhaps because of the perception that wildlife had little value, minimal knowledge of the role of wildlife in disease transmission and inappropriate colonial policies. Examples include black rhinoceros south of the Ngorongoro Crater because they were a menace for farmers (Prins, 1996; Stanley, 2000), wildebeest in Botswana because they were thought to compete for grazing and spread malignant catarrh (Spinage, 1992), elimination of all game to keep disease-free corridors along the border between Tanzania and Zambia (Plowright, 1982) and of lions and wild dog because they were thought to prevent the recovery of game species (Stevenson-Hamilton, 1974). They all have in common the objective of extermination based on assumptions, beliefs, or proof that wild species were interfering with desired management goals.

In arid land ecosystems, these seasonal movements of many ungulate species play a critical role in their survival. Erecting fences for the protection of livestock against contagious diseases has often disrupted these migration routes, sometimes resulting in local extinction of wildlife species. For example, fences have caused massive mortality in wildebeest in Botswana that were prevented from migrating by fences during droughts (Spinage, 1992). From 1960 to the present, veterinary control fences have been built in Botswana, Namibia, South Africa, Zambia and Zimbabwe. The early fences were mainly directed at the control of foot and mouth disease but, as veterinary research progressed in the latter half of the 20th century, it became apparent that numerous other diseases affecting cattle had to be considered (Morkel, 1988). The fence along the international boundary between Botswana and Namibia was constructed in the early 1960s and disrupted wildlife movement between the two countries.

Perhaps the most important habitat modification results from the use of fire and tree felling to create an environment unsuitable for tsetse flies (Ford, 1971) or the felling of trees to increase grass production. The use of fire especially causes a decrease of woody species and an increase in grass cover (Norton-Griffiths, 1979; Van Wijngaarden, 1985; Buss, 1990; Dublin, 1995). The combination of cattle, small-stock, and fire over hundreds of years probably has had a profound effect on wildlife herbivores by creating habitats suitable to livestock keeping (Smith, 1992; Stutton, 1993; Marshall, 1994) although some wildlife species have benefited and their numbers have increased. Also, by the provision of water and locating available fodder, domestic stock become extreme generalists and can dominate the rangelands with human assistance (Homewood and Rodgers, 1991). In arid drought-prone rangelands, low-input animals, such as zebu cattle are physiologically adapted to 'track' available feed and water supply by shifting their metabolic rate by season and between years, thus permitting optimum exploitation of highly variable environments. In the semi-arid and arid areas, wildlife species such as the oryx or the addax have developed mechanisms to cope with low and erratic water supply and, therefore, can use areas that livestock cannot access. Unfortunately, in western

Africa, most of the ungulates adapted to arid conditions are close to extinction, perhaps because of the combined effects of hunting using modern weapons and provision of artificial water-points and expansion of livestock into areas where there was competition with arid-adapted ungulates such as addax and Scimitar-horned oryx.

Livestock are threatened by a number of diseases to which wildlife species have developed a high degree of resistance, such as trypanosomiasis, although the wildlife-livestock-human disease interface is still not well understood. According to Grootenhuis (2000) the main cost of wildlife to livestock appears to accrue from disease interaction, though at the same time, diseases harboured by livestock also pose an enormous threat to wildlife. For example, the population of greater kudu in Namibia, a very important trophy animal, was reduced by some 40,000 individuals by rabies originating from dogs (Mcdonald, 1993; Swanepoel et al, 1993) and the number of lions in the Serengeti was recently reduced by 30 per cent due to canine distemper, also harboured by domestic dogs (see also Alexander and Appel, 1994).

Complementarity of wildlife and livestock can also refer to forage use. As an example, giraffe that browse up to a height of 5m or more can keep the savanna open and thus create more room for shrubs, smaller trees and herbaceous vegetation, accessible to livestock. Grazing by livestock can also be beneficial for wildlife. Moderate grazing favours plant biodiversity. Livestock grazing around wetlands brings additional nutrients into these areas, which can then support higher populations of water birds or fish (Brouwer, 2001), and the livestock and birds can coexist without disturbing each other (Touré et al, 2001) as long as the population density of livestock and their owners is not too high. In the moister semi-arid and sub-humid rangelands, range management practices such as controlled burning can be beneficial for both livestock and wildlife.

Another aspect of complementarity is in terms of food security. Hunting of wildlife tends to be intensified in dry years when crop yields are low and when livestock may also not perform well, although under such conditions there is always a danger that hunting pressures may be acute and unsustainable in the long-term. According to Kreuter and Workman (1997) the skills required to manage cattle and wildlife operations also differ. While successful cattle ranchers may be insular, safari operators require good public relations skills in order to attract and entertain hunting clients. Both should have good business and range management skills, but cattle ranchers also require animal husbandry skills while safari operators require tracking and hunting experience and knowledge of the habitats and behaviour of wildlife species favoured by commercial clients. Thus, the land and management needs for cattle and wildlife ranching do not readily coincide. While cattle can be managed independently by each rancher, landowners could pool their ranches for the limited purpose of safari hunting. In the case of community-based natural resources management, the local people may still keep livestock while benefiting from wildlife revenues from safari hunting. A summary of the advantages and disadvantages of wildlife and livestock is given in Table 8.1.

Table 8.1 *Comparative analysis of advantages and disadvantages of wildlife and livestock land uses*

Advantages	*Disadvantages*
Livestock are easily controlled and bred.	Livestock can be an expensive investment for poor farmers; if the animal dies, the loss can be devastating.
Ownership and tenure are well defined.	Livestock are prone to disease, especially in remote, 'wild' areas.
State support and subsidies are often offered.	Livestock are not as resilient as wildlife to local environmental changes such as droughts (certain animals excluded).
Livestock are easily traded for cash, goods and services.	Access to grazing is often controlled by local elites.
The benefits are immediate when livestock are sold or consumed, and transaction costs tend to be minimal.	Environmental costs result if ranges are poorly managed.
Livestock can be used for work.	
The preference for meat of wildlife over that of domestic animals.	Wildlife is a mobile resource and difficult to control.
Strong cultural sentiment or religious significance of wildlife.	There is rarely individual ownership (unless the animal is dead).
Strong link to wildlife hunting in sport and culture.	Tenure over wildlife rests with the state or, in some cases, the community but not with the individual unless the land title is freehold.
Wildlife's superior disease resistance and tolerance of local environmental change.	Wildlife resources usually require a collective management system, often even where land title is held individually.
Generally (although not always) better use of and impact on habitat by wildlife than by domestic stock (an exception being large elephant populations in southern Africa, which confer negative impact).	Wildlife often poses a threat to other livelihoods through direct competition or disease transmission.
Income or other benefits to the community if there is a community-based natural resource management (CBNRM) programme present.	In communally managed situations, direct consumptive use is often discouraged and sometimes made illegal.

Source: Based on Murphree (2003)

Rangelands and their Management

Characteristics of rangelands

In Africa, rangelands are estimated to cover about 13.4 million km^2 or 60 per cent of the continent. In the drier part of western Africa receiving less than 800mm annual rainfall, grasses are predominantly annuals, except on run-on sites that accumulate more soil moisture, whose productivity is almost entirely determined by rainfall. Perennial grasses dominate in eastern and southern Africa even at lower annual rainfall of up to 400mm, especially in eastern Africa, perhaps because rainfall is bimodal, but even here the proportion of ephemeral annuals increases in dry years (Njoka, 1994; Herlocker et al, 1993).

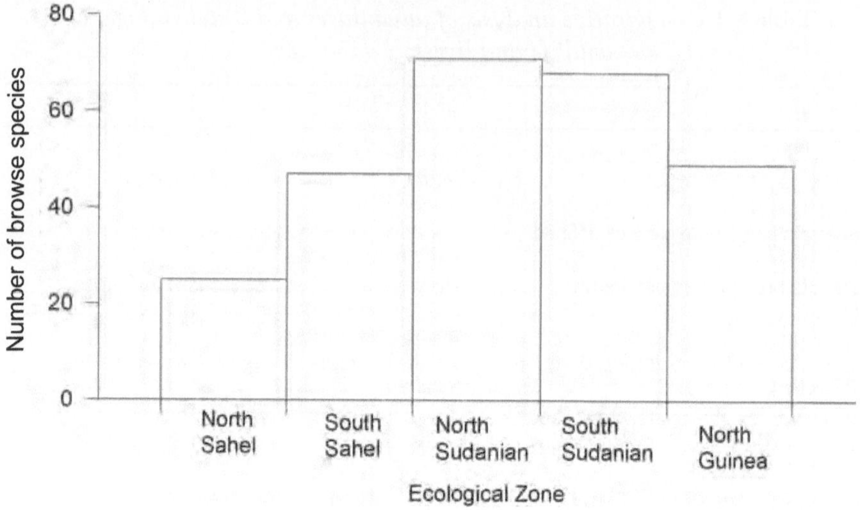

Figure 8.5 *Abundance of browse species in different ecological zones of western Africa*

Source: Based on Le Houérou (unpublished)

The browse resource is a key food for grazing animals, especially in the dry season, because of the diversity of species, the length of the production cycle, the variety of feed components (fresh and dry leaves, flowers and fruits/pods) and the high protein content and minerals (Sanon, 2007). Rangeland production in the Sahel is highly seasonal, with a rainy season occurring from June to September. The herbaceous layer is composed almost exclusively of annual plants, the composition and growth rate of which are strongly influenced by the pattern and amount of seasonal rainfall. Most of the trees and shrubs are deciduous but have longer leaf longevity and production cycles than the herbaceous plants. Wickens (1980) estimated that the flora of tropical Africa, including dry forests and woodlands, contains more than 7000 species of trees or shrubs of which at least 75 per cent are browsed to a greater or lesser extent. In western Africa peak abundance of browse species is in Sudanian woodlands and declines northwards in sahel and southwards in the Guinea dry forest (Figure 8.5). Hood (1972) identified a total of 14 browse species in a 1.24ha wet miombo in northern Zambia although only eight of these were palatable.

Grazing systems

Within the context of African dry forests and woodlands two livestock management systems are recognized (Upton, 2004). These are (i) rangeland-based systems and (ii) mixed crop-livestock systems. The systems vary in terms of the land required for production and the degree of intensification. Rangeland-based systems depend on feed from extensive areas of rangelands mainly in arid and semi-arid regions. Feed mainly consists of herbaceous species and browse. The degree of intensification ranges from relatively low in traditional subsistence systems to high in commercial systems. The livestock species kept in rangeland-based systems include cattle, goats, sheep and camels. According to Upton (2004) mixed crop-livestock production systems are the most important source of ruminant livestock production globally. In Africa, these systems are common in regions where rainfall is adequate to support the cultivation of crops in some years. As highlighted earlier, crop and livestock production complement each other in the smallholder sector (Rukuni, 1994). Livestock provide draught power and manure that are used for crop production. Crop residues and other plant materials are in turn fed to livestock. Upton (2004) has argued that livestock and crops compete for land as livestock numbers and intensity of production increase. Generally the same livestock species are kept as in the rangeland-based livestock production systems but the degree of intensification is higher.

Herdsmen practising varying degrees of transhumance have traditionally managed natural steppe and savanna rangelands. While using these rangelands, they try to strike a balance between grazing grass in its optimum stage, seeking water (ponds, rivers, wells), minimizing the inconveniences and risks from insects (flies, horseflies, tsetse flies) and moving around avoiding conflict with the farmers (Bellefontaine et al, 2000). In western Africa during the rainy season, the herds use the pastures in the drier parts of the Sudanian zone and drink at temporary ponds and/or wells and are generally spread out widely. At the beginning of the dry season, after the crops are harvested, the herds can move into the wetter Sudanian zone and utilize greener and hence more nutritious grass on natural vegetation, fallow-lands, river flood-plains, post-fire regrowth or crop residues (Bellefontaine et al, 2000).

Thus in semi-arid rangelands where rainfall variability is large, opportunistic rangeland exploitation appears to be more advantageous than a sedentary utilization of rangelands. According to Bayer and Waters-Bayer (1994) pastoralists in both sedentary and nomadic herd systems have developed a variety of survival strategies. These strategies include:

- Keeping herds with a mixture of animal species which feed on different components of the vegetation (see Figure 8.3).
- Maximizing herd size during favourable periods so that animal losses during drought do not reduce herd size below a viable level.
- Using adapted breeds that have a lower demand on forage during periods of low supply.
- Adjusting herd composition so that animals with lower nutrient requirements are kept during dry periods.
- Opportunistic movement of herds to track sporadic rainfall events in arid zones.
- Opportunistic or regular movement of herds to use more productive key forage resources, such as wetlands, together with less productive upland range in the same agro-ecological zone.
- Regular movement of herds between different agro-ecological zones.

This was also observed by De Leeuw and Tothill (1993) who noted that traditional livestock owners in semi-arid and Sudanian zones in western Africa have adopted several strategies to reduce fodder quality constraints in the dry season that include herd mobility and reliance on feed sources other than dry standing grass.

Transhumance patterns in western Africa can involve long-distance treks to better feed sources in floodplains, such as the Niger river and its tributary, the Benue river, or along the shores of Lake Chad. In northern Nigeria, sedentary stock-owners herd their animals to cultivated land to feed on crop residues after harvest in the early dry season. In central Mali rice straw and grass regrowth in fields that were irrigated during the wet season are grazed by cattle during the dry season, while small stock are taken to upland grazing grounds to browse.

The desirable criteria for forage are provision of palatable non-toxic nutritious foliage and fruit, ability to withstand browsing, lopping, pruning and coppicing, and quick growing especially during the early stages of growth (Wickens, 1995). In this respect, trees provide a valuable browse for game and livestock including nutritious pods. Depending on the species, a flush of growth is often available at the end of the dry season before the grasses start to grow. Woody vegetation therefore provides a high quality food source at the critical period. Browse is essential to all herbivores in arid and semi-arid zones since grasses alone are unable to supply maintenance requirements for more than a few months of the year. For example, in Niger cattle deprived of dry season browse suffered vitamin A deficiency.

In some cases encroachment onto grazing land, including creation of national parks and game reserves and by squatter farmers, is listed high as a cause of increasing grazing pressure. It can be argued, however, that converting land from grazing to crops increases overall feed supplies, in particular in sub-humid areas with high fire frequencies. Fire hazards are lessened by greater herbage and woody biomass removal and a greater heterogeneity and patchiness of the vegetation. Defoliation by livestock tends to increase tillers, shoots at the base of tussock grasses, although few tillers may reach the flowering stage. Under extreme continuous grazing pressure shorter cycle and lower yielding annuals replace longer cycle species and although these are palatable they are very resistant to grazing due to their short growing period (De Leeuw and Reid, unpublished). The transhumant herdsmen seasonally utilize large geographical areas to ensure a fairly balanced diet for their livestock. Seasonal transhumance between lowlands and highlands is practised in parts of eastern Africa whereas in western Africa transhumance is between drier areas during the wet season and more humid areas during the dry season. In eastern and southern Africa woodland trees produce new leaves two months before the onset of the rainy season (Lawton, 1980), and similar observations have been made for some rare species in the Sudanian zone of western Africa. This phenological behaviour of woody plants in dry forest and woodlands is widely exploited by the herdsmen. Yields of deep-rooted woody species generally vary less than those of herbaceous plants but they also are affected by water-stress. Although some woody species are well adapted to severe defoliation, most trees and shrubs suffer under heavy browsing and cutting (Bayer and Waters-Bayer, 1994).

Rangeland management practices

There are a number of traditional and conventional practices applied in the management of rangeland resources. These include controlled grazing levels, resting or deferred grazing, burning and a variety of silvopastoral practices.

There is evidence to support the view that light or moderate grazing by livestock increases rangeland productivity in many grazing systems. For example, removal of coarse, dead stems by fire permits the regrowth of nutritious new shoots in grasses such as *Themeda triandra* in African savannas.

Resting, a term used when livestock are excluded from a grazing area for a certain period of time, is a comparatively simple way of countering the effects of overgrazing for what is generally only a short period of time. In extreme cases, resting is needed for several years (see also Box 10.1). It may also be accompanied by some light regeneration operations. Short resting may be one element of grassland management. In the Sahelian zone, during the rainy season, it permits optimum development of annual grasses, a good dissemination of seeds and may help to regenerate woody plants. In the Sudanian zone, it guarantees greater grass production that, once burnt, makes it possible to properly control brush encroachment. The two notions of resting and deferred grazing are not the same but in practice resting coincides most of the time with deferred grazing of an area (Bellefontaine et al, 2000). In western Africa the period of protection ranges from 1.5 to 5 years. The main reason for such a protection regime is to protect all highly palatable natural regeneration from browsing and to prevent soil compaction following exposure and trampling by livestock.

A silvicultural system is a set of rules applied to a rangeland in order to ensure its renewal. One such system is the coppice system, which in a broad sense involves different patterns of tree cutting for biomass harvesting and regeneration. This includes pruning, pollarding, topping and thinning, as well as using livestock to promote tree regeneration.

In the African woodlands pruning is a practice used to supply fodder and/or firewood and to reduce crop shade. In Niger, the pruning of the *Acacia nilotica* bottomland formations is recommended to produce wood and fodder and to use the small spiny branches in order to protect the natural regeneration in the clearings (Peltier et al, 1995).

Pollarding or trimming-out consists of cutting off tips of branches or young adventitious shoots on the main stem of a tree. Pollarding is widely used in some areas to keep the browse accessible to livestock (Wickens, 1995), although continuous browsing should be avoided because browsers tend to overexploit preferred species that may be killed (Figure 8.6). It is therefore necessary to practice rotational grazing to reduce mortality of preferred browse species.

Topping in rangelands often involves the virtual removal of all the branches for fodder production at the end of the dry season, and fruit production by

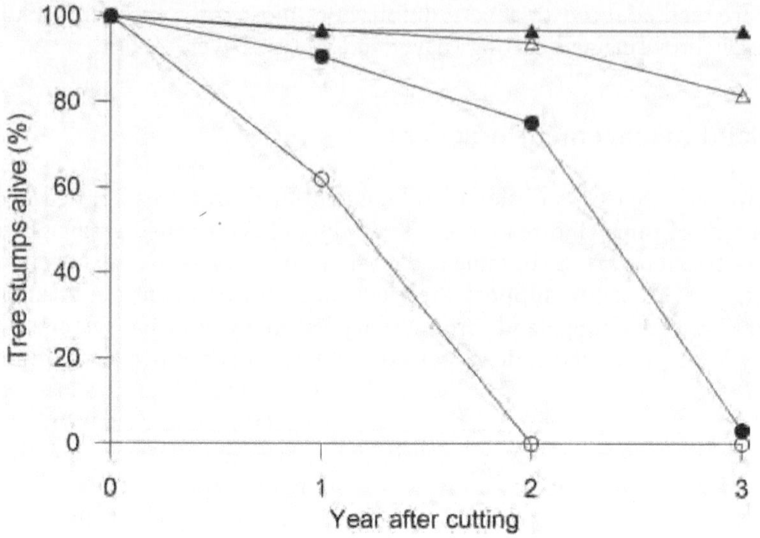

Figure 8.6 *Tree stump mortality caused by browsing by cattle in wet miombo in northern Zambia*

Note: The species disappeared in order of browsing preference as follows: *Baphia bequaertii* (open circles), *Brachystegia spiciformis* (closed circles), *Julbernadria paniculata* (open triangles) and *Isoberlinia angolensis* (closed triangles).
Source: Based on Hood (1972)

controlling tree height to facilitate fruit gathering. It also enables sprouts to occur at a safe height, out of reach of browse or fire damage.

Thinning is usually carried out in order to encourage the development of certain trees considered to be valuable. However, thinning can be applied in rangeland management to promote herbage production. For example, in miombo woodlands of eastern and southern Africa, tree shading depresses herbage production that is required for grazing. Opening up the canopy by thinning or clear-cutting can promote herbage production (Hood, 1972) and therefore improve grazing. When thinning is done by cutting trees at stump level, then the trunks produce resprout shoots that can provide additional forage for livestock.

Seed dispersal by both wild and domestic ruminants is known to encourage tree regeneration in rangelands. For example, the seeds of some plant species are spread efficiently by being carried in cattle guts, then deposited in dung in favourable seed beds or trampled into the soil. This favours the recruitment of tree species under certain conditions by grazing animals. Reid and Ellis (1995) found that goats play a key role in enhancing recruitment reliability of *Acacia tortilis* in South Turkana, Kenya. The passage of herbage through the gut and out as faeces modifies the nitrogen cycle, so that grazed pastures tend to be richer in nitrogen than ungrazed ones.

Wildlife-based management systems

Most African dry forest and woodland ecosystems are rich in wild large mammal species. Best known are the eastern and southern African woodland/savanna systems that teem with unparalleled populations of large mammals. The trees and grasslands provide forage, browse and habitat to many wildlife animals. In turn a large number of carnivorous animals also depend on the herbivores forming an intricate food chain, which ultimately depends on the dry forests and woodlands/savannas. The increasing demand for nature-based recreation has induced a dynamic private sector involvement in the management of game reserves and parks in Kenya, Namibia and South Africa.

Many wildlife-based tourism activities revolve around woodland/savanna ecosystems, establishing a strong cause-and-effect relationship between eco-tourism development and woodland and savanna use. This is particularly important for the woodlands and savannas of Africa since these support many animal species. The capacity of dry forests and woodlands/savannas to support wildlife depends on how they are protected and managed. Most countries with dry woodlands and savannas have a vibrant wildlife sector based on networks of national parks, game reserves, game management areas, state hunting concession areas, communal lands and private lands. Different management systems are imposed on the land types with a more intensive management on private lands. Controlling fires is perhaps the most common woodland management strategy in the wildlife management areas. For example, ranchers use fire to stimulate new grass growth and to control encroachment or recruitment of woody plants in rangeland.

Nowadays schemes which involve local communities in managing natural resources and gaining financial benefits are common throughout the world and more so in developing countries (e.g. CAMPFIRE in Zimbabwe: Bond and Frost, 2005). Community-based wildlife management schemes in woodland and savanna systems have become an important channel for the capture of international funding and are generating good financial returns for communities (Barnes, 2001). For example, hunting safaris have effectively doubled the areas used for wildlife conservation in Africa through the provision of incentives for setting aside area for wildlife and habitat protection. The areas include state owned concessions, wildlife management areas where local communities live but where wildlife is the primary land-use such as are found in Botswana, Namibia and Tanzania, and Game Management Areas where communities live and wildlife is not a primary land use, such as in Zambia and Communal Areas Management Programs for Indigenous Resources (CAMPFIRE) in Zimbabwe (Lindsey et al, 2007). Nature-based tourism is also being advocated as a way of stopping illicit networks of poachers and traffickers through empowering local communities (Duffy, 2006). Box 8.1 shows some examples of community-based wildlife management programmes in selected African countries and regions.

Box 8.1 Community-based natural resource management programmes in selected countries and regions in Africa

- CAMPFIRE (Communal Areas Management Program for Indigenous Resources) in Zimbabwe
- ADMADE (Administrative Management Design Programme) in Zambia
- ZICGC (Zones d'Interêt Cynegetique à Gestion Communautaire) in Cameroon
- GEPRE-NAF (Gestion Participative des Ressources Naturelles et de la Faune) in western Africa
- ECOPAS (Ecosystèmes Protégés en Afrique Sudano-Sahélienne)

Source: Lindsey et al (2007)

Contribution to Livelihoods and Economy

Contribution to livelihoods

Given the large herbivore assemblages of African woodlands and savannas and their high productivity (Prins, 1994, 1996), the question that arises is why would people adopt livestock rearing and give up a hunter-forager lifestyle? Indeed, elsewhere it has been shown that the potential meat yields for traditional hunting range from 250kg per km^2 per year (in a system dominated by migratory game: Serengeti National Park in northern Tanzania) to 500kg per km^2 per year (in a system dominated by sedentary game: Lake Manyara National Park, also in northern Tanzania) (Prins, 1994). This compares favourably with present-day African pastoralism, which yields approximately 800kg per km^2 per year (Prins, 1994). The biggest benefit of pastoralism in comparison to hunting-gathering appears to be the reduction in uncertainty of food procurement. Indeed, Wetterstrom (1993) also gave this as the most important explanation for the adoption of domesticates in Egypt around 7000 BP, and Muzzolini (1993) gave a comparable explanation for the transition in the Sahara around 5500 BP. Most likely, the biggest advantage of the domestic species is their potential to produce milk (and blood) that can be collected quite easily. Thus, when production is not market-oriented but is aimed at self-sufficiency, the exploitation of woodlands and savannas through herding livestock appears to be superior to hunting not so much because secondary productivity is higher under herding but especially because it is less risk-prone than hunting. Whatever the exact mechanism, eventually pastoral societies emerged in eastern, western and southern African woodlands and savannas.

Rangelands, including woodlands and savannas occupy about 90 per cent of the agricultural land in Africa and sustain the livelihoods of 25 million people. Grazing systems are usually geared to the production of multiple outputs, including meat, milk, blood, hides and skins, dung fuel, transportation and flexible household capital reserves, while ranching systems are generally geared more narrowly towards meat production (Mearns, 1996). The livelihoods of millions

Table 8.2 *Effect of herd size (number of animals) on maize production factors in the communal agricultural sector in Zimbabwe*

Production factor	*Explanatory power (r^2) of herd size on production factor*
Area under maize (ha)	96%
Area manured (ha)	99%
Manure applied (tonnes)	97%
Maize yield (kg per ha^{-1})	97%

Source: Based on Rukuni (1994)

of people in Africa are therefore dependent on livestock (Shackleton et al, 2001, 2005; Dovie et al, 2006). Cattle, goats and sheep are the dominant livestock species kept. They are kept for multiple purposes resulting in both direct and indirect benefits. Direct benefits include provision of meat, milk, blood, draught power, manure, cash and many cultural functions. Important indirect benefits are demonstrated by the interaction between crop production and livestock ownership. For example, Rukuni (1994) found an important positive relationship between livestock ownership and crop production factors in Zimbabwe: planted area, manure applied and maize yield increased as cattle herd size increased (Table 8.2). This was attributed to larger cultivated areas, improved soil fertility and structure due to use of manure, and timely ploughing and planting in households with large cattle herds. Thus livestock is an important component of agricultural production contributing to food security and human well-being.

Crop-livestock integration is the main avenue for intensification in sub-humid dry forests and woodlands where external inputs are not available, and can support higher rural populations than grazing systems alone. Integrated crop-livestock systems can be environmentally beneficial, as by-products from one production component (e.g. crop residues, manure) serve as inputs to the other component. Livestock play a key role in energy and nutrient cycling, as well as providing a diverse range of outputs.

Animal draught power and nutrient cycling through manure compensate for lack of access to modern inputs, such as tractors and fertilizers. In the Sahelian zone about 22 per cent of the total nitrogen fertilizer and 38 per cent of phosphate is estimated to be of animal origin from the manure and represents US$1.5 billion worth of commercial fertilizer (Steinfeld et al, 1997). In addition the use of livestock as gifts, for dowry or slaughtering for traditional feasts or religious ceremonies reinforces family and social links.

In southern and eastern Africa wildlife management is an important complement to livestock keeping on rangelands. Income from game viewing and/or trophy hunting can exceed the income from livestock, and a combination of both provides higher income than livestock or wildlife alone (Kiss, 1990), especially under community-based management systems. Although large game animals are now rare in western Africa, smaller game animals, such as duikers, grass cutters (cane rats) and giant rats, contribute substantially to local meat supply (Caspary, 1999) of the western African woodlands and savannas.

The main economic advantage of wildlife over cattle in marginal lands is generally considered to be their potential for providing multiple products of high value while at the same time reducing ecological pressure (Johnstone, 1975; Muir, 1988; Child, 1988). For instance, safari hunting and photo-tourism have been found to be lucrative in semi-arid savannas with diverse wildlife communities, depending less on high population densities than does meat production (Muir, 1988).

However, despite claims that African wildlife can generate greater profits than cattle (Joubert et al, 1983; Clarke et al, 1986; Hopcraft, 1986; Child, 1988), the relative profitability of extensive cattle and wildlife has not been well established for semi-arid African savannas with a limited diversity of wildlife.

The wildlife and the cultural landscapes of the dry forests and woodlands of Africa provide a rich and diversified basis of eco-tourism, which includes game safaris, trophy hunting and wildlife photography. Lindsey et al (2007) estimated that sub-Saharan Africa receives about US$201 million per year from trophy hunting, making trophy hunting an important driver of conservation. According to Mearns (1996) many developing countries have adopted policies to encourage ecotourism but only limited success has so far been achieved owing to:

- lack of analysis of the real economic potential of ecotourism;
- failure to capture a greater proportion of ecotourism's benefits locally and nationally;
- failure to regulate visitor environmental and cultural impacts, arising from the often false perception that total numbers of visitors will yield greater revenues than efforts to increase fees and charges from a smaller number of visitors;
- failure to provide additional funds for biodiversity conservation.

Contribution to national economies

Livestock production makes a significant contribution to GDP although its importance in the economy differs among countries (Figure 8.7). The highest contribution to GDP is in eastern Africa (6.0± 1.3 per cent) followed by western Africa (4.2± 1.1 per cent) and is least in southern Africa (2.6 ± 0.5 per cent). The contribution of livestock to economic growth is likely to increase because of the projected increase in livestock population (see Figure 8.1) and demand for livestock products due to global growth in the human population (Upton, 2004; AGL, 2005; Pica-Ciamarra, 2005). Furthermore, livestock is expected to play an important role in fulfilling the Millennium Development Goal of reducing poverty by 2015, especially given that most poor people in Africa keep livestock and live in rural areas. However, this development goal

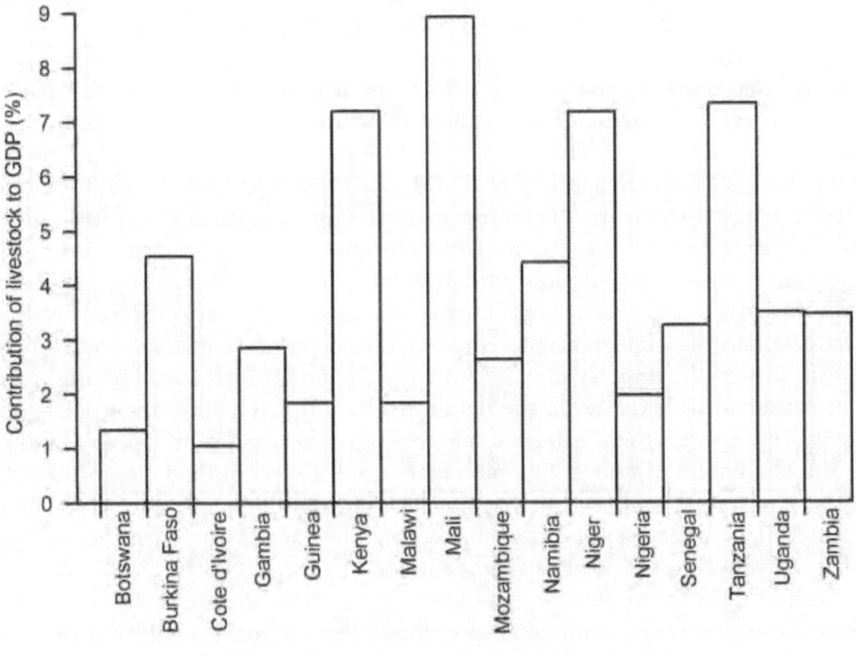

Figure 8.7 *Contribution of livestock to national agricultural gross domestic product*

Source: Based on FAO (2005)

should be weighed against the processes of human demographic dynamics and the role of livestock in the climate change debate. Nevertheless, the livestock sector contributes substantially to the agricultural GDP in sub-Saharan Africa, especially when manure and draught power are included: its contribution to GDP is estimated to be as high as 35 per cent (Winrock, 1992) and the livestock sector alone accounts for 12.3 per cent of the GDP in Burkina Faso (FAO, 2003).

In Kenya the opportunity cost of wildlife biodiversity conservation in protected areas measured in terms of forgone livestock and agricultural production has been estimated at about US$203 million per year or 2.8 per cent of total GDP, while revenues from wildlife tourism and forestry contribute only about US$42 million per year to the national economy (Norton-Griffiths and Southey, 1995). This is partly because, as noted above, grazing systems under livestock are usually geared to the production of multiple outputs, while ranching systems are generally geared more narrowly towards meat production (Mearns, 1996). Crop-livestock integration is the main avenue for intensification in sub-humid rangelands where external inputs are not available, and can support higher rural populations than grazing systems alone.

Key Issues and Challenges

Surveys suggest that over 65 per cent of the original wildlife habitat in Africa has been lost (Kiss, 1990) as a result of agricultural expansion, deforestation, and overgrazing, which have been fueled by rapid human population growth and poverty (Newmark and Hough, 2000). As a result protected areas for biodiversity conservation are becoming increasingly ecologically isolated while wildlife on adjacent lands is actively eliminated for various reasons. This phenomenon, in combination with the small size of most protected areas, indicates that in the absence of intensive management, most protected areas in Africa will not be large enough to conserve many species (Newmark and Hough, 2000). Given the underlying determinants of habitat loss, it has been argued that conservation activities in the field must be intimately linked with development. This approach has gained some popularity because it provides an option to the challenges of conserving biological diversity within existing protected areas. The trans-boundary conservation areas in southern Africa are also aimed at achieving conservation and livelihood objectives within large landscapes that include many forms of land use and countries.

Additionally, rural poverty and external markets will continue to encourage both subsistence and commercial poaching of many species within protected areas. Analysis in Zambia suggests that it costs US$200 per km^2 per year to effectively control commercial poaching of species, such as elephant and rhinoceros in protected areas (Leader-Williams and Albon, 1988). Unfortunately, few, if any, African countries have such financial resources, and central governments are unlikely to allocate significantly more funds for wildlife management in the future, given the many other competing demands for governmental resources. Recognition of these problems has led many workers to argue that the only way to enlarge and link existing protected areas (Newmark, 1996) and control commercial poaching (Owen-Smith and Cumming, 1993) is to develop cooperative relationships with adjacent communities through community-based or joint natural resources management approaches. This is particularly essential because protected areas have adversely affected many indigenous people in Africa. For example, all of the large savanna parks of eastern Africa have been established on former Masai rangelands (Århen, 1985; Parkipuny and Berger, 1993) although the Masai have retained some degree of control in a few areas, such as Mara.

However, implementation of a conservation-development approach in Africa is faced with difficulties arising from external forces. For example, sources of potential revenues in community-based or joint natural resources management initiatives are usually unreliable and insufficient. The challenge for wildlife conservation therefore is not simply to replace domesticated livestock production with domesticated wildlife but to keep ecosystems healthy. In this way ecosystems can maintain their capacity to provide goods and services while ensuring that people have incentives to use and conserve wildlife resources. Because exchange rate fluctuations and political turmoil often make tourist revenues unreliable, basing cash inducements to communities on tourism needs to be reviewed (Barrett and Arcese, 1995). The dramatic decline in tourism in recent years in Uganda, Kenya and Zimbabwe highlights the high vulnerability of this industry to political unrest and economic downturns. Additionally, as Barrett and Arcese (1995) noted, there are few protected areas in Africa where the revenues from gate receipts exceed the cost of management; thus, it is unlikely that many communities will ultimately benefit from such revenue-sharing practices. Furthermore, as Norton-Griffiths and Southey (1995) have pointed out, if opportunity costs are taken into account, protected areas and their buffer zones may impose economic penalties on their surrounding communities that far outweigh any potential financial advantages from revenue-sharing arrangements and pursuing such initiatives may only worsen the poverty situation in many African dry forest and woodland/savanna countries.

Sustainable rangeland development in semi-arid Africa therefore must focus on reducing poverty by increasing the value of products derived from marginal land. However, further studies on the spatial and temporal variation in browse production, including a proper evaluation of production of fruits from various browse species, are needed to generate new knowledge that can form the basis for improving the management of rangeland resources. Increased fodder production through planting local and improved browse species that are not invasive and their integration in the production systems should be tested. Seed germination and plant growth characteristics of selected local species should also be studied.

The new thinking in rangeland ecology suggests an urgent need for training of a new generation of range managers able to combine technical insight with socio-economic analysis and with the ability to think holistically and link with a multi-disciplinary team. They need to have the ability to focus on the needs of multiple, competing users, facilitate participatory planning processes and negotiate during conflicts, among many other technical skills.

REFERENCES

AGL (2005) 'Responding to the "Livestock Revolution" – the case for livestock public policies', Livestock Information, Sector Analysis and Policy Branch, Animal Production and Health Division, FAO, Rome

Alexander, K. A. and Appel, M. J. G. (1994) 'African wild dogs (*Lyacon pictus*) endangered by a canine distemper epizootic among domestic dogs near the Masai Mara National Reserve, Kenya', *Journal of Veterinary Diseases*, vol 30, pp481–485

Århen, K. (1985) 'Pastoral man in the Garden of Eden: The Maasai of the Ngorongoro Conservation Area, Tanzania', *Uppsala Research Reports in Cultural Anthropology* vol 3, pp1–123

Arsenault, R. and Owen-Smith, N. (2008) 'Resource portioning by grass height among grazing ungulates does not follow body size relation', *Oikos*, vol 117, pp1711–1717

Barnes, R. D. (2001) 'The African acacias – a thorny subject', *Southern African Forestry Journal*, vol 190, pp9–18

Barrett, C. B. and Arcese, P. (1995) 'Are integrated conservation-development projects (ICDPs) sustainable? On the conservation of large mammals in sub-Saharan Africa', *World Development*, vol 23, pp1073–1084

Bayer, W. and Waters-Bayer, A. (1994) 'Forage alternatives from range and field: Pastoral forage management and improvement in the African rangelands', in I. Scoones (ed) *Living with Uncertainty: New Directions in Pastoral Development in Africa*, IIED and Intermediate Technology Publishing Ltd, London

Bellefontaine, R., Gaston, A. and Petrucci, Y. (2000) *Management of Natural Forests of Dry Tropical Zones*, FAO, Rome

Bond, I. and Frost, P. G. H. (2005) 'CAMPFIRE and the payments for environmental services', Paper prepared for the workshop: Payments for Environmental Services (PES) – Methods and Design in Developed and Developing Countries, Titisee, Germany, 15–18 June, CIFOR and IIED

Brouwer, J. (2001) 'Analyse des interactions entre le bétail et la faune sauvage au Niger, en faisant l'accent sur les zones humides', Contribution to workshop 'Faune sauvage et bétail: Complémentarité, coexistence ou competition?', Niamey, 16–19 January

Buss, I. O. (1990) *Elephant Life: Fifteen Years of High Population Density*, Iowa State University Press, Ames

Caspary, H. U. (1999) 'Utilisation de la faune sauvage en Côte d'Ivoire et Afrique de l'Ouest: Potentiel et contraintes pour la cooperation au développement', GTZ, Eschborn, Germany

Child, B. (1988) 'The role of wildlife utilization in the sustainable development of semi-arid rangelands in Zimbabwe', PhD thesis, Worcester College, University of Oxford, Oxford

Child, B. and Child, G. (1986) 'Wildlife economic systems and sustainable human welfare in semi-arid rangelands in southern Africa', Paper presented at FAO/Finland Workshop on Watershed Management in Semi-Arid Zones of the SADCC Countries, April, Maseru, Lesotho

Clarke, V. J., Cumming, D. H. M., Martin, R. B. and Peddie, D. A. (1986) 'The comparative economics of African wildlife and extensive cattle production', in *Proceedings of FAO African Forestry and Wildlife Meeting, January 1986, Bamako, Mali*, FAO, Rome

Cumming, D. H. M. (1982) 'The influence of large herbivores on savanna structure in Africa', in B. H. Walker (ed) *Ecology of Tropical Savannas*, Springer-Verlag, Berlin

De Leeuw, P. N. and Tothill, J. C. (1993) 'The concept of rangeland carrying capacity in subSaharan Africa – myth or reality', in R. H. Behnke Jr, I. Scoones and C. Kerven (eds) *Range Ecology at Disequilibrium: New Models of Natural Variability and Pastoral Adaptation in African Savannas*, Overseas Development Institute, London

Dekker, B. (1997) 'Calculating stocking rates for game ranches: Substitution ratios for use in the mopani veld', *African Journal of Range and Forage Science*, vol 14, pp62–67

Dovie, D. B. K., Shackleton, C. M. and Witkowski, E. T. F. (2006) 'Valuation of communal area livestock benefits, rural livelihoods and related policy issues', *Land Use Policy*, vol 23, pp260–271

Dublin, H. T. (1995) 'Vegetation dynamics in the Serengeti-Mara ecosystem: The role of elephants, fire and other factors', in A. R. E. Sinclair and P. Arcese (eds) *Serengeti II, Dynamics, Management and Conservation of an Ecosystem*, University of Chicago Press, Chicago

Duffy, R. (2006) 'The potential and pitfalls of global governance: The politics of transfrontier conservation area in southern Africa', *Political Geography*, vol 25, pp89–112 du Toit, J. T. and Cumming, D. H. M. (1999) 'Functional significance of ungulate diversity in African savannas and the ecological implications of the spread of pastoralism', *Biodiversity and Conservation*, vol 8, pp1643–1661

du Toit, P. C. V., Blom, C. D. and Immelman, W. F. (1995) 'Diet selection by sheep and goats in the arid Karoo', *African Journal of Range and Forage Science*, vol 12, pp16–26

FAO (2003) 'Livestock sector brief. Burkina Faso', Livestock Information, Sector Analysis and Policy Branch, FAO-AGAL, Rome

FAO (2005) 'Livestock Sector Briefs', www.fao.org/ag/againfo/resources/en/pubs_sap.html

FAO (2006) 'Country Pasture Profiles', www.fao.org/ag/AGP/AGPC/doc/Counprof/regions/africa.htm

Field, C. R. (1972) 'The food habits of wild ungulates in Uganda by analyses of stomach contents', *East African Wildlife Journal*, vol 10, pp17–42

Ford, J. (1971) *The Role of the Trapanosomiases in African Ecology: A Study of Tsetse Fly Problem*, Oxford University Press, Oxford

Frost, P., Medina, E., Menaut, J.-C., Solbrig, O., Swift, M. and Walker, B. (1986) *Responses of Savannas to Stress and Disturbance, Biology International* (Special Issue 10), IUBS and Unesco MAB, Paris

Grootenhuis, J. G. (2000) 'Wildlife, livestock and animal disease reservoirs', in H. H. T. Prins, J. G. Grootenhuis and T. T. Dolan (eds) *Wildlife Conservation by Sustainable Use*, Kluwer Academic Publishers, Norwell, MA

Grootenhuis, J. G. and Prins, H. H. T. (2000) Wildlife utilization: a justified option for sustainable land use in African savannas', in H. H. T. Prins, J. G. Grootenhuis and T. T. Dolan (eds) *Wildlife Conservation by Sustainable Use*, Kluwer Academic Publishers, Boston

Herlocker, D. J., Dirsehl, H. J. and Frame, G. (1993) 'Grasslands of East Africa', in R. Couplanel (ed) *Natural Grasslands: Eastern Hemisphere (Ecosystems of the World, vol 8B)*, Elsevier, Amsterdam, Netherlands

Homewood, K. M. and Rodgers, W. A. (1991) *Maasailand Ecology: Pastoralist Development and Wildlife Conservation in Ngorongoro, Tanzania*, Cambridge University Press, Cambridge

Hood, R. J. (1972) 'The development of a system of beef production for use in the *Brachystegia* woodlands of northern Zambia', PhD thesis, University of Reading (Department of Agriculture), Reading, UK

Hopcraft, D. (1986) 'Wildlife land use: A realistic alternative', in S. MacMillan (ed) *Wildlife/Livestock Interfaces on Rangeland*, Inter-African Bureau of Animal Resources, Nairobi, Kenya

Johnstone, P. A. (1975) 'Evaluation of a Rhodesian game ranch', *Journal of the South African Wildlife Management Association*, vol 5, pp43–51

Joubert, E., Brand, P. A. J. and Visagie, G. P. (1983) 'An appraisal of the utilization of game on private land in South West Africa', *Madaqua*, vol 13, pp197–219

Kiss, A. (ed) (1990) 'Living with wildlife: Wildlife resource management with local participation in Africa', *World Bank Technical Paper no 130*, World Bank, Washington, DC

Kreuter, U. P. and Workman, J. P. (1997) 'Comparative profitability of cattle and wildlife ranches in semi-arid Zimbabwe', *Journal of Arid Environments*, vol 35, pp 171–187

Lamprey, H. F. (1963) 'Ecological separation of large mammal species in the Tarangire Game Reserve, Tanganyika', *East African Agriculture and Wildlife Journal*, vol 1, pp63–93

Lawton, R. M (1980) 'Browse in miombo woodlands', in H. N. Le Houerou (ed) *Browse in Africa: The current state of knowledge*, ILCA, Addis Ababa

Leader-Williams, N. and Albon, S. D. (1988) 'Allocation of resources for conservation', *Nature*, vol 336, pp533–535

Lindsey, P. A., Roulet, P. A. and Ramanach, S. S. (2007) 'Economic and conservation significance of the trophy hunting industry in sub-Saharan Africa', *Biological Conservation*, vol 134, pp455–469

Marshall, F. (1994) 'Archaeological perspectives on East African pastoralism', in E. Fratkin, K. A. Galvin and E. A. Roth (eds) *African Pastoralist Systems: An Integrated Approach*, Lynne Riener, Boulder, CO

Mcdonald, D. W. (1993) 'Rabies and wildlife: a conservation problem?', *Onderstepoort Journal of Veterinary Research*, vol 60, pp351–355

Mearns, R. (1996) 'When livestock are good for the environment: Benefit-sharing of environmental goods and services', Paper presented at the World Bank/FAO workshop 'Balancing livestock and the Environment', 27–28 September, Washington, DC

Mizutani, F. (1999) 'Biomass density of wild and domestic herbivores and carrying capacity on a working ranch in Laikipia District, Kenya', *African Journal of Ecology*, vol 37, pp226–240

Mott, J. J., Williams, J., Andrew, M. H. and Gillison, A. N. (1985) 'Australian savanna ecosystems', in J. C. Tothill and J. J. Mott (eds) *Ecology and Management of the World's Savannas*, Australian Academy of Sciences, Canberra

Morkel, P. (1988) 'Introduction of disease-free African buffalo (*Syncerus caffer*) to game farms in Namibia', Republic of Namibia, Ministry of Environment and Tourism, Windhoek, Namibia

Muir, K. (1988) 'The potential of indigenous resources in the economic development of the arid environments in sub-Saharan Africa', Department of Agricultural Economics and Extension Working Paper AEE 9/88, University of Zimbabwe, Harare, Zimbabwe

Murphree, M. J. (2003) 'Counting Sheep: The Comparative Advantages of Wildlife and Livestock – A Community Perspective 1', Institute of Natural Resources/SASUSG, Scottsville, South Africa

Muzzolini, A. (1993) 'The emergence of a food-producing economy in the Sahara', in T. Shaw, P. Sinclair, B. Andah and A. Okpoko (eds) *The Archaeology of Africa: Food, Metals and Towns*, Routledge, London

Newmark, W. D. (1996) 'Insularization of Tanzanian parks and the local extinction of large mammals', *Conservation Biology*, vol 10, pp1549–1556

Newmark, W. D. and Hough, J. L. (2000) 'Conserving wildlife in Africa: Integrated conservation and development projects and beyond', *BioScience*, vol 50, pp585–592

Njoka, T. J. (1994) 'Evaluation of seasonal and drought effects on range trend analysis in Kajiado District, Kenya', in W. Siderius (ed) *Proceedings of the Workshop on Land Evaluation for Extensive Grazing (publication no 6)*, International Institute of Land Reclamation and Improvement, Wageningen, Netherlands

Norton-Griffiths, M. (1979) 'The influence of grazing, browsing and fire on the vegetation dynamics of the Serengeti', in A. R. E. Sinclair and M. Norton-Griffiths (eds) *Serengeti: Dynamics of an Ecosystem*, University of Chicago Press, Chicago

Norton-Griffiths, M. and Southey, C. (1995) 'The opportunity costs of biodiversity conservation in Kenya', *Ecological Economics*, vol 12, pp125–139

Olff, H., Ritchie, M. E. and Prins, H. H. T. (2002) 'Global environmental controls of diversity in large herbivores', *Nature*, vol 415, pp901–904

Owen-Smith, N. and Cumming, D. H. M. (1993) 'Comparative foraging strategies of grazing ungulates in African savanna grasslands', *Proceedings of the XVII International Grasslands Congress*, New Zealand

Parkipuny, M. S. O. and Berger, D. J. (1993) 'Maasai rangelands: Links between social justice and wildlife conservation', in D. Lewis and N. Carter (eds) *Voices from Africa: Local Perspectives on Conservation*, World Wildlife Fund, Washington, DC

Peltier, R., Lawali, A. and Djibrilla, M. (1995) 'Amémagement sylvo-pastoral des foréts inondables a *Acacia nilotica* du Niger', Groupement SEED-CIRAD-Forêt. Rép. Du Niger (Project Energie II)

Pica-Ciamarra, U. (2005) 'Livestock polices for poverty alleviation: Theory and practical evidence from Africa, Asia and Latin America', www.fao.org/ag/againfo/projects/en/pplpi/docarc/wp27.pdf

Plowright, W. (1982) 'The effects of rinderpest and rinderpest control on wildlife in Africa', *Symposium of Zoological Society of London*, vol 50, pp1–28

Prins, H. H. T. (1994) 'Productivity of tropical herbivore assemblages: Nature's basis for exploitation by man', in E. A. Huisman, J. W. M. Osse, D. van der Heide, S. Tamminga, B. J. Tolkamp, W. G. P. Schouten and H. H. T. Prins (eds) *Ecology and Behaviour of the African Buffalo: Social Inequality and Decision Making*, Chapman and Hall, London

Prins, H. H. T. (1996) *Ecology and Behaviour of the African Buffalo: Social Inequality and Decision Making*, Chapman and Hall, London

Prins, H. H. T. and Olff, H. (1998) 'Species richness of African grazer assemblages: Towards a functional explanation', in D. M. Newbery, H. H. T. Prins and N. D. Brown (eds) *Dynamics of Tropical Communities*, Blackwell Scientific, Oxford

Rannestad, O. T., Danielsen, T., Moe, S. R. and Stokke, S. (2006) 'Adjacent pastoral areas support higher densities of wild ungulates during the wet season than in the Lake Mburo National park in Uganda', *Journal of Tropical Ecology*, vol 22, pp675–683

Rees, W. A. (1974) 'Preliminary studies into bush utilization by cattle in Zambia', *Journal of Applied Ecology*, vol 11, pp207–214

Reid, R. S. and Ellis, J. E. (1995) 'Impacts of pastoralists on woodlands in south Turkana, Kenya: Livestock-mediated tree recruitment', *Ecological Applications*, vol 5, pp978–992

Rukuni, M. (1994) *Report of the Commission of Inquiry into Appropriate Agricultural Land Tenure Systems. Volume II: Technical Reports*, Government Printer, Harare

Sanon, H. O. (2007) 'The importance of some Sahelian browse species as feed for goats', PhD thesis, Swedish University of Agricultural Sciences, Uppsala, Sweden

Shackleton, C. M., Shackleton, S. E. and Cousins, B. (2001) 'The role of land based strategies in rural livelihoods: The contribution of arable production, animal husbandry and natural resource harvesting in communal areas in South Africa', *Development Southern Africa*, vol 18, pp581–604

Shackleton, C. M., Shackleton, S. E., Netshiluvhi, T. R. and Mathabela, F. R. (2005) 'The contribution and direct-use value of livestock to rural livelihoods in the Sand River catchment, South Africa', *African Journal of Range and Forage Science*, vol 22, pp127–140

Smith, A. B. (1992) *Pastoralism in Africa: Origins and Development Ecology*, Hurst and Co, London

Spinage, C. A. (1992) 'The decline of the Kalahari wildebeest', *Oryx*, vol 26, pp147–150

Stanley, J. (2000) 'The Machakos Wildlife Forum: The story from a woman on the land', in H. H. T. Prins, J. G. Grootenhuis and T. T. Dolan (eds) *Wildlife Conservation by Sustainable Use*, Kluwer Academic Publishers, Norwell, MA

Steinfeld, H., de Haan, C. and Blackburn, H. (1997) 'Livestock-environment interactions, issues and options', FAO, WRENmedia, Suffolk, UK

Stevenson-Hamilton, J. (1974) *South African Eden: From Sabi Game Reserve to Kruger National Park*, Collins, London

Stutton, J. E. G. (1993) 'Becoming Masaailand', in T. Spear and R. Waller (eds) *Being Masaai: Ethnicity and Identity in East Africa*, James Currey, London

Swanepoel, R., Barnard, B. J. H., Meredith, C. D., Bishop, G. C., Brockner, G. K., Foggin, C. M. and Hobschle, O. J. B. (1993) 'Rabies in southern Africa', *Onderstepoort Journal of Veterinary Research*, vol 60, pp325–346

Touré, I., Ickowicz, A., Sagna, C. and Usengumuremyi, J. (2001) 'Étude de l'impact du bétail sur les resources du Parc National de Djoudj (PNOD-Senegal)', Contribution to workshop 'Faune sauvage et bétail: Complémentarité, coexistence ou compétition?', Niamey, 16-19 January

Upton, M. (2004) 'The role of livestock in economic development and poverty reduction', www.fao.org/ag/againfo/projects/en/pplpi/docarc/wp10.pdf

Van Wijngaarden, W. (1985) 'Elephants-trees-grass-grazers: Relationships between climate, soils, vegetation and large herbivores in a semi-arid savanna ecosystem (Tsavo, Kenya)', *ITC Publication 4*, ITC, Enschede, Netherlands

Walker, B. H. (1976) 'An assessment of the ecological basis for game ranching in southern Africa', *Proceedings of the Grassland Society of Southern Africa*, vol 1, pp125–130

Walker, B. H. (1979) *Management of Semi-arid Ecosystems*, Elsevier, Amsterdam

Wetterstrom, W. (1993) 'Foraging and farming in Egypt: The transition from hunting and gathering to horticulture in the Nile valley', in T. Shaw, P. Sinclair, B. Andah and A. Okpoko (eds) *The Archaeology of Africa: Food, Metals and Towns*, Routledge, London

Wickens, G. E. (1980) 'Alternative uses of browse species', in H. N. Le Houérou (ed.) *Browse in Africa: The Current State of Knowledge*, International Livestock Centre for Africa (ILCA), Addis Ababa

Wickens, G. E. (1995) *Role of Acacia Species in the Rural Economy of Dry Africa and the Near East*, FAO, Rome

Winrock (1992) *Assessment of Animal Agriculture in sub-Saharan Africa*, Winrock International Institute for Agricultural Development, Morrilton, Arkansas

9

Plantations and Woodlots in Africa's Dry Forests and Woodlands

Shabani Chamshama, Patrice Savadogo and Crispen Marunda

Introduction

Africa's dry forests and woodlands have been exploited for centuries. Firstly, by local iron-smelting industries, which from 2000 years ago resulted in selective felling of hardwoods that produced good quality charcoal. Secondly, trade in African blackwood (*Dalbergia melanoxylon*) to China from eastern African miombo woodlands. Thirdly, during the colonial period when hardwoods were cut to supply wood for railway sleepers, timber for the mining industries (see Chapter 6) and specialist woods (particularly African blackwood) for niche markets such as musical instruments. By the mid-20th century timber was being exported to Europe (Buckle, 1959). The growing population of Africans during the same period started exerting pressure on the natural forests for timber for construction purposes. The growth in population also led to an increase in the demand for agricultural land leading to the clearing of large swaths of land. Finally there was, during the post-colonial period, commercial exploitation of high value hardwoods for export to Asia, during the construction of the Tanzania-Zambia railway and, more recently, a major surge in hardwood use and export during the past decade, especially to China (Mackenzie, 2006). These factors led to the impetus towards a tree growing culture in the dry forest regions of Africa, which inherently have low timber yields when compared to the tropical rain forests.

The use of dry forests and woodlands by people is well documented (Murphy and Lugo, 1986), and consequently are most threatened and less protected than other ecosystems (Mertz et al, 2007). Recent studies indicated that the dry forest and woodlands regions of Africa are undergoing unprecedented forest and land degradation as a result of climate change and overharvesting by human beings (Lambin, 1999; Stephenne and Lambin, 2001; Huang et al, 2003; Wardell et al, 2003; UNEP, 2006). The causal factors of degradation include: clearing trees for agricultural expansion; firewood gathering and charcoal production; uncontrolled fires; overgrazing; human settlements; infrastructural and industrial developments; and trade policies (Baumer, 1990; Blay et al, 2004). Additionally, there are various underlying causes which include high population growth, rural poverty and poor policies (Palo, 1999; Palo et al, 2000). The main consequences of land degradation which impact negatively on human livelihoods and the environment include: shortages of firewood and timber; shortages of non-wood forest products; loss of biodiversity; increased sediment deposits, siltation of dams, floods and land slides; drying of springs and water bodies; increased incidence of waterborne diseases; climate change and desertification (Palo, 1999; Blay et al, 2004; Lamb et al, 2005). All these reduce land productivity and influence ecological services and the means of existence for forest-dwelling people (Perrings, 2000; Darkoh, 2003; Zegeye et al, 2006).

The establishment and management of plantations and woodlots in the dry regions of Africa can contribute towards availability of wood and non-wood products as well as rehabilitation of degraded areas, and relieve some of the pressure on natural forests and rangelands. It is because of this need that African governments in the 1960s and 1970s, in most cases with donor support, established plantations and assisted communities with establishment of woodlots. Little has been documented about these plantations and woodlots especially in the dry forest and woodland regions of Africa. The overall objective of this chapter is to give an overview of histories of plantations and woodlots for wood and non-wood forest products, the critical and common management practices, and the common barriers to plantation and woodlots expansion in dry areas of Africa. Whilst the rest of the chapters in this book focus on naturally occurring dry forests and woodlands, this chapter focuses on plantations and woodlots. This focus is important in that tree planting still remains a viable option to increasing woody biomass and tree products to meet the long-term needs of the people. In more recent times, trees have also been planted to produce environmental services, such as soil stabilization and amelioration, windbreaks and shade, and carbon sequestration. Plantations and woodlots, if well established and protected from fire and livestock, are much more productive than natural forests and allow economic management in a sustainable manner (FAO, 2001a). Thus the inclusion of this chapter in the book completes the whole picture of forests, woodlands and forestry in the dry regions of Africa.

Plantations and Woodlots: Context

Plantations are defined as those forest stands established by planting or/and seeding in the process of afforestation or reforestation. They are either of introduced species or indigenous species which meet a minimum area requirement of 0.5ha; tree crown cover of at least 10 per cent of the land cover; and total height of adult trees above 5m (FAO, 2001b). Woodlots are small plantings or clumps of trees near villages, as well as larger plantings which are intended for firewood, building materials, poles, laths and droppers for local villages, but not for industrial purposes, such as production of sawn timber, mining timber or pulpwood. These woodlots are usually associated with a community.

Development of both plantations and woodlots has not been without challenge. Particularly challenging has been the need to understand the relationship between tree and site conditions and the effects of weather, especially rainfall. Both factors called for the need to conduct thorough research into species and seed source trials on different sites. The research has yielded a significant body of literature on species and site productivity, and management techniques that are employed for various species and on different sites for different products. Furthermore, plantations have expanded to occupy a significant portion of the landscape. Consequently, the plantations and woodlots have a landscape role in hydrology, biodiversity and carbon sequestration. Thus the plantations and woodlots have to be seen not only for their role in supplying wood products, but their service, support and regulative functions in the environment.

Establishment of plantations and woodlots in Africa has also happened in a complex policy environment. Land tenure and security are some of the policy areas that have had a significant bearing on the success of tree planting. The net economic and environmental impacts of plantations and woodlots have to be assessed within the complex policy environment. Despite the technical challenges and policy constraints, tree planting still remains an important activity, especially in the dry forest regions of Africa where timber and other forest products from natural forests and woodlands may not meet demand.

History of Plantations

Whilst the dry forests and woodlands of Africa produce some of the most valuable natural timbers, the productivity of the forests is low. The establishment of trees in many African countries is a viable land management option capable of achieving economic, environmental and social benefits. Tree planting using exotic species was mainly driven by the need to secure round-wood supply for construction and pulp industries. Plantation forestry in various countries of Africa was preceded by species and provenance trials mainly of exotic tree species in the period between the mid-1800s and early 1900. Successful results from species and provenance trials led to large-scale planting mainly using exotic tree species, although in some parts of Africa trials of indigenous species, such as *Afzelia quanzensis* and *Millettia stuhlmannii* in Mozambique and *Baikiaea plurijuga* in Zambia (Saramaki et al, 1986) and Zimbabwe (Calvert and Timberlake, 1993) were also planted. There were several objectives of plantation establishment depending on the ecological zone. In the humid and sub-humid zones, the emphasis has been on high value industrial plantations while for the semi-arid and arid areas, emphasis has been on woodfuel production (peri-urban plantations and woodlots), as well as to improve environmental conditions, including desertification control and sand dune fixation (FAO, 2003). However data on planted areas have not been disaggregated into ecological zones or type of forests, whether plantations or woodlots.

The first extensive plantings of industrial tree crops in Africa occurred during the period 1900–1945, mostly in countries with little utilizable natural forest and where there had been an early influx of European settlers (Evans, 1992). In 1938, for example, South Africa had 520,000ha of plantations (SAIF, 2000). For many countries, most species were introduced during this period and were planted in trial plots (Evans, 1992). There has been a steady expansion in forestry plantations since 1945. In 2000 it was estimated that dry forest and woodland countries, excluding South Africa, had plantations covering a total of 3570km^2 (FAO, 2005) but with large variations between phytoregions and countries (Figure 9.1). South Africa alone had 15,540 km^2 under plantations in 2000 (FAO, 2005). According to FAO (2001b) Africa's total plantation area and the annual planting area are the lowest among all the continents. Across many regions of Africa, most plantations are restricted to a few countries. In southern Africa, 51 per cent are in South Africa, in western and central Africa, 71 per cent are in Côte d'Ivoire, Nigeria, Rwanda and Senegal, and in eastern Africa, 96 per cent are in Ethiopia and Sudan (FAO, 2005). *Eucalyptus* is the most widely planted genus covering 22.4 per cent of all planted area, followed by *Pinus* (20.5 per cent), *Hevea* (7.1 per cent), *Acacia* (4.3 per cent) and *Tectona* (2.6 per cent) (FAO, 2001b). The area covered by other broadleaved and other conifers is respectively 11.2 per cent and 7.2 per cent, while the unspecified species cover 24.7 per cent of the total area (FAO, 2001b). Drought tolerant species, such as *Eucalyptus* and *Acacia*, are the common species planted in the dry parts of Africa for timber, firewood, tanbark and gums. Other species planted in dry areas include *Azadirachta indica*, *Cupressus* spp, *Casuarina equisitilifolia* and *Casuarina senegalensis*.

The ownership of plantation forests and woodlands in Africa extends from governments and large industrial corporations to individual farmers, and their management varies considerably, from relatively simple and low-input to highly sophisticated and intensive systems (FAO, 2005). Industrial plantations are 52 per cent publicly owned, 34 per cent privately owned and 14 per cent other or unspecified (FAO, 2001b). For non-industrial plantations, 62 per cent are publicly owned, 9 per cent privately owned and 29 per cent other or unspecified (FAO, 2001a). Because of increasing land pressure for plantation expansion, there is now a trend for large forestry companies to go into out-grower partnership schemes with communities (Mayers, 1999). In South Africa, for example

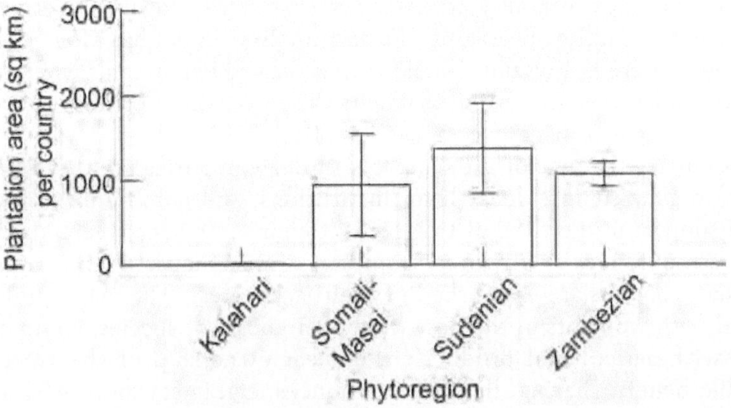

Figure 9.1 *Area under plantations in dry forest and woodland countries in sub-Saharan Africa in 2000*

Note: The value for the Zambezian phytoregion excludes South Africa.
Source: Based on FAO (2005)

smallholders grow trees on a contract basis with credit and fertilizer support from large forestry companies (Cairns, 2000; Nawir et al, 2007). These out-grower schemes are increasing the plantation forestry base and local communities are getting involved and benefiting from plantation forestry development.

History of Woodlots

Investment in forest plantations in the 1960s to 1980s was in most countries done with donor support, with emphasis on industrial plantations. In the 1970s emphasis was shifted to establishment of woodlots for social or community forestry and in the 1980s the environmental side of social forestry was reinforced by increasing public concern regarding rapid deforestation in many tropical countries (Person, 2003). For example, in Zimbabwe, the Rural Afforestation Project was funded by the World Bank and its major objective was to meet the demand for firewood and poles by setting up nurseries, small-scale plantations, and demonstration and trial plots of high-yielding exotic species. The planting of woodlots was largely driven by demand from communities to meet household needs for firewood, timber for construction and other forest products. The woodlots were also established for environmental reasons, such as providing a stop-gap measure to natural woodland deforestation through the supply of wood. In most countries, the woodlots were established using mainly eucalypt species because of their fast growth and high wood productivity. However, in spite of the many dry forest and woodland countries in sub-Saharan Africa, statistics on woodlots established and areas involved are difficult to find. The exception, although lying largely outside the geographical coverage of this book, is South Africa, where small-scale growers are estimated at 19,000 with woodlots averaging 2ha and totaling 43,000ha. Most of these woodlot owners are in Kwazulu-Natal province near the pulp mills where there is a traditional land tenure system that suits the allocation of plots to individual households for tree growing (Mayers et al, 2001).

Cross cutting features of the woodlots throughout Africa are that communities are organized into local level institutions; the communities receive technical support in the form of tree seeds; training in nursery techniques and tree culture practices is provided from state forest departments and local NGOs; and woodlots are established in open access areas. The major problems associated with woodlots in the past were poor choice of species, competition for land with agricultural production, unclear ownership of the trees and inequitable benefit sharing. In certain instances, e.g. in southern Africa, the products from the trees (e.g. poles) were found to be more valuable than the original intended product (firewood) leading to changes in the original objective of tree planting. These problems led to most of the projects failing to achieve their goals. Eucalypts were and are still the main species planted for community woodlots in dry parts of Africa, and over the years there has been growing concern over the high water usage of the trees leading to drying of catchments. But still, eucalypts yield high returns. For example, in a village level survey in Tigray in Ethiopia, Jagger and Pender (2003) illustrated that planting eucalypts resulted in high rate of return, almost 20 per cent above the baseline scenario where there are no plantations. So whilst there are problems with woodlots, the potential for economic returns and supplementary income for rural households still exists.

Management of Plantations in the Dry Areas of Africa

The performance of plantations and woodlots in the dry forest and woodland regions of Africa has generally been very variable. In some situations, outstanding performance (up to 30m^3 per hectare per year) has been observed while poor performance (1–2m^3 per hectare per yr) has been noted in other situations (Zobel et al, 1987; Tiarks et al, 1998; FAO, 2001a, 2003). A recent study has shown that ownership pattern has significant impact on the quality of management and productivity (Chamshama and Nwonwu, 2004). Overall, government owned industrial or energy plantations are characterized by planting and replanting backlogs, low intensity site preparation techniques, poor quality trees due to use of unimproved seed, low survival due to poor species-site matching and delayed or low intensity weeding. It is also noted that they are generally neglected or have irregular pruning and thinning, constant fire, disease and pest attack, and generally suffer from illegal felling and encroachments. Nearly all the public sector plantations were initially established with donor support. After expiry of the donor support, funding of the plantation activities reverted to the governments. It is then that the many problems of mismanagement cropped up. On the other hand, privately owned plantations have been found to be of high productivity due to careful site selection, intensive cultural practices, selection of best species/provenances and genetic improvement through research. For the semi-arid areas, drought has also been shown to reduce both survival and growth (Zobel et al, 1987).

With regard to individual and community woodlots, Arnold (1984), Jackson (1984) and Taylor and Soumare (1984) have observed that in addition to some of the above factors, variability in performance has also been shown to be due to:

- Projects being based on the perceptions and priorities of planners and foresters followed by project implementation in a socio-cultural context where perceptions and priorities are often quite different.
- Donor initiatives and priorities which were often not forestry department priority.
- Limited political commitment, as a result of which the level of funding provided to the forestry sector has been meagre.
- Limited/lack of trained technicians and extension staff, leading to poor species choice and management techniques.
- Tree planting and management occurring when farm labour is most in demand resulting in neglect of these activities.
- In some cases, adequate market studies have not preceded plantation establishment and the sale of resulting products may be considerably more difficult than originally estimated.
- Communal planting efforts have sometimes been plagued by lack of adequate attention to detail of the eventual distribution of the products.

Tree establishment and growth is challenging in the dry forest and woodland regions of Africa as the areas are characterized by low soil moisture and nutrient reserves and pests such as termites (Box 9.1). Adoption of best management practices is key to successful tree establishment. Practices include good site species matching, access to improved seed, good nursery practices and silvicultural practices. The best management practices are briefly discussed in the following sections. Detailed explanations of these practices are not the objective of this chapter – they are described in publications such as those by Goor and Barney (1968), FAO (1974), Zobel et al (1987), Evans (1992) and SAIF (2000).

Good site–species matching

Due to their evolutionary development, most tree species are generally site specific and careful consideration must be given to matching species to overall site characteristics (Jackson, 1984; Zobel et al, 1987; Evans, 1992). Most planted species (e.g. *Eucalyptus* spp, *Pinus* spp, *Tectona grandis* and *Gmelina*

Box 9.1 Mortality of tree seedlings in rural woodlots in Zambia

Emmanuel N. Chidumayo

Chidumayo (1988) observed a large variation in tree survival rate among rural woodlots in eastern and central Zambia; the variation was larger in *Eucalyptus grandis* than in *Gmelina arborea*, at least during the first two years of establishment. Tree mortality declined with increasing age of woodlots in both species: *G. arborea* mortality declined from 1.0–1.3 per cent per month during the first 6–8 months following planting, to 0.4–0.5 per cent per month during the subsequent 9–18 months. Average mortality of *E. grandis* declined from 4.6–6.1 per cent per month during the first 6–8 months, to 0.3 per cent per month during the subsequent 9–18 months.

Woodlot owners indicated that the major cause of mortality among *E. grandis* trees was root damage by termites and 71 per cent of the dying seedlings in 6 to 8 month old woodlots in Chongwe district were damaged by termites. Kwesiga et al (1999) noted that the main problems cited by small scale farmers participating in growing trees in agroforestry schemes are insect pests, browsing by livestock, drought and poor seed. Indeed drought does contribute greatly to seedling mortality. For example, the 1986–1987 drought resulted in total mortality of *E. grandis* seedlings in six of the eight newly established woodlots in Chongwe district (Chidumayo, 1988); the other two woodlots were irrigated and survival of seedlings one year later was 87 per cent in one woodlot and 100 per cent in the other.

Fire is another potential cause of tree mortality. A one-and-half year old *E. grandis* woodlot in Chongwe district was destroyed by a bush fire during the 1987 dry season and 57 per cent of the trees died, while the root-stocks of the remainder survived and coppiced (Chidumayo, 1988). Bohlin and Larsson (1983) also reported that a woodlot in Chadiza district of Eastern Province was completely destroyed by fire.

In Chongwe district frost is reported to have damaged all but one tree among one-and-half year old *Gmelina* trees although these eventually coppiced and developed into multi-stemmed plants (Chidumayo, 1988).

arborea) are site-specific, therefore require special consideration for site selection in order to maximize plantation survival and productivity (Bekker et al, 2004). Soil type, altitude, slope gradient, natural vegetation and environment are the main parameter considerations in site selection and site classification. Selection of sites for planting also requires adequate knowledge of the climate, edaphic and topographic factors both in the natural habitat of the species (for exotic species) and in the proposed country of introduction. The low productivity of various tree species in plantations and woodlots has in some cases been attributed to 'off-site' planting (Jackson, 1984; Zobel et al, 1987; FAO, 2001a, 2001c, 2001d, 2002b). This arises from starting plantations or woodlots (sometimes due to political pressure) without species or provenance trials or if trials were carried out, they were of very short duration i.e. less than the recommended half rotation period. Of all the factors considered in various reviews, species/provenance-site matching is considered to have the greatest and often longest effect on differences in productivity of various plantations and woodlots (Vichnevetskaia, 1997; FAO, 2002b). Off-site planted trees grow under stress, and this may increase their susceptibility to pests and diseases (Zobel et al, 1987; FAO, 2001d).

Access to improved seeds

When large-scale establishment of plantations and woodlots started in various countries of sub-Saharan Africa, seed requirements were initially met by importation from other countries. Seed for eucalypts and pines were imported from Australia and Central America, respectively. Over the years, countries in Africa with tree breeding programmes, such as Zimbabwe (Marunda, 1997) and South Africa supplied large quantities of seed. With time, local seed sources i.e. seed stands (essentially an interim source) and seed orchards were established for the major tree species. Local seed sources continued to be supplemented by importation to meet domestic demand, especially for species which fail to flower or produce a reasonable number of viable seeds in exotic environments (Zobel et al, 1987).

While some countries in sub-Saharan Africa continued with tree improvement efforts to produce advanced generation seed orchards or clonal material, others neglected both the established seed stands and un-rogued first generation seed orchards. As a result, improper seeds from unregistered sources are used, thus limiting the productivity and quality of the plantations and woodlots (Zobel et al, 1987; Chamshama et al, 1996; Bekele, 2001; FAO, 2002a). As a consequence, in some sub-Saharan African countries, forest plantations and woodlots constitute a significant proportion of trees of low productivity and quality.

In order to ensure high productivity and quality of forest plantation species, proper seed source is imperative. The main sources of improved tree seeds are seed orchards, usually established and managed by the forestry service, although some private companies have their own seed sources. In cases where there are no seed orchards seed can be imported from other countries with tree breeding programmes.

Silvicultural management

A critical part of plantation or woodlot establishment is the quality of planting stock. Chavasse (1980) summarized the factors that affect seedling quality as: nursery site (soil fertility, moisture, climate, shelter); genetic make-up of the stock; seed (size, variability, storage, germination treatments); methods of production (bare-rooted, rooted cuttings, container seedlings); seedbed density; time of the year seedlings are lifted; weed control methods and effectiveness;

seedling nutrition; methods of seedling conditioning; and insect and disease control. Nitrogen and potassium are two of the important nutrients that encourage good seedling growth and post nursery survival.

In the semi-arid regions, drought, livestock, fire and termites are major factors that determine establishment and growth of seedlings (Veenendaal et al, 1995; see Box 9.1). The actual performance of newly planted seedlings depends not only on inherent seedling performance but also on the extent to which the environmental conditions allow this potential to be expressed (Folk and Grossnickle, 1996; Zida et al, 2008). Root pruning or wrenching is one way of hardening seedlings against shock during lifting and planting; and to ensure high field survival and growth. This practice results in reduced plant dry weight, increased root-shoot ratio, high relative turgidity of leaves, high root growth capacity and increased translocation of photosynthates to roots (Rock, 1969, 1971; Bacon and Hawkins, 1977; Bacon and Bachelard, 1978; Burdett et al, 1983; Chamshama and Hall, 1987). Top pruning is sometimes used as a cultural technique with the aim of creating a favourable shoot-root ratio (Chamshama and Hall, 1987). However, excessive pruning may be lethal due to reduced photosynthetic surface.

Planting site preparation

The literature on the effects of mechanical site preparation on tree survival and growth is voluminous. Most reports generally agree that seedling establishment and growth is improved though initial responses may not persist. For the dry areas, preparation of the planting sites using methods that conserve moisture is important. For plantation forests, the methods for site preparation are well developed. Treatments that directly modify the soil are of primary interest because in addition to reducing competition from unwanted vegetation, they improve infiltration and reduce surface run-off (Perrow and Davy, 2003). Poor site preparation may result in reduced site productivity. The timing of site preparation is important. For example, mechanical site preparation should not be carried out when the soil water content is at field capacity as this may result in a significant soil compaction, associated with increased bulk density and decreased macro-pore space.

The foregoing practice, which requires intensive management by elaborate site preparation, may be difficult to obtain under village situations in the absence of ox-ploughs or tractors. However, observations made in semi-arid areas of Gairo, Morogoro in Tanzania have shown that deep cultivation using hoes ensures the necessary management intensity (Chamshama, unpublished). A critical issue in tree establishment is to maximize seedling establishment and early growth from available moisture. A number of water harvesting and conserving techniques are available in many parts of Africa. For example, in parts of western Africa, the zaï techniques, half-moons and stone cords or bunds are used extensively to harvest rainwater and help crop and tree establishment. In Burkina Faso, Niger and Mali the zaï approach for restoration of degraded areas has resulted in increased crop yield. The zaï is a micro-catchment with a diameter of 20 to 30cm and a depth of 15 to 20cm and is dug primarily to increase water infiltration and to reduce erosion. The zaïs are fertilized with animal dung. The zaï forestier technique is used to encourage tree growth and regeneration. In contrast to the traditionally narrow spacing (2.4 × 2.4m) used in forest practice in moist areas, wider spacing (3 × 3m to as much as 7 × 7m) is recommended for arid and semi-arid areas (Goor and Barney, 1968). The wide spacing ensures that the low soil moisture reserves are made available to fewer trees thus increasing their survival prospects.

Tree protection

Control of weeds and pests are important practices that affect survival and growth of trees. In dry sites, the generally low soil moisture levels dictate complete weeding. The weeding can be done manually or mechanically and in most woodlots where trees are intercropped with agricultural crops (rotational woodlot arrangement), the weeding carried out to maintain the crop will simultaneously provide adequate weeding to the trees. The latter has the added advantage of ensuring that trees are not neglected during the peak season when labour is directed to farming.

Because of the open access of the areas where woodlots are planted, there could be problems with browsing animals. Protection from animals is sometimes costly. For example, in Zimbabwe, reforestation and afforestation projects using eucalypts faced the problems of browsing, and the Forestry Commission of Zimbabwe had to provide fencing material as support and incentive to tree planting by communities. In Mali, the Joliba Trust working with the Dogon people, reports losses of young *Faiherbia albida* trees to browsing by animals. Protecting the trees still remains a major bottleneck to successful tree planting. Various methods have been tried which include protecting the young trees with brush fence and sprinkling the seedlings with water mixed with cow dung as a repellant (cordon sanitaire method).

In the dry parts of Africa, eucalypts are the most widely planted species and the young trees are very susceptible to termites. Trees between three and four years are particularly susceptible and termiticides have to be applied to control the pests. For eucalypt plantations, tree mortality due to termite attack can be 30–50 per cent but can reach 100 per cent in the absence of any control. For many years, organo-chlorines, such as dieldrin, chlordane and aldrin, were used to protect trees from termites but their persistence in the environment means that they are now not recommended. Carbosulfan, in the controlled-release granule formula (Marshall suSCon) is one of the widely used chemicals applied around individual plants in many parts of sub-Saharan Africa (Atkinson et al, 1992). The best means found to overcome the problem is to use eucalypt species naturally resistant to the disease. For example, *Eucalyptus maculata* seems to be more resistant than other species, while *E. citriodora*, *E. Saligna* and *E. maideni* are very sensitive (FAO, 1958). Dry forests and woodlands have few really serious insect pests or fungal diseases of trees. Perhaps this is due in part to the fact that dry climates do not favour dense stands or growth of underbrush, environments where harmful insects and pathogens often breed in humid climates (Goor and Barney, 1968; FAO, 1974). However, during years of severe drought, many trees become weak and may be severely attacked by pests and diseases. Vigilance is required to detect any attacks and in the absence of biological control, replanting might be necessary.

Pruning

Pruning is a deliberate removal, preferably while still alive, of some of the branches from the lower trunk (bole) of a tree, with an objective of reducing knots in sawn timber and similar finished products (SAIF, 2000). Branches form knots, which are the most common defects of timber, especially those formed by dead branches. The lateral grain distortion around knots leads to reduced timber strength. Pruning is generally due when crowns touch and the pruning schedules vary according to management objectives. Most growers of sawlogs prune 40–50 per cent of the living crown three to five times (Zobel et al, 1987). The decision to prune or not to prune must almost be entirely based on the consideration of economic factors. High pruning is associated with price differentiation between pruned and unpruned timber.

In countries of sub-Saharan Africa, pruning schedules have been based on research results and/or adapted with modifications from countries with longer experience in growing the various species. Despite the presence of the pruning schedules, many countries of sub-Saharan Africa have pruning backlogs, mainly in public industrial plantations, and this is often attributed to budgetary constraints (see e.g. MENR, 1994; Nshubemuki et al, 2001; Kenya FOSA, 2001; FAO, 2002a). Additionally, the absence of price differentiation between pruned and unpruned timber serves as a disincentive to prune and also narrows the prospects for exporting high valued timber. Pruning schedules should be adhered to and mechanisms for pushing price differentiation for pruned and unpruned timber both for the local and export market have to be developed.

Thinning

Artificial thinning is the removal of a proportion of individual living trees from a stand before clear felling (SAIF, 2000). It is generally understood to take place after the onset of competition. The major objectives of thinning are: to reduce the number of trees in a stand so that the remaining ones have more space for crown and root development to encourage stem diameter increment and so reach a utilizable size sooner; to remove trees of poor form; to prevent severe stress which may induce pests, diseases and stand instability; and to provide an intermediate financial return from sale of thinnings (Evans, 1992; SAIF, 2000). More trees are initially established than the required final crop mainly to ensure sufficient trees from which the final crop can be selected, enhance early canopy closure to suppress weed growth and to utilize the site better (SAIF, 2000). As indicated for pruning schedules, thinning schedules have also been based on research. Thinning schedules generally have initial stocking in the range 1111–1680 stems per hectare, two to four thinnings and stocking at clear felling of 220–370 stems per hectare.

While thinning is an important silvicultural operation, which must be done at the right time, in the right way and at the right intensity, various reports and the authors' observations show that thinning operations in many public industrial plantations in sub-Saharan Africa do not follow the prescribed schedules (Zobel et al, 1987; Åhlback, 1988; MENR, 1994; FAO, 2002a). Where thinnings have been carried out, they have been fewer and lighter than recommended, resulting in the standing volume being distributed to too many small trees rather than fewer ones of greater value per m^3. The main reasons given for the neglect of thinnings have been shortage of funds, lack of markets for unsawn thinnings, lack of plantation management skills and experience, foresters' traditional attitude against waste and lack of processing plants (Åhlback, 1988; FAO, 2002a).

Productivity of Plantations

The yields from forest plantations, which are rapidly expanding in sub-Saharan Africa (FAO, 2001b), vary considerably across ecosystems (Table 9.1). There is a large body of empirical field evidence indicating that, with appropriate species–site matching and silvicultural management, plantations can remain productive (Tiarks et al, 1998). Intensive silvicultural operations that increase the availability of water and nutrients in forest stands, management during the inter-rotational (harvesting) period, removal of competing vegetation, slash

Table 9.1 *Yield of the most important timber and fuelwood species*

Species	*Source*	*Country*	*Rotation length (years)*	*Mean annual increment (m^2 per ha per yr)*
Gmelina arborea	Onyekwelu (2004)	Nigeria	5	16.6
Pinus patula for sawn timber	van Vuuren et al (1978)	South Africa	25	14.8
Pinus patula for pulp	van Vuuren et al (1978)	South Africa	25	14.8
Eucalyptus urophylla	van Vuuren et al (1978)	Cameroon	8	30
E. grandis	van Vuuren et al (1978)	South Africa	25	31.8
Tectona grandis	Nunifu and Murchison (1999)	Ghana	12	11.05
Tectona grandis	Nunifu and Murchison (1999)	Ghana	20	13.67

management and harvesting operations, have a potentially great impact on the productivity and long-term sustainability of forest stands.

Because the dry forests and woodlands of Africa do not produce high volumes of merchantable timber, the establishment of forest plantations guarantees a sustainable supply of wood. The main products from wood from plantations are industrial round wood for sawn timber, veneer, particle-boards, pulpwood, firewood and charcoal. South Africa is the largest producer of wood from industrial plantations in Africa whilst countries such as Ethiopia, Kenya and Rwanda have large firewood plantation resources. A large portion of wood in most of the countries is consumed as firewood showing the large dependence of the countries on traditional energy sources. Most of the firewood however is sourced from natural forests, but plantations are increasingly becoming important and the preferred source of firewood. In countries, such as Ethiopia, peri-urban eucalypt plantations are the major sources of firewood (Gebrehiwot, 1997). In Malawi, plantations meet about 11 per cent of the firewood demand (Jumbe and Angelsen, 2006). These plantations provide a source of livelihoods to many people. However, due to overexploitation and lack of clear protection strategies, the extent of the plantations is declining at a very fast rate. With increasing rates of urbanization throughout Africa, peri-urban plantations are going to remain as one of the most important sources of wood energy.

Trade in Forest Products

The development of the plantation-grown wood trade from Africa, particularly western Africa, has been very much dependent on old colonial links and the demand for timber by the building and construction industry in Europe after the Second World War (Attah, 2007). Today, 96 per cent of Africa's plantation-grown wood is exported to Japan, the EU and other markets. With the increasing requirement that timber comes from legal and sustainable sources, there is a gradual decline in demand from environmentally aware markets (Hillring, 2006), such as the UK and Germany, which were key traditional markets for timber from western and central Africa. However, there is a growing demand in China. Canby (2007) reported that 13 per cent of Africa's pulp exports were sent to China in 2006. Swaziland and South Africa are the top suppliers of African plantation-grown wood to China. The potential for plantations to meet growing domestic and international demand for wood is now recognized and countries in Africa with suitable growing conditions and suitable areas should invest in expanding the forest plantation resource base.

Contribution of Plantations and Woodlots to National Economies

From the socioeconomic perspective, plantations and woodlots generate revenue and foreign exchange for national governments. At the local level, they provide jobs offering economic opportunities for rural residents (Whiteman and Lebedys, 2006). In addition, there may be opportunities for local residents to use the residues and by-products left behind after trees have been harvested for firewood or timber. Indirect benefits may include government reinvestment of the revenue generated from plantations into education, medicine and infrastructure development in local communities (Morrison and Bass, 1992). The social dimensions associated with plantation forestry are not necessarily positive and, indeed, may be quite adverse. Afforestation of agricultural land leads to a decline in the rural population through voluntary or forced removal of existing land users (Evans and Turnball, 2004). In South Africa, the impacts of plantations on water resources is a source of conflict between the forestry sector and other sectors dependent on water as the plantations intercept rainfall and use up a lot of water thereby reducing run-off from planted areas.

Non-wood Products from Plantations and Woodlots

The previous section considered aspects of plantation forestry whose primary objective is to supply timber for commercial use. Plantations and woodlots also provide non-wood forest products that are used in industries and consumed at a household level (FAO, 2005). Traditionally local communities and households have depended on natural forests and woodlands for the supply of timber and non-wood forest products (see Chapters 4 and 5 on non-timber forest products). But due to loss of forests and woodlands and increasing demand for non-timber products, most African countries particularly those with dry forests and woodlands, have invested in the growing of woodlots for supplying wood and trees to provide non-timber forest products. Most of these investments have been channeled through rural development initiatives and donor funded projects to help rural communities.

In order to replace the diminishing harvests of non-wood forest products, there are increasing tree-planting schemes to supply some products. Firewood is perhaps the most important product and many initial firewood plantations were in response to the perceived firewood crisis in many African countries. However, this perception changed as some of the introduced species produced more valuable products, such as poles, and hence firewood became a by-product. The main non-timber products that are harvested from planted trees include gum arabic, tannins, oils, resins, fruits and medicines. This section discusses non-timber forest products with a special focus on gums, oils, tannins and indigenous fruits from countries in the dry regions of Africa.

Gums, resins and essential oils from plantations

Acacia gums

Acacia gums are harvested in the 'gum belt' of Sudan, Ethiopia and Eritrea (Giffard, 1975). From the Horn of Africa, this region extends southwards through Tanzania to the southern African countries of Angola, Namibia, Zimbabwe, Botswana and South Africa. The main product is gum arabic, obtained from the stems and branches of *Acacia senegal* and *A. seyal*. The major source (95 per cent) is *A. senegal* (*hashab*), with the remaining 5 per cent derived from *A. seyal* and sold as an entirely separate product (gum *talha*). In southern Africa, the main source of gum is *A. karroo*.

It has long been observed that *A. senegal* trees are disappearing from many areas in the northern range of their distribution in Sudan. This was largely attributed to increased human and livestock pressure that converted most of the stands to grasslands (Obeid and Seif El Din, 1971). This, together with other factors such as the need for improved gum and increasing international demand, led to the establishment of artificial plantations of gum in arid areas of Africa. *A. senegal* is the most widely planted species. Countries that have gum plantations include Burkina Faso, Ethiopia, Mali, Nigeria, Senegal and Sudan. Sudan has the largest plantations of *A. senegal*, with well over 100,000ha, although weak land tenure appears to affect the viability of these plantations.

Most of the gum production is done in agroforestry systems known as 'gum gardens' (Rahim, 2005), in which gum trees are grown on farm plots at a spacing of 4m × 4m. During the first four to five years, agricultural crops are planted between the lines of trees, thereby supplying the farmers with food. The trees are nitrogen-fixers, especially during their first few years. A significant amount of the fixed nitrogen is assumed to be transferred to adjacent non-nitrogen-fixing trees or crops, probably as a result of below-ground turnover of roots and nodules (Njiti and Galiana, 1996; Dean et al, 1999; Raddad et al, 2006). Gum production begins at around year four and continues annually until the trees are 20 to 25 years old. The practices employed by growers to improve gum yield include tapping intensity and timing of tapping. Other factors that affect yield include rainfall and temperature (Ballal et al, 2005). Faye et al (2006) also showed that inoculating ten-year old trees with rhizobia generally resulted in increased gum yield. There are still problems with identifying high yielding varieties and vegetative propagation technologies. In South Africa and Zimbabwe, a network of *A. senegal* and other gum producing species, such as *A. karroo* and *A. seyal*, provenance trials were established around 1996 and these will provide valuable information on the genetics and productivity of the species (Barnes et al, 1999).

Oleo-resin

Another key non-wood forest product is oleo-resin or resin from pine plantations. This is produced in South Africa, Kenya, Zimbabwe and Uganda (Coppen and Hone, 1995). *Pinus elliottii*, *P. caribaea* and *P. radiata* are the three species of pine that are tapped in Zimbabwe, Kenya and South Africa. Management of pine species for resin production has economic advantages at both the community (providing employment) and national (import substitution or foreign exchange) levels. Resin tapping also provides an incentive for sustainable forest management based on the principle of multiple uses as long as strong tenure exists. The major constraints in meeting the demands of the market and in raising productivity include the limited yield from the varieties grown, limiting environmental conditions and tapping methods. Thus, long-term solutions are needed and current research on *Pinus* hybrids in South Africa holds particular promise for the region (FAO, 1995a).

Essential oils

Eucalyptus is an important source of essential oil; production is based on either a short-rotation coppice system (harvesting intervals of 6–16 months) or the conversion of waste leaf material, available after trimming, with the felled trees being destined for pulp or sawmills. Eucalypt oils are found in the leaves, fruits, buds and bark of the tree. The oil has roles in medicine, perfumery and flavourings (FAO, 1995b). The principal species used are *Eucalyptus smithii*, *E. cloeziana* and *E. radiata*, which produce medicinal oils containing high levels of cineole. The species *E. globulus*, *E. citriodora* and *E. camaldulensis* are good sources of both medicinal oils for a number of ailments (arthritis, bronchitis, catarrh, cold sore, cold, cough, fever, flu, poor circulation and sinusitis remedies) and perfumery oils (Lawless, 1995). African production of medicinal oil amounts to approximately 250 tonnes per year and is split between South Africa and Swaziland (FAO, 1995b). Yields of oil from leaves vary somewhat between species, but on a commercial scale are in the order of 1 per cent on a 'fresh' weight basis. For example, production from *E. smithii* in Swaziland yields approximately 15 tonnes per hectare of leaf, corresponding to about 150 litres per hectare of oil (FAO, 1995b).

Tannins

Plantations of *Acacia mearnsii* were planted in South Africa, Kenya, Morocco and Zimbabwe for the production of tanbark. The total estimated area is 325,000ha with South Africa having 130,000ha. The tannin is used in the tanning industry and the wood used for firewood and charcoal making.

Indigenous fruit trees in the dry forests and woodlands of Africa

Indigenous fruits are consumed by many people in Africa and sold to raise supplementary incomes (Schreckenberg et al, 2006). Traditionally fruits were harvested from natural forests and woodlands, but due to high rates of deforestation in many parts of Africa, fruit trees are disappearing at a fast rate

Table 9.2 *Priority indigenous fruit species from three phytoregions of Africa*

Somali-Masai phytoregion	*Zambezian phytoregion*	*Sudanian phytoregion*
Adansonia digitata	*Uapaca kirkiana*	*Adansonia digitata*
Tamarindus indica	*Strychnos cocculoides*	*Tamarindus indica*
Ziziphus mauritiana	*Parinari curatellifolia*	*Vitellaria paradoxa*
Sclerocarya birrea	*Ziziphus mauritiana*	*Ziziphus mauritiana*
Balannites aegyptica	*Adansonia digitata*	*Parkia biglobosa*

Source: Adapted from Akinnifesi et al (2007)

(Simons and Leakey, 2004). The supply of fruit and other resources is worsened by commercialization, which is resulting in a decline of the resource and the livelihoods that have become dependent upon it (e.g. Cunningham and Milton, 1987). Domestication of indigenous fruit trees emerged as a farmer-driven, market-led process and has become an important agroforestry initiative in the tropics to supply fruit and other products (Akinnifesi et al, 2007; Leakey et al, 2005). The World Agroforestry Centre (ICRAF) has been conducting research on domestication and commercialization of indigenous fruit trees in order to increase productivity and conserve genetic resources of some of the endangered species.

Most of the dry forest and woodland countries have priority lists of species for domestication (Table 9.2). The planting of most of the species listed in Table 9.2 is still limited to research plots, but individual planting of the trees around homestead boundaries and gardens is common. The research still focuses on selection of priority species, germplasm collection and tree genetic improvement, propagation systems and field management, harvesting and post-harvest technology, economic analysis and market research. It is envisaged that planting of indigenous fruit trees is going to be second to planting eucalypts. There are of course barriers to the expansion of these plantings, some of which are similar to those hindering the expansion of eucalypt woodlots. Research results so far indicate that there is potential for growing the different indigenous fruit trees. The main consideration now at national policy level is for governments to give greater recognition of the potential of indigenous fruit trees to contribute to poverty reduction as a component of more diversified, sustainable and environmentally friendly livelihood options (Schreckenberg et al, 2006).

Barriers to Development of Plantations and Woodlots

The main barriers to the successful expansion of plantations and woodlots in sub-Saharan Africa include limited land base, competing land-use demands, tenure insecurity and inadequate technical and institutional capacities. These are briefly described below.

Diminishing land resources for plantation establishment

Most plantations in Africa are established in areas that receive above average rainfall. Availability and access to suitable land is one of the cornerstones in developing and maintaining a steady planting programme (Chamshama and Nwonwu, 2004). The diminishing land base has also forced plantations and woodlots to be established on marginal sites often with poor soils and increased susceptibility to drought. This has raised the need for further research on site–species matching and tree culture techniques that promote tree growth in the marginal sites.

Tenure insecurity

Land tenure and land law is the greatest hindrance to promoting plantations and woodlots. Most African tenure systems are characterized by the existence of multiple tenants, with several users having access to different resources on the same piece of land (Peters, 2007). In general, there is a duality between customary and statutory land rights in many African countries (Chimhowu and Woodhouse, 2006). The main difficulty associated with customary ownership is that the specific act of tree planting is associated with staking a claim to land ownership and the reluctance of owners to lose their land for a long-term period. This often causes confusion as to whether the land belongs to the group or to specific individuals who planted the trees. With community woodlots, this confusion can also lead to management problems (who does what and who is accountable?) and to inequitable benefit sharing when the trees are harvested.

In most countries of western Africa, even though all land has been officially nationalized, there continues a system of inheritance and hereditary rights. This can lead to tenure insecurity, a significant impediment for long-term forestry projects. Clearly, any sustainable plantation or woodlot establishment plans or activities will need to begin with a clear understanding of local land and resource tenure and access rights (Ruitenbeek and Cartier, 1998; FAO, 2001a; Chamshama and Nwonwu, 2004). Where customary owned land is acquired for planting, Evans and Turnball (2004) recognized two principles that help to overcome difficulties:

1. every effort should be made to meet the wishes of the local people regarding land use proposals;
2. landowners must be involved in the project as key partners.

The failure to involve local people and have tangible benefits accrue to them can disassociate people from the planted forests and even make them co-exploiters and co-destroyers of the forests.

Governance

The governance processes surrounding land and forests are critical for local communities and the sub-Saharan African countries. Good governance lies at the heart of sound environmental management. Considering that most plantation projects have a long gestation period, any investment is liable to be risky unless backed by long-term economic and political stability. Many African countries face political volatility and unpredictable governance systems making long-term forest investments risky. Therefore, it is necessary to have good governance practices in place at national and local levels. This goes from providing scope for meaningful participation in the forest decision-making process to improving the transparency and accountability of forest institutions (Christy et al, 2007). In recent years much progress in forest governance has been made in many countries, such as South Africa and Swaziland, where national policy has opened up to more stakeholders and rights of local people have been strengthened.

Inadequate technical capacity

Forest plantations are long-term investments and require good research and planning. A key success factor in plantation establishment is identifying the right species for the right site. This requires investment in research activities that include site–species matching, tree genetic improvement and protection against pests and diseases. Over the years, investment in such types of research has been on the decline in most African countries. The effects of such low investment in forest research will manifest themselves in low tree survival rates and low plantation forestry productivity.

Institutional barriers

Key functions of most forest institutions in Africa were to conduct research and generate information for forest planning and decision-making. However, many of the countries in sub-Saharan Africa have forestry institutions that are weak and inadequately equipped to implement their functions. These shortcomings stem from diminishing funding, shortage of skilled staff, the absence of adequate training and research facilities, and the lack of integration and cooperation among the major institutions involved in the management of plantations and woodlots.

Macro-economic conditions

Increased economic activity can lead to increased demand for forest products leading to a higher level of demand for round wood and ultimately more plantations and efficient wood processing. Therefore, the development of plantations depends on the macro-economy of a country. For example, Kowero and Mabugu (2006) demonstrated a relationship between macro-economic policies and the development of the forest sector in Zimbabwe.

Conclusion

The preceeding sections have highlighted the importance of plantations and woodlots in meeting demands for wood and non-wood forest products in the dry parts of Africa. Plantation and woodlots remain an important aspect of national development and provide security to wood supplies. At a community level, plantations are an important component of rural development and are sources of energy, non-timber forest products and other environmental services. Many of the countries have put emphasis on reforestation activities in their forest policies, legislative frameworks and national forestry action programmes (Chamshama and Nwonwu, 2004). However, there are constraints that need to be addressed to promote tree establishment. The main threats and constraints to the promotion of plantations and woodlots in sub-Saharan Africa have been extensively reviewed (Evans and Turnball, 2004) and addressing these constraints can contribute greatly to the successful implementation of plantation and woodlot programmes in sub-Saharan Africa.

References

Åhlback, A. J. (1988) 'Forestry for development in Tanzania', Working Paper no 71, IDRC, Swedish University of Agricultural Sciences, Uppsala, Sweden

Akinnifesi, F. K., Ajayi, O. C., Sileshi, G., Kadzere, I. and Akinnifesi, A. I. (2007) 'Domesticating and Commercializing Indigenous Fruit and Nut Tree Crops for Food Security and Income Generation in Sub-Saharan Africa', Paper presented at the New Crops International Symposium, 3–4 September, Southampton, UK

Arnold, J. E. M. (1984) 'Forestation for local community development', in K. F. Wiersum (ed) *Strategies and Designs for Afforestation, Reforestation and Tree Planting*, Proceedings of an international symposium on the occasion of 100 years of forestry education and research in the Netherlands, Wageningen, Netherlands

Atkinson, P. R., Nixon, K. M. and Shaw, M. J. P. (1992) 'On the susceptibility of *Eucalyptus* species and close to attack by *Macrotermes natalensis* Haviland (Isoptera: Termitidae)', *Forest Ecology and Management*, vol 48, pp15–30

Attah, A. N. (2007) 'Market competitiveness of West and Central African forest products and services and the influence of market conditions on forest investments', Paper presented at Issues and Opportunities for Investment in Natural Tropical Forests, the West and Central Africa Tropical Forest Investment Forum, Accra, Ghana

Bacon, G. J. and Bachelard, E. P. (1978) 'The influence of nursery conditioning treatments on some physiological responses of recently transplanted seedlings of *Pinus caribaea* var *hondurensis* B. & G.', *Australian Forest Research*, vol 8, pp171–183

Bacon, G. J. and Hawkins, P. J. (1977) 'Studies on the establishment of open root Caribbean pine planting stock in Southern Queensland', *Australian Forestry*, vol 40, pp173–191

Ballal, M. E., El Siddig, E. A., Elfadl, M. A. and Luukkanen, O. (2005) 'Relationship between environmental factors, tapping dates, tapping intensity and gum arabic yield of an *Acacia senegal* plantation in western Sudan', *Journal of Arid Environments*, vol 63, pp379–389

Barnes, R. D., Filer, D. L., Milton, S. J. (1999) 'Acacia karoo: Monograph and annotated bibliography', *Trop. For. Rap*, vol 32, Oxford Forestry Institute, UK

Baumer, M. (1990) *Agroforestry and Desertification*, Technical Centre for Agricultural and Rural Cooperation, Wageningen, Netherlands

Bekele, M. (2001) 'Forestry outlook study for Ethiopia (FOSA)', Addis Ababa, Ethiopia

Bekker, C., Rance, W. and Monteuuis, O. (2004) 'Teak in Tanzania: II. The Kilombero Valley Teak Company', *Bois et forêts des Tropiques*, vol 279, pp11–21

Blay, D., Bonkongou, E., Chamshama, S. A. O. and Chikamai, B. (2004) 'Rehabilitation of degraded lands in sub-Saharan Africa: Lessons from case studies', Forestry Research Network for Sub-Saharan Africa (FORNESSA) and the International Union of Forestry Research Organisations' Special Programme for Developing Countries (IUFRO-SPDC)

Bohlin, F. and Larsson, G. (1983) 'Village tree planting study carried out for IRDP, Eastern Province, Zambia', Swedish University of Agricultural Sciences, Uppsala, Sweden

Buckle, D. H. (1959) 'Timber operations in West Africa', *Unasylva*, vol 13, p1

Burdett, A. N., Simpson, D. G. and Thompson, C. F. (1983) 'Root development and plantation establishment success', *Plant and Soil*, vol 71, pp103–110

Cairns, R. I. (2000) *Outgrower Timber Schemes in KwaZulu-Natal: Do They Build Sustainable Rural Livelihoods and What Interventions should Be Made?*, Instruments for Sustainable Private Sector Forestry, South Africa Series, IIED/CSIR, London/Pretoria

Calvert, G. M. and Timberlake, J. R. (1993) 'Burning trials in Baikiaea woodland, Gwayi Forest, Zimbabwe', in G. D. Piearce and D. J. Gumbo (eds) *The Ecology and Management of Indigenous Forests in Southern Africa*, pp347–365, Proceedings of an international symposium, Victoria Falls, Zimbabwe, 27–29 July 1992, Zimbabwe Forestry Commission and SAREC, Harare

Canby, K. (2007) 'China and the global trade for forest products: Transforming trade to benefit forests and livelihoods', Forest Trends presentation, 8 March

Chamshama, S. A. O. and Hall, J. B. (1987) 'Effects of nursery treatments on *Eucalyptus camaldulensis* field establishment and early growth at Mafiga, Morogoro, Tanzania', *Forest Ecology and Management*, vol 21, pp91–108

Chamshama, S. A. O. and Nwonwu, F. (2004) 'Case studies on forest plantations in Africa', AFORNET/KSLA/FAO, Nairobi

Chamshama, S. A. O., Madoffe, S. S., Sabas, E., Msanga, H. P. and Shehaghilo, I. M. (1996a) 'Seed supply for forest plantations in Tanzania', *Faculty of Forestry, Sokoine University of Agriculture Record*, no 63, pp58–64

Chavasse, C. G. R. (1980) 'Planting stock quality: A review of factors affecting performance', *New Zealand Journal of Forestry*, vol 25, pp144–171

Chidumayo, E. N. (1988) 'A re-assessment of effects of fire on miombo regeneration in the Zambian Copperbelt', *Journal of Tropical Ecology*, vol 4, pp361–372

Chimhowu, A. and Woodhouse, P. (2006) 'Customary vs private property rights? Dynamics and trajectories of vernacular land markets in sub-Saharan Africa', *Journal*

of Agrarian Change, vol 6, pp346–371
Christy, L. C., Di Leva, C. E. and Lindsay, J. M. (2007) *Forest Law and Sustainable Development: Addressing Contemporary Challenges Through Legal Reform*, Herndon, VA
Coppen, J. J. W. and Hone, G. A. (1995) 'Gum naval stores: Turpentine and rosin from pine resin', *NWFPs*, no 2, FAO, Rome
Cunningham T. and Milton, S. J. (1987) 'Effects of the basket weaving industry on the mokola palm (*Hyphaene petersiana*) and on dye plants in northwestern Botswana', *Economic Botany*, vol 42, pp386–402
Darkoh, M. B. K. (2003) 'Regional perspectives on agriculture and biodiversity in the drylands of Africa', *Journal of Arid Environments*, vol 54, pp261–279
Dean, R. M., Rivers, R. L., Zeidel, M. L. and Roberts, D. M. (1999) 'Purification and functional reconstitution of soybean nodulin 26: An aquaporin with water and glycerol transport properties', *Biochemistry*, vol 38, pp347–353
Evans, J. (1992) *Plantation Forestry in the Tropics* (second edition), Clarendon Press, Oxford
Evans, J. and Turnball, J. W. (2004) *Plantation Forestry in the Tropics: The Role, Siviculture and Use of Planted Forests for Industrial, Social, Environmental and Agroforestry Purposes* (third edition), Oxford University Press, Oxford
FAO (1958) 'Principal pests and diseases of eucalypts outside Australia', *Unasylva*, vol 12
FAO (1974) *Tree Planting Practices in African Savannas*, FAO, Rome
FAO (1995a) 'Non-wood forest products for rural income and sustainable forestry', *NWFPs*, no 7, FAO, Rome
FAO (1995b) 'Flavours and fragrances of plant origin', *NWFPs*, no 1, FAO, Rome
FAO (2001a) *Global Forest Resources Assessment 2000*, FAO, Rome
FAO (2001b) 'Mean annual increment of selected industrial plantation species', FAO, Rome
FAO (2001c) 'The impact of forest policies and legislation on forest plantations', FAO, Rome
FAO (2001d) 'Protecting plantations from pests and diseases', FAO, Rome
FAO (2002a) *Hardwood Plantations in Ghana*, FAO, Rome
FAO (2002b) 'Tropical forest plantation areas 1995 data set', FAO, Rome
FAO (2003) 'Planted forest data base: Analysis of annual planting trends and silvicultural parameters for commonly planted species', FAO, Rome
FAO (2005) *Global Forest Resources Assessment 2005*, FAO, Rome
Faye, A., Sarr, A. and Lesueur, D. (2006) 'Effect of inoculation with rhozobia on the gum-arabic production of 10-year-old *Acacia senegal* trees', *Arid Lands Research and Management*, vol 20, pp79–85
Folk, R. S. and Grossnickle, S. C. (1996) 'Determining field performance potential with the use of limited environmental conditions', *New Forests*, vol 13, pp121–138
Gebrehiwot, A. (1997) 'Fuel-wood as a source of urban household energy in Ethiopia: A supply perspective', *Boiling Point*, no 39, p9
Giffard, P. L. (1975) 'Gum trees for the reforestation of Sahelian regions', *Bois et Forêts des Tropiques*, vol 161, pp3–21
Goor, A. Y. and Barney, C. W. (1968) *Forest Tree Planting in Arid Zones*, The Ronald Press Company, New York
Hillring, B. (2006) *Status of Recovered Wood in Europe*, Swedish University of Agricultural Sciences (SLU), Uppsala, Sweden

Huang, W., Pohjonen, V., Johansson, S., Nashanda, M., Katigula, M. I. L. and Luukkanen, O. (2003) 'Species diversity, forest structure and species composition in Tanzanian tropical forests', *Forest Ecology and Management*, vol 173, pp11–24
Jackson, J. K. (1984) 'Why do forest plantations fail?', in K. F. Wiersum (ed) *Strategies and Designs for Afforestation, Reforestation and Tree Planting*, Proceedings of an international symposium on the occasion of 100 years of forestry education and research in the Netherlands, Wageningen, Netherlands
Jagger, P. and Pender, J. (2003) 'The role of trees for sustainable management of less-favoured lands: The case for eucalypts in Ethiopia', *Forest Policy and Economics*, vol 5, pp 83–95
Jumbe, C. B. L. and Angelsen, A. (2006) 'Household's choice of fuel-wood source in Malawi: A multinomial probit analysis', Paper prepared for the International Association of Agricultural Economists Conference, Gold Coast, Australia, 12–18 August
Kenya FOSA (2001) *Kenya Forestry Outlook Study*, Forest Department, Nairobi
Kowero, G. and Mabugu, R. (2006) 'Macroeconomic policies and industrial wood processing and trade in Zimbabwe', *Forest Policy and Economics*, vol 8, no 1, pp22–34
Kwesiga, F. R., Franzel S., Place F., Phiri D. and Simwanza, C. P. (1999) '*Sesbania sesban* improved fallows in eastern Zambia: Their inception, development and farmer enthusiasm', *Agroforest Systems*, vol 47, pp49–66
Lamb, D., Erskine, P. D. and Parrotta, J. A. (2005) 'Restoration of degraded tropical forest landscapes', *Science*, vol 310, pp1628–1632
Lambin, E. F. (1999) 'Monitoring forest degradation in tropical regions by remote sensing: Some methodological issues', *Global Ecology and Biogeography*, vol 8, pp191–198
Lawless, J. (1995) *The Illustrated Encyclopaedia of Essential Oils*, Element Books, Shaftsbury, UK
Leakey, R., Shackleton, S. and du Plessis, P. (2005) 'Domestication potential of marula (*Sclerocarya birrea* subspecies *caffra*) in South Africa and Namibia: 1. Phenotypic variation in fruit traits', *Agroforestry Systems*, vol 64, pp25–35
Mackenzie, C. (2006) *Forest governance in Zambezia, Mozambique: Chinese takeaway!*, Final report for the Forum of NGOs in Zambezia (FONGZA), Mozambique
Marunda, C. T. (1997) 'Forest genetic resources and the seed production system in Zimbabwe', *Zimbabwe Science News*, vol 30
Mayers, J. (1999) 'Company–community forestry partnerships: A growing phenomenon', *Unasylva*, vol 200, pp33–41
Mayers, J., Evans, J. and Foy, T. (2001) *Raising the Stakes*, IIED, London
MENR (1994) *Kenya Forestry Master Plan Development Programme*, Ministry of Environment and Natural Resources, Nairobi
Mertz, O., Ravnborg, H., Lövel, G., Nielsen, I. and Konijnendijk, C. (2007) 'Ecosystem services and biodiversity in developing countries', *Biodiversity and Conservation*, vol 16, pp2729–2737
Morrison, E., and Bass, S. M. J. (1992) 'What about the people', in C. Sargent and S. M. J. Bass (eds) *Plantation Politics: Forest Plantations in Development*, Earthscan Publications, London
Murphy, P. G. and Lugo, A. E. (1986) 'Ecology of tropical dry forest', *Ecology and Systematics*, vol 17, pp67–88
Nawir, A. A., Kassa, H., Sandewall, M., Dore D., Campbell, B., Ohlsson, B. and Bekele, M. (2007) 'Stimulating smallholder tree planting – lessons from Africa and

Asia', *Unasylva*, vol 228, pp53–58

Njiti, C. F. and Galiana A. (1996) 'Symbiotic properties and rhizobium requirements for effective nodulation of five tropical dry zone acacias', *Agroforest. Syst.*, vol 34, pp265–275

Nshubemuki, L., Chamshama, S. A. O. and Mugasha, A. G. (2001) 'Technical specifications on management of forest plantations in Tanzania', Forestry and Beekeeping Division, Ministry of Natural Resources and Tourism, Dar es Salaam

Nunifu, T. K. and Murchison H. G. (1999) 'Provisional yield models of Teak (*Tectona grandis* Linn F.) plantations in northern Ghana', *Forest Ecology and Management*, vol 120, no 1, pp171–178

Obeid, M. and Seif El Din, A. (1971) 'Ecological studies of the vegetation of the Sudan: The germination of seeds and establishment of seedlings of *Acacia senegal* (L.) Willd. under controlled conditions in the Sudan', *Journal of Applied Ecology*, vol 8, pp191–201

Onyekwelu, J. C. (2001) *Growth Characteristics and Management Scenarios for Plantation-grown* Gmelina arborea *and* Nauclea diderrichii *in South-western Nigeria*, HieronymusVerlag, Berlin

Onyekwelu, J. C. (2004) 'Above-ground biomass production and biomass equations for even-aged *Gmelina arborea* (ROXB) plantations in south-western Nigeria', *Biomass and Energy*, vol 26, no 1, pp39–46

Palo, M. (1999) 'No end to deforestation?', in M. Palo and J. Uusivuori (eds) *World Forests, Society and Environment*, Kluwer Academic Publishers, Dordrecht, Netherlands

Palo, M., Lehto, E. and Uusivuori, J. (2000) 'Modelling causes of deforestation with 477 subnational units', in M. Palo and H. Vanhanen (eds) *World Forests from Deforestation to Transition*, Kluwer Academic Publishers, Dordrecht, Netherlands

Perrings, C. (2000) *Economics of Biodiversity Conservation in Sub-Saharan Africa*, Edward Edgar Publishing Ltd, Cheltenham

Perrow, M. R. and Davy, J. A. (2003) *Handbook of Ecological restoration, vol 1, Principles of Restoration*, Cambridge University Press, Cambridge

Person, R. (2003) *Assistance to Forestry: Experiences and Potential for Improvement*, CIFOR, Bogor

Peters, P. E. (2007) 'Challenges in land tenure and land reform in Africa: An anthropological perspective', Working Paper no 141, Center for International Development, Harvard University, Harvard

Raddad, E. Y., Luukkanen, O., Salih, A. A., Kaarakka, V. and Elfadl, M. A. (2006) 'Productivity and nutrient cycling in young *Acacia senegal* farming systems on Vertisol in the Blue Nile region, Sudan', *Agroforestry Systems*, vol 68, no 3, pp193–207

Rahim, A. H. (2005) 'Adoption and abandonment of gum arabic agroforestry in Sudan', Conference paper, AFTA 2005

Rock, D .A. (1969) 'Water relations of wrenched and unwrenched *Pinus radiata* seedlings on being transplanted into condition of water stress', *New Zealand Journal of Forestry*, vol 14, pp50–58

Rock, D. A. (1971) 'Effect of undercutting and wrenching on growth of *Pinus radiata* D. Don seedlings', *Journal of Applied Ecology*, vol 8, pp477–490

Ruitenbeek, J. and Cartier, C. (1998) 'Rational exploitations: Economic criteria and indicators for sustainable management of tropical forests', *Occasional Paper no 17*, CIFOR, Bogor

SAIF (2000) *South African Forestry Handbook*, South African Institute of Forestry and

V&R Printers, Pretoria
Saramäki, J., Sekeli, P. M. and Kamwi, E. M. (1986) 'Early development of mukusi in plantation trials', in G. D. Piearce (ed) *The Zambezi Teak Forests*, Forest Department, Ndola, Zambia, pp186–195
Schreckenberg, K., Awono, K., Degrande, A., Mbosso, C., Ndoye, O. and Tchoundjeu, Z. (2006) 'Domesticating indigenous fruit trees as a contribution to poverty reduction', *Forests, Trees and Livelihoods*, vol 16, pp35–51
Simons, A. J. and Leakey, R. R. B. (2004) 'Tree domestication in tropical agroforestry', *Agroforestry Sytems*, vol 61, pp167–181
Stephenne, N. and Lambin, E. F. (2001) 'A dynamic simulation model of land-use changes in Sudano-Sahelian countries of Africa (SALU)', *Agriculture Ecosystems & Environment*, vol 85, pp145–161
Taylor, G. F. and Soumare, M. (1984) 'Strategies for forestry development in semi-arid tropics: Lessons from Sahel', in K. F. Wiersum (ed) *Strategies and Designs for Afforestation, Reforestation and Tree Planting*, Proceedings of an international symposium on the occasion of 100 years of forestry education and research in the Netherlands, Wageningen, Netherlands
Tiarks, A., Nambiar, E. K. S. and Cossalter, C. (1998) 'Site Management and Productivity in Tropical Forest Plantations', *Occasional Paper no 16*, CIFOR, Bogor
UNEP (2006) *Environmental Outlook of Africa 2*, UNEP, Nairobi
van Vuuren, N. J. J., Banks, C. H. and Stohr, H. P. (1978) *Shrinkage and Density of Timbers Used in the Republic of South Africa*, bulletin no 57, South African Forestry Research Institute, Pretoria
Veenendaal, E. M., Swaine, M. D., Agyeman, V. K., Blay, D., Abebrese, I. K. and Mullins, C. E. (1995) 'Difference in plant and soil water relations in and around a forest gap in West Africa during the dry season may influence seedling establishement and survival', *Journal of Ecology*, vol 83, pp83–90
Vichnevetskaia, K. (1997) 'Factors affecting productivity of tropical forest plantations: Acacia, eucalypt, teak and pine', *Working Paper GFSS/WP/02*, Global Fibre Supply Study Working Paper Series, FAO, Rome
Wardell, D. A., Reenberg, A. and Tettrup, C. (2003) 'Historical footprints in contemporary land use systems: Forest cover changes in savannah woodlands in the Sudano-Sahelian zone', *Global Environmental Change – Human and Policy Dimensions*, vol 13, pp235–254
Whiteman, A and Lebedys, A. (2006) 'The contribution of the forestry sector to African economies', *International Forestry Review*, vol 8, no 1, pp31–44
Zegeye, H., Teketay, D. and Kelbessa, E. (2006) 'Diversity, regeneration status and socio-economic importance of the vegetation in the islands of Lake Ziway, south-central Ethiopia', *Flora – Morphology, Distribution, Functional Ecology of Plants*, vol 201, pp483–498
Zida D., Tigabu, M., Sawadogo, L. and Oden, P. C. (2008) 'Initial seedling morphological characteristics and field performance of two Sudanian savanna species in relation to nursery production period and watering regimes', *Forest Ecology and Management*, vol 255, pp2151–2162
Zobel, B. J., Van Wyk, G. and Stahl, P. (1987) *Growing Exotic Forests*, John Wiley & Sons, New York

10

Environmental Services from the Dry Forests and Woodlands of Sub-Saharan Africa

Crispen Marunda and Henri-Noël Bouda

INTRODUCTION

The Millennium Ecosystem Assessment described three categories of environmental services as:

1. Provisional services which are renewable resources such as wood, fibre, fruits and other non-timber products that have been traditionally harvested and used directly by communities.
2. Cultural services that include the spiritual, religious, aesthetic, educational, scientific, inspirational and recreational (e.g. nature-based tourism).
3. Supporting services which underpin all other services and include biodiversity conservation, soil formation, nutrient and water cycling, and regulatory services which include air quality regulation, climate regulation, water regulation, erosion regulation, water purification, disease regulation and natural hazard regulation.

In Africa, the main environmental services from dry forests and woodlands are conservation of biological diversity, protection of watersheds and regulation of water flow, desertification control and soil amelioration, and climate stabilization through carbon sequestration (Nair and Tieguhong, 2004). Some of the environmental services in Africa are becoming commercialized and therefore provide benefits to the communities and create new incentives for sustainable forest management (Jindal, 2004). This re-configured approach to the way forests and woodlands are viewed underpins a fundamental change from treating the forests and woodlands as sources of free and inexhaustible products, to seeing them as a source of services that enhance the economic value of the forests and woodlands and their contribution to livelihoods and the environment. Payments for environmental services (PES) are emerging market-based instruments that can be used to create sources and markets for such services (Engel et al, 2008; Wunder, 2005; Ferraro and Kiss, 2002 cited in Wunder, 2005; Lindell-Mills and Porras, 2002). The fact that the majority of the people in Africa are the customary custodians of the dry forest and woodland resources, and PES is about people managing natural resources for which they are compensated, makes environmental services an important service that can be tapped, marketed and the benefits used to improve people's livelihoods.

This chapter describes the environmental services from the dry forests and woodlands of Africa focusing on two services: carbon sequestration and watershed services. The other major environmental services from forests and woodlands, such as biodiversity conservation are described in detail in Chapter 3. Environmental services that support agricultural production, such as soil conservation, windbreaks and nutrient recycling, are only briefly described. These environmental services are used locally and have no markets or if available the markets are extremely thin and highly uncertain.

The chapter also explores the development and promotion of payments for environmental services beyond the boundaries of traditional forestry as a new way of providing market-based incentives for sustainable forest management in Africa. An analysis of barriers and recommendations to adopting environmental payment schemes within the context of Africa is provided. More importantly, the chapter aims to improve our understanding of the importance of environmental services in enhancing the value the dry forests and woodlands beyond production of timber and non-timber forest products.

Environmental and Ecosystem Services

In a general sense, environmental services, or ecosystem services (ES) are the functions and services that biologically driven systems provide which contribute to the overall functioning of the environment, but that are not generally considered to be owned by individuals. In short, environmental services are the natural functions of an ecosystem that can be secondarily used for the benefit of humans (Barbier and Swanson, 1992; Daily, 1997). The four most recognisable environmental services (Wunder, 2005; Grieg-Gran et al, 2005; Katila and Puustjärvi, 2003; Pagiola et al, 2002) are:

1. carbon sequestration;
2. regulation of fresh water and river flows;

3. forest landscape beauty;
4. biodiversity.

These four main environmental services have received a lot of attention because of the international nature of the services, previous valuation studies and the potential for markets for these services. In the dry forests and woodlands of Africa, some of the minor environmental services (e.g. soil improvement, water and nutrient circulation, wind-break and shelter) are very localized and do not lend themselves to market mechanism (i.e. there are no buyers). These are particularly important in the dry forest of Africa since a large percentage of the population depends on natural resources for livelihoods. Unlike the physical products from dry forests and woodlands that are easily recognized and valued at a local scale and often within individual land units managed specifically for production of the goods, environmental services are somewhat less obvious and may appear over large spatial scales and extending over multiple sectors and landscapes.

In order to demonstrate the growing importance of the environmental services concept, we need to consider how it emerged and how it is being used as an incentive for sustainable forest management. The concept of environmental or ecosystem services emerged from the conservation and development imperative and from the need to provide new incentives and mechanisms for sustainable forest management, community participation in nature conservation, collaborative forest management and successful implementation of integrated conservation and development projects. Most of these practices are anchored on the premise that the communities (or other stakeholders) can protect and manage forests and woodlands and benefit directly from products such as timber, fruits, fodder and non-wood forest products. The dry forests and woodlands of Africa are well recognized for many products (e.g. Campbell, 1996; Malaise, 1997) but the commercial benefits to the state are meagre, making contemporary forms of forest management practices unattractive (Matta and Kerr, 2006). Environmental services from forests and woodlands are now providing an alternative mechanism for optimizing investment and creating new forms of incentives in environmental protection and conservation through valuation of services and linkages to markets (Costanza et al, 1997).

The matching of sources of ES (different land-use systems) and markets (usually carbon-emitting sectors in the industrialized countries) has given rise to what are now termed payment for environmental services (Lindell-Mills and Porras, 2002; Pagiola et al, 2002; Wunder, 2005). Sven Wunder (2005) defined PES as 'a voluntary transaction in which a well-defined environmental service (ES), or a land use likely to secure that service, is being bought by at least one ES buyer from at least one ES provider if, and only if, the ES provider secures ES provision'. In Africa, examples of PES are few and these have concentrated on carbon sequestration projects in eastern Africa (Jindal, 2004). In 2008, the Katoomba Group commissioned a study that inventoried PES projects across eastern and southern Africa (Kenya, Tanzania, Uganda, Malawi and South Africa) and 68 PES and PES-like initiatives split between bio-diversity, carbon, water and other were identified (Bond, 2008). Watershed services have also received some attention within the broader context of integrated catchment management. Cases for watershed services in Africa are rare largely because this is often a site-specific service (Dillaha et al, 2008). Current work on the Rovuma in Tanzania shows some promise. Other environmental services, such as soil erosion regulation, nutrient and water cycling have not been analysed within the context of PES but more as approaches that enhance agricultural productivity at the household level. By drawing attention of the policy-makers and international community to the ES from the dry forests and woodlands, their value can be influenced and swing the balance towards conservation. As Wunder (2005) noted 'might work in marginal areas... such as dry forests where modest attention (payments) might favour conservation'. African dry forests and woodlands could now be viewed as more valuable for the environmental services they provide, although in the absence of a system for valuation and their inclusion in national economic accounting, there is often much less recognition of their overall contribution (Nair and Tieguhong, 2004).

Carbon Sequestration in the Dry Forests and Woodlands of Africa

Global context of carbon sequestration

The Intergovernmental Panel on Climate Change (IPCC, 2001) described carbon sequestration as the process of removing carbon dioxide (CO_2) and other greenhouse gases from the atmosphere and storing it in other potential reservoirs, such as land, forests and oceans. Increasing concentration of greenhouse gases in the atmosphere is expected to warm the Earth and change the climate. Carbon sequestration lends itself to transfer value techniques in which the value estimates obtained for sequestration in one location can be transferred to other regions (UNFCCC, 2003). Carbon dioxide (one of the six greenhouse gases) mixes relatively rapidly and completely in the global atmosphere and thus the contribution of carbon sequestration to the general circulation system is independent of where it takes place (Sohngen and Brown, 2006). This transferability of carbon sequestration makes it a suitable candidate for climate change mitigation under the UN Framework Convention on Climate Change (UNFCCC) Kyoto Protocol through the Clean Development Mechanism (CDM). Afforestation and reforestation create effective sinks for absorbing carbon from the atmosphere and these two activities are currently eligible for funding under the CDM (UNFCCC, 2003).

At the 13th Bali Conference of Parties (CoP13) of the UNFCCC held in Indonesia in December 2007, a new strategy on Reduced Emissions from Deforestation and forest Degradation (REDD) took centre stage as a new strategy to reduce harmful gases from the atmosphere. The REDD policy entails that developing countries which are experiencing deforestation may on a voluntary basis receive compensation if they reduce national deforestation rates (Skutsch and Trines, 2008). Tackling deforestation has been a priority developmental and environmental issue for most countries in the dry zones of Africa. The REDD deal gives more opportunities and scope for such countries to reduce deforestation rates through new incentives. For dry forest and woodland countries with lower carbon stocks but higher deforestation rates compared to wet forest countries, the challenge remains as to what approaches to adopt to reduce deforestation and, more importantly, to identify the markets for carbon credits created by such strategies.

Forests and woodlands play an important role in carbon sequestration and by investing in forest development and conservation, countries in the dry forest and woodland regions of Africa can benefit from carbon trading. A number of corporate institutions in Europe are already benefiting from carbon trading by investing in tree planting in some parts of Africa. The market for environmental services from forests and woodlands is growing rapidly around the world, often facilitated by national and regional policies as well as international conventions and agreements. Certain segments of society that are able and willing to pay for these services are creating opportunities for the forest owners. Markets for carbon sequestration based on dry forest and woodland management and protection have been adopted in Mozambique, Senegal, Uganda, Tanzania, Malawi and Madagascar (Jindal, 2004; Bond, 2008). Authorities responsible for managing dry forests and woodlands in Africa need to articulate the contribution of the forests and woodlands in sequestrating carbon and demonstrate that they have management strategies in place that enhance net carbon assimilation and attract buyers. Unfortunately, Africa contributed less than 3 per cent of the total global trade in carbon offsets in 2003–2004 (Jindal, 2006, citing Lecocq and Cupoor, 2005). The carbon projects in Africa are expected to sequester about 26.85 million tonnes CO_2 (Jindal et al, 2008: the 2006 international prices for carbon credits ranged from US$3.50 per tonne CO_2 at Chicago Climate Exchange to US$15.80 per tonne CO_2 in various European markets).

Carbon stocks and balance in dry forests and woodlands of Africa

The potential of dryland ecosystems to sequester carbon has been estimated to be up to 0.4–0.6 billion tonnes of carbon per year (UNFCCC, 2006). Lal (2004) gave estimates of carbon sequestration in dryland ecosystems and concluded that such areas have a huge potential as carbon sinks. The large surface area of African dry forests and woodlands, albeit fragmented in places, gives CO_2 sequestration in these systems some global significance. To demonstrate the potential of dry forests and woodlands to sequester carbon, there is need first to assess the carbon stock currently held and deforestation rates. The amount of carbon stored in the soil and in the woody biomass depends on the soil and vegetation type. Recognizing that little is known of the belowground carbon stocks in dry forests and woodlands, Tables 10.1 and 10.2 show the

Table 10.1 *Trends in carbon stocks in forest biomass 1990–2005*

Region/sub-regions	*Carbon in living biomass (Giga tonnes)*		
	1990	*2000*	*2005*
Eastern and southern Africa	15.9	14.8	14.4
Northern Africa	3.8	3.5	3.4
Western and central Africa	46.0	43.9	43.1
Total Africa	65.8	62.2	60.8

Note: Giga tonne = 1 billion tonnes.
Source: Based on FAO (2005)

trends in carbon stocks in above ground living biomass from the major regions and selected African countries with dry forests and woodlands. It is quite clear that the eastern, southern and northern subregions of Africa that are covered by the dry forests and woodlands, have less above ground living woody biomass compared to western and central African tropical forests. This is due to the inherently slow growth rates of dry forest and woodland species and the high levels of utilization (leading to high levels of deforestation). Because of the low above ground carbon stocks in the dry forests and woodlands, not much interest has been given to these areas compared to rainforests (Zahabu et al, 2007). However, new opportunities for compensating reduced emission through reversing deforestation now exist under REDD and dry forest and woodland countries need to take advantage of the global focus on reduced emissions from deforestation.

Table 10.2 *Carbon stock in forests and woodlands (living woody biomass) and annual forest area change (%) for selected African countries with dry forests and woodlands*

Region/Country	*Carbon stock (Megatonnes)*	*Annual forest areas change (%)*
Southern Africa		
Botswana	141.5	–0.9
Malawi	161.0	–0.8
Mozambique	606.3	–0.2
Namibia	230.9	–0.8
South Africa	823.9	0
Eastern Africa		
Ethiopia	252.0	–0.9
Kenya	334.7	–0.3
Sudan	1530.7	–0.8
Tanzania	2254.0	–1.0
West Africa		
Burkina Faso	298.0	–0.3
Chad	236.0	–0.6
Mali	241.9	–0.7
Niger	12.5	–2.3
Senegal	371.0	–0.4

Source: FAO (2005, 2006, 2007)

Table 10.3 *Carbon stock values for selected vegetation types*

Woodland types	*Carbon stock (tonnes carbon per ha)*	*Reference*
Congo-Guinea and Congo-Zambezian	160–209	Chapter 2, this volume
Zambezian warm dry forests	88–97	Chapter 2, this volume
Sudanian Savanna	56–78	Chapter 2, this volume
Kalahari Highveld	22–34	Chapter 2, this volume
Sahel with isolated shrubs	12–31.2	Woomer et al (2004)
Somali-Masai	13–78	Chapter 2, this volume
Degraded savanna and remnant forests	9–113	Woomer et al (2004)

Table 10.2 also shows the annual change in forest cover in the selected countries. It is quite clear that there is a net loss of forest cover, implying a net loss of woody carbon. Most of the countries (except South Africa) have a net loss in forest cover. It also has been reported that dryland ecosystems also store huge amounts of soil organic carbon (SOC) with the world's dryland soils containing 241Pg (pentagram) of SOC (Eswaran et al, 2001). Estimates for SOC stored in Africa's dry forests and woodlands are very few. As an example, the miombo woodlands in Southern Africa have a capacity of storing more than 100 tonnes of carbon per hectare of soil (Williams et al, 2008). Tschakert and Tappan (2004) gave an estimate of 11.3 tonnes per hectare for Senegal whilst Woomer et al (2004) gave similar figures of 11.6–25.3 for carbon per hectare for that country's Sahel Transition Zone. Soil organic carbon is also subject to loss due to soil erosion. For example, Lal et al (1999) estimated that erosion in the drylands leads to emissions of 0.21 to 0.26Pg of carbon per year. Table 10.3 shows the carbon stocks in different woodland types. It is quite clear that there is a wide variation in the estimates of the amount of carbon within woodland type and between different types.

Deforestation, land clearing for agriculture and uncontrolled burning has resulted in aggregate loss of terrestrial carbon from all vegetation types in the drylands of Africa. The net change of forest area in Africa is the highest among the world's regions, with an annual net loss, based on country reports, estimated at –5.3 million hectares annually, corresponding to –0.78 per cent annually (FAO, 2000). Table 10.2 also shows that for most countries with dry forests and woodlands, there is a net loss of forest area. The dry forests and woodlands of Africa have a history of disturbance due to human activities through land clearance for agriculture, fire, charcoal production and firewood collection. Such activities impact on the carbon stock for woodlands such as miombo (Chidumayo, 2002) and result in net loss of carbon into the atmosphere. Williams et al (2008) showed that clearing for agricultural land in miombo woodlands in Mozambique resulted in the loss of 19 tonnes of carbon per hectare. Woomer et al (2004) estimated that Senegal lost 292 megatonnes of carbon over 35 years from 1965 to 2000. Apart from deforestation, the dry regions of Africa are also prone to desertification which exacerbates the loss of soil carbon through exposure of carbonaceous material to climatic elements caused by soil erosion (Lal et al, 1999). Managing this imbalance in dry forest and woodland landscapes is a key strategy for removing atmospheric CO_2.

Managing natural woodlands for carbon sequestration in Africa

A key element for payment for environmental services is the ability to clearly define 'a land use system likely to secure that service' (Wunder, 2005). Payment for environmental services systems presuppose the existence and integration of technical capacity to enhance carbon storage in production systems and that resource users (or communities) have the capacity to adopt and maintain land resource practices that sequester carbon. This is the main challenge facing dry forest countries and they will need to articulate the CO_2 sequestration roles of woodlands to attract carbon finances. Whilst the dry forests and woodlands of Africa have low carbon stocks per unit area but not in total amounts, they present a huge potential to act as sinks provided appropriate woodland management strategies are developed and implemented. The new developments under UNFCCC (e.g. REDD) provide countries with high rates of deforestation an opportunity to reduce deforestation and offset carbon at a global level. A country like Tanzania, for example, with 34 million hectares of forest land, can potentially earn US$630 million per year or around US$119 per rural household from the REDD policy if woodland management strategies to reduce deforestation and degradation are developed and implemented (Zahabu et al, 2007).

Forestry climate change mitigation activities may be grouped into three categories. The first includes activities that avoid the release of emissions from carbon stocks, such as forest conservation and protection. The second includes activities that store carbon, for example, afforestation, reforestation and agroforestry. The third involves substituting the use of carbon-intensive products and fuels with sustainably harvested wood products and wood fuel, for example, wood substituting for concrete or steel and bio-electricity substituting for fossil fuel electricity. Within the context of dry forests and woodlands of Africa, the first two options are plausible as they provide other goods and services for immediate consumption by both local and international communities. In sub-Saharan Africa there are a number of carbon offsetting projects based on managing forests and woodlands and Jindal (2006) identified 19 such projects.

Avoided deforestation through forest conservation and protection

The capacity of dry woodlands of southern Africa to regenerate is well documented (Piearce, 1993). The species commonly found in these woodlands are vigorous resprouters and if left undisturbed, can accumulate huge amounts of carbon. For example, regrowth miombo woodlands in Mozambique are estimated to accumulate 0.7 tonnes of carbon per hectare per year (Williams et al, 2008). Most of the miombo species are able to regenerate through stump resprouts and root suckers after destruction of the above ground parts (Backéus et al, 2006; Chidumayo, 1997, 2004; Frost, 1996). For other dry forest vegetation types, such the Sudanian savanna, Ky-Dembele et al (2007) found that disturbed savanna woodlands in Burkina Faso regenerated mainly from coppice, root suckers and water sprouts. *Acacia*-dominated woodlands regenerate well from seeds scarified as they pass through the alimentary canal of browsing animals and this has been documented as the major source of seedlings compared to fires. In southern Africa, the resilience of *Acacia*, *Dichrostachys* and *Commiphora* species in certain cases can even cause extensive bush encroachment (Wiegand et al, 2005). Consequently, if savannas were to be protected from fire and grazing, most of them would accumulate substantial carbon and the carbon sink would be large (San Jose et al, 1998).

A number of interventions have been tried and tested in the dry woodlands of Africa to improve the condition of woodlands and their productivity and service function. In Tanzania, the *Ngitili* concept has gained international recognition as a management system for miombo woodlands practised by a community for production of goods and services (Barrow and Mlenge, 2003; Monela et al, 2005). Box 10.1 briefly describes the *Ngitili* practice while Box 10.2 is another example of a community based woodland management project for carbon sequestration in the Nhambita area of Mozambique. This project involves communities managing natural woodlands and planting trees for products and services (Zohlo, 2005; Jindal, 2004). In some cases the private sector has taken the lead in establishing forests and woodlands for carbon sequestration although concerns are emerging about the negative impacts on local people of some such initiatives (Box 10.3). In arid lands of western Africa, the Sequestration of Carbon in Soil Organic Matter (SOCSOM) project (Box 10.4) traced the fate of soil carbon over long periods of time and simulated management strategies to increase carbon stock (Touré, 2006). Over periods of time these practices could stabilize and the carbon in the woodlands and soil will act as a critical adjunct to reducing emissions. In such management systems, the carbon stocks are at a very low risk of loss and thus represent a pool of sequestered carbon that is effectively permanent in the dry forest and woodland areas.

Silvicultural practices to increase carbon stocks in dry forests and woodlands

At the field level, woodland management practices to enhance biomass production have been documented by a number of researchers and these are summarized in Table 10.4. The management systems are designed to increase the woody biomass of different woodland systems. These management approaches can present a suite of practices that resource managers can adopt to enhance the value of environmental services and fulfil one of the conditions of PES, to engage in management practices that ensure the production of the services by the service provider. The response of the woodlands in terms of biomass accumulation varies from place to place as it is a function of a number

Box 10.1 Ngitili: An indigenous natural resource management system of Tanzania

The *Ngitili* natural resource management system is an indigenous practice used by the Sukuma people of the Shinyanga and Mwanza districts of Tanzania. The objective of the practice was to provide fodder for livestock during the dry season and droughts, restore soil fertility and produce wood products. The success of the practice in restoring vegetation and woody biomass has increased its value as a carbon sequestrating strategy. The main trees in *Ngitili* are *Acacia*, *Brachystegia*, *Albizia*, *Commiphora* and *Dalbergia*.

The practice involves conservation of fallow areas and rangelands to restore vegetation, especially browse species and grasses. *Ngitili* is a Sukuma term meaning enclosure. A household can have its own *Ngitili* (about 5ha) and over the years as livestock herds increased, communal *Ngitilis* became common; these can be as large as 50ha. The practice is based on excluding livestock from grazing in the protected areas and prohibiting cutting trees for a period of time. During the wet season, the *Ngitilis* are closed off to animals in order to allow vegetation to regenerate. During the dry season the animals are allowed to graze in the *Ngitilis* at a very low intensity level to start off but increase as the other sources of dry season fodder get depleted. The success of the project depended very much on the local ownership rights of the local communities through local community law (village law) enforced by local scouts called *sungusungu* or *wasalama*. Penalties for breaking local by-laws such as causing forest fires would attract a penalty of US$10 which has increased over time to US$40.

The main management strategies include formulating local by-laws and institutions, developing forest management plans that cover site selection, species selection, responsibility, conservation management, protection, preparation of fire guards around degraded woodlands, promoting early burning and promoting more benign uses, such as beekeeping and honey production.

So far over 300,000ha have been put under a *Ngitili* form of management and the communities are benefiting from goods and services. Records show increased species diversity in some areas, increase of biomass (carbon stock) and increased ecosystem integrity. The value of the *Ngitilis* is estimated to be US$14 per person per year, US$1190 per household per year and US$89.6 million per district per year (Monela et al, 2005). The *Ngitili* natural resource management practice was recognized by the UN in 2002 and was awarded a prize under the Equator Initiative. The challenge now is to expand the *Ngitili* practice to other areas, prevent selling of 'restored areas' to outsiders and create new markets for other environmental services from the areas e.g. carbon sequestration.

Source: Based on Kamwenda (2002)

of factors which include rainfall, soil type, temperature and management practices. Because of the low returns from the dry forests and woodlands in terms of physical products, some of the management practices are deemed uneconomic. The management practices are designed to achieve specific tangible products, and examples of management of forest for carbon sequestration

Box 10.2 Miombo community land-use and carbon management in Mozambique – Nhambita Pilot Project

Nhambita is a small community located near Gorongosa National Park in the Sofala province of Mozambique. The community was resettled after initially being displaced from the National Park. There are three main land-use systems: protected area, buffer zone and communal area in which the local communities are engaged in subsistence farming, hunting, fishing, livestock rearing and charcoal production (Zohlo, 2005). The dominant tree species are *Brachystegia boehmii*, *B. spiciformis* and *Julbernardia globiflora*.

The Miombo Community Land-Use and Carbon Management project aims to develop forestry and land-use practices that promote sustainable rural livelihoods in partnership with rural communities in a way that raises living standards, and to assess the potential of these activities to generate verifiable carbon emission reductions.

The project was developed as a result of the increasing concern about global climate change, and the recent evolution of carbon markets. The Nhambita project was launched in 2003 as a collaboration between the environmental company Enviro-trade Ltd and the University of Edinburgh. The project is supported by the European Commission. The project is a collaborative effort between several different organizations, which include the University of Edinburgh, the Edinburgh Centre for Carbon Management, Enviro-trade (UK), International Centre for Research in Agroforestry (Kenya) and the Park Administration of the Gorongosa National Park (Mozambique).

Local farmers and forest communities manage the planting and growth of trees in return for proceeds from the sale of carbon offsets to customers in the developed world using the Plan Vivo methodology developed by the Edinburgh Centre for Carbon Management. The Plan Vivo is a carbon management system that was developed for small farmers under the Scolel Te Project in Mexico in 1996. The Plan Vivo is a Trust Fund which provides technical and financial assistance to local farmers to take up forestry/agroforestry activities and then, on their behalf, sells carbon offsets that are generated.

By May 2007 the project had planted 230,000 trees as a combination of agroforestry and woodland restoration and has over 500 farmers involved who have benefited from the payments and have been encouraged to become involved in other micro-finance initiatives, such as beekeeping and carpentry using miombo tree species planted by the project. The project will pay US$242.60 per hectare to farmers who agree to undertake carbon sequestration activities on their farms, such as planting of trees, promoting agroforestry, etc. The project will also pay US$40.50 per hectare to a community fund on the basis of the number of hectares that are brought under carbon sequestration.

The Mozambique Carbon Livelihoods Trust was launched in 2007 to ensure that the community's and individual farmers' proceeds of carbon offset sales from Carbon Livelihoods projects were safeguarded. Approximately one-third of the proceeds of any carbon sale goes directly to this fund and is paid out to individual farmers over seven years; other payments are reserved for forest management and conservation activities.

Source: Chapter authors

Box 10.3 Concern with carbon trading forests in Tanzania

In its Bulletin no 35, the World Rainforest Movement highlighted the growing concern with proposed carbon trading forests in Tanzania. The Escarpment Forestry Company, which is a subsidiary of Norwegian company, Tree Farms, has planted 1900ha of *Pinus patula* and *Eucalyptus saligna* in Sao Hill, Mufindi and Kilombero districts, being the beginning of proposed carbon trading forests. The company is also supporting the Tanzania Greenhouse Gas Action Trust. However, serious concerns have been expressed, locally and internationally, about the socio-political, ecological and economic benefits of the carbon forests. For one thing, the company has paid a paltry land rent of US$1.9 per hectare. The negotiated 99-year land lease has important sovereignty implications. However, the most serious concern is that the operations will exploit cheap labour and will contribute to further marginalization of the rural poor.

Source: Owino (no date)

are very few. Perhaps some of these practices might have a greater impact if applied at a larger landscape scale for the production of environmental services.

There are a number of barriers to adopting some of these management practices at the local level. Campbell et al (2007) and Kokwe et al (2005) analysed some common barriers applicable to open access dry forests and woodlands with special reference to the miombo woodlands of Zambia. Key barriers include tenure rights which in many parts of Africa are not defined adequately and hence it is difficult to allocate responsibility and benefits. This

Table 10.4 *Summary of common silvicultural practices in the dry woodlands, savanna and parklands of Africa*

Woodland type	*Management practice*	*Reference*
Miombo woodlands	1. Clear felling, coppice with standard, selective cutting and thinning 2. Silvicultural thinning and pruning 3. Controlled burning to reduce fuel load 4. Grazing to reduce fuel load 5. Coppicing and pollarding 6. Fertilizing miombo fruit trees	1. Shackleton and Clarke (2007) 2. Mbwambo and Nshubemuki (2007) 3. Frost (1996); Campbell (1996); Chidumayo (1988) 4. Gambiza et al (2008) 5. Chidumayo et al (1997) 6. Akinnifesi et al (2006)
Sahelian *Acacia-Vitellaria* parklands	7. Selective thinning to improve productivity	7. Baumer (1994); Lovett and Haq (2000)
Gum arabic *Acacia* savanna	8. Exclusion and grazing to reduce fuel load	8. Egadu et al (2007)
Sudanian and Zambezian woodlands	9. Agroforestry	9. Unruh et al (1993)

Box 10.4 The SOCSOM project in Senegal

Sequestration of Carbon in Soil Organic Matter (SOCSOM) is a project funded by USAID to investigate carbon sequestration potential in Senegal. The project used remotely sensed imagery and biogeochemical computer modelling to develop carbon sequestration strategies that could eventually be applied across the country and the rest of western Africa.

The semi-arid and sub-humid regions of Africa have the highest potential for carbon sequestration in the world due to their severely degraded soils. But by increasing irrigation, reduced tillage, restoring grasslands and savannas, and extending fallow periods, coupled with applications of compost and organic waste and the adoption of erosion control strategies, countries could replenish their soil carbon stocks. In the process, this would benefit farmers through increased agricultural productivity.

Researchers estimated the patterns of carbon stocks and changes using a computer simulation model. Models on the impact of human activities and climate change were developed and included information on changes in land cover and land use. Additionally, data on crop composition, crop rotation patterns, grazing, fire, fertilizer use and irrigation from the literature and censuses were incorporated into the simulations. The model results showed the actual and potential soil carbon gains, and the economic and ecological costs and benefits. Between 1900 and 2000 the total carbon stock in soils and vegetation decreased from 141 to 89 tonnes per hectare – a reduction of 37 per cent. In the world of carbon sequestration this means that Senegal has much more 'space' left to fill with carbon. The potential revenue from managing the dry landscape for CO_2 sequestration was about €70 million (US$95 million) a year, or just under €2.5 billion (US$3.4 billion) over the 35 years that it will take to reach a new carbon steady-state.

The project has expanded to include other western African countries, such as Burkina Faso, Ghana, Mali and Niger.

Source: Touré (2006)

has been addressed through community based natural resource management but there are still a number of lingering problems (elite capture, definition of community, governance and transparency issues, inequitable benefit sharing). Participation of community members in woodland management is usually very low unless they have immediate and direct benefits. The other barrier is lack of incentive to encourage resource users (and owners) to provide their inputs (labour, time and technical capacity) in management of dry forest (Nhira, 1996). The low value of dry forests and woodlands in terms of revenues from products and services also serves as a disincentive to managing dry forests and woodlands. According to Inamadar et al (1999) local communities or user groups do not participate in forest management programmes whose transaction costs are perceived to be higher than benefits. This is true for dry forests and woodlands because there is a perception that the natural capital and value of the resource is low. As the global community recognizes the value of dryland ecosystems as sinks for carbon, new opportunities might emerge from REDD programmes and the described management practices will be useful in increasing carbon stocks provided the barriers discussed above are addressed.

Afforestation and forestation activities to create carbon sinks

Under the Clean Development Mechanism, forestation and afforestation projects are eligible for funding under the Kyoto Protocol of the UNFCCC. In principle, tree planting has been the major driver of forests as sinks for atmospheric carbon (Murdiyarso and Skutsch, 2006). Whilst tree planting in the dry forests and woodlands has been mainly for timber products (plantation), woodfuel (woodlots) and products (e.g. gums from acacia), there has been a growing need to recognize the multiple functions of the plantation to include carbon sequestration. Because dry forests and woodlands have relatively low carbon stocks, tree planting for timber has historically been the main source of additional carbon stock. There is now growing evidence that plantations are a huge source of carbon even after they have been harvested and most of the carbon locked away in buildings, furniture and paper products (Forestry Tasmania, unpublished reports). Most of the commercial plantations in Africa are located in the relatively wetter parts of the dry forest regions (e.g. the pine and eucalypt plantations in Kenya, South Africa, Mozambique and Zimbabwe).

Agroforestry

Carbon sequestration through changes in land use and management is one of the important strategies to mitigate the global greenhouse effect (Unruh et al, 1993; Tan and Lal, 2005). Agroforestry has been described as an important land-use system suitable for carbon sequestration because of the carbon storage potential in its multiple plant species and soil. Whilst reforestation and afforestation projects are plausible as a means of stabilizing CO_2, there are a lot of other competing land-use systems. This is particularly so in the dry forest areas of Africa where there is competition for land for crop production and livestock grazing. Agroforestry provides an alternative land-use system that combines tree planting and agriculture (Sanchez, 1999). In many parts of Africa, agroforestry has been introduced to improve small-holder agricultural productivity through nutrient input by trees, fodder for livestock, wind-breaks and erosion control. Essentially agroforestry increases woody biomass per unit agricultural area. In the dry forest area, the potential of agroforestry to sequester carbon has been reported by Unruh et al (1993) who estimate that nearly 1549 million hectares are under agroforestry and possibly contain about 11,540 million tonnes of above ground woody biomass. Agroforestry can also indirectly help CO_2 sequestration by decreasing pressure on natural forests (avoided deforestation) and plantations that are the largest sink of CO_2 (Montagnini and Nair, 2004).

The amount of carbon accumulated in agroforestry systems depends on the existing land use and environmental conditions. In Senegal, for example, the capacity to sequester carbon increases with rainfall and generally decreases with temperature provided soil conditions are not limiting (Batjes, 2001). From the studies by Unruh et al (1993), agroforestry systems that included a firewood production component had the greatest potential to accumulate carbon. However, this presents a problem in that all the firewood is converted to energy increasing the 'leakage of carbon' from the systems. Even with fodder trees that are fed to animals, the carbon is eventually lost to the atmosphere. The residence time for the carbon is short but the advantage of the system is that pressure is temporarily removed from natural forests.

Success of agroforestry in the carbon budget will depend on the availability of suitable tree species, secure tenure system, critical number of farmers and suitable climatic conditions. Countries with dry forest and woodland areas need to invest in research to identify suitable agroforestry species and quantify their carbon stocks. Such capacity and information will be crucial if countries in the dry regions of Africa are to promote the carbon sequestration service function of agroforestry systems.

Urban tree planting

Urban tree planting schemes also have a potential to sequester carbon, apart from adding aesthetic value to cities and towns. Such schemes may qualify for carbon credits under 'agriculture, forestry and other land uses', which is increasingly gaining momentum as part of the post-Kyoto climate change regime. In South Africa, the carbon storage of *Jacaranda mimosifolia* street trees in the City of Tshwane (Pretoria) was estimated to be 41,978 tonnes of carbon, valued at US$839,560 (Stoffberg and van Rooyen, 2006). Tree planting in or near urban centres has been practised in many urban centres in Africa with the objective of supplying firewood to urban dwellers. Whilst most schemes did not succeed due to a number of factors (ill-defined ownership, inappropriate species), they may become an important stock of carbon in the future. The high level of urbanization in most African countries is driving up the demand of traditional energy (from firewood and charcoal) in the cities. The ultimate effect is increased rates of tree cutting (and deforestation) to meet the energy demands in the growing cities. Thus tree planting near towns using the right species and with clearly defined ownership rights can help to reduce pressure on natural forests and woodlands.

WATERSHED ENVIRONMENTAL SERVICES IN THE DRY FORESTS AND WOODLANDS OF AFRICA

A sufficient and reliable water system is essential for supporting livelihoods through food production, preserving ecosystems, sustaining economic development and providing vital goods and services, such as hydroelectricity. The major river basins in Africa provide habitat to the majority of the population and are the main areas of economic activities. The yield and quality of water in the river

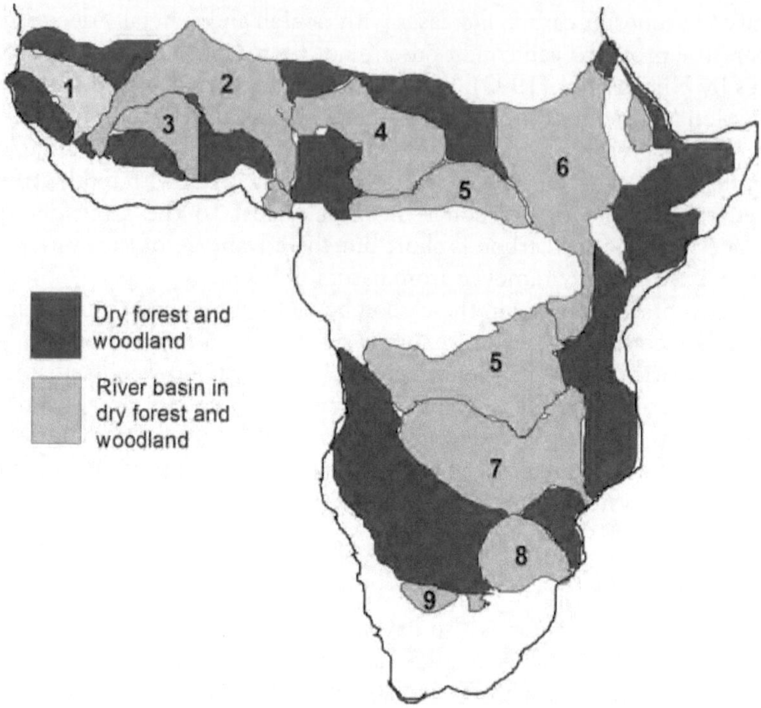

Figure 10.1 *Extent of major river basins in dry forests and woodlands of Sub-Saharan Africa*

Note: Senegal (1), Niger (2), Volta (3), Chad (4), Congo (5), Nile (6), Zambezi (7), Limpopo (8) and Orange (9)

systems is largely a function of rainfall input and land-use patterns within the catchment areas. There is now a growing recognition of the importance of upstream land management on downstream activities that are dependent on water (e.g. water supply to municipal areas, irrigation and hydroelectricity). This basic relation has given rise to the recognition of watershed services. Good upstream forest management provides a host of watershed services, including water purification, groundwater and surface flow regulation, erosion control and stream-bank stabilization. The importance of these watershed services is increasing as water yield and quality become critical issues around the world. The financial value of watershed services becomes particularly apparent when the costs of protecting an ecosystem for improved water quality are compared with downstream benefits, access to water and water security issues. Innovative market-based mechanisms for watershed services include self-organized private payments, public payments or incentives, and trading schemes

Box 10.5 Relationship between forests and production of water service

- Forests slow rate of runoff.
- Forests reduce soil erosion and sedimentation in waterways.
- Forests filter contaminants and influence water chemistry.
- Forests reduce the total annual water flow in a watershed.
- Forests can increase or decrease groundwater recharge.
- Forest loss shifts aquatic productivity.
- Forests may influence precipitation at a larger regional scale, but the effect of forest cover in most areas is limited.

Source: Adapted from Johnson et al (2001)

Major river basins in Africa containing dry forests and woodlands

All the major river basins in sub-Saharan Africa are either located or have most of their headwaters in dry forests and woodlands (Figure10.1). Of these river basins, five (the Congo, Nile, Niger, Chad and Zambezi) occupy about 42 per cent of the geographical area and sustain more than 44 per cent of the African population (UNEP, 2006; Singh et al, 1999). The Volta river basin covers large parts of western Africa and flows through the dry forest areas in that part of Africa. The river basins of Africa support many economic activities that include agricultural production, fishing, forestry, transport and tourism, and contain the largest areas of dry forests and woodlands. Apart from rainfall input, the land-use systems and human activities within the basins affect the quality and quantity of water in the basin. The forested or wooded areas are generally superimposed over watershed areas, and thus the status and management of these forest areas may have a significant effect on the hydrology of the rivers.

The growing population in the fertile river basins is exerting pressure on the forests and woodlands, soil and water resources. Most land use practices have unintentionally altered the hydrological function of most of the watershed areas. This major threat has been exacerbated by extreme and temporal variability in climate and rainfall. It is generally well known, however, that the relationship between forest cover and the production of water services is variable, complicated and poorly understood. Box 10.5 shows some of the general biophysical relationships between forests and water.

Watershed services

The watershed services in the dry forests and woodlands of Africa can be viewed from three levels. The first is rather modest and local, often involving watersheds that supply rural settlements and individual rural households and their livestock. The second is the national water supply to dams for urban water supply, industries and hydroelectricity generation. The third is an extension of the second at a regional level where watercourses may be shared and thus invariably the services offered, e.g. the Zambezi. Whilst it is difficult to establish the actual management practices that produce watershed services, the fact that woodland and forest form the dominant land-use system in the headwaters of most rivers implies their importance and function in the hydrology of the rivers. Currently, the main drivers behind the generation of the watershed services are the protection practices and degradation on upstream areas of the catchment areas, and these become more complicated by level.

The basic notion is that where upstream action generates downstream benefits in terms of water and other aesthetic benefits, and the consumption (or enjoyment) of the service by beneficiaries, this creates the watershed service function of forested/wooded catchments. With regards to the buyers of watershed services, the market is rather small and is usually within the specific countries or between neighbouring countries. Studies by IIED found little evidence of the existence of, or demand for, market-based mechanisms, either by governments or potential 'buyers' of watershed services (Geoghegan, 2005). Public budgets are the main source of funding for watershed services (Gouyon, 2002). Dillaha et al (2008) listed some of the potential buyers of water services and these include hydroelectricity companies, municipal water suppliers, irrigation schemes, industrial companies and tax revenue collection institutions. Within the African context, the lack of investment in such sectors and low value of capital makes payments for water services difficult or uneconomic. The general approach is to compensate those people in the upper reaches of drainage basins who refrain from land uses that exacerbate flooding, periodic water shortages, water quality problems and other problems at lower elevations (Landell-Mills and Porras, 2002; Pagiola et al, 2002). In fact Dillaha et al (2008) suggested that PES in Africa are 'essentially public works programmes'. The Working for Water Program in South Africa is one such programme. The programme aims to control invasive alien species in catchments for the protection of water resources and ensuring water supplies (van Wilgen et al, 2001) resulting in socioeconomic benefits, such as creating labour-intensive tree clearing programmes in the catchments (Le Maitre et al, 2001).

It would appear perhaps that for African dry forest and woodland countries, economic sectors benefiting from watershed services should create funding to invest in forest management in water catchments or pay land users and owners who actively participate in land management. In the case of most countries where the state government still controls and manages forest resources, the payments for water services could be via tax revenues that are specifically collected and allocated to the natural resource management ministries or agencies. In most dry forest countries, state-led systems of protection and regulation of water are giving way to more decentralized ones that emphasize community-based and co-management approaches. The trend is the creation of new institutions for water catchment management variously called water catchment authorities, councils or boards. These institutions may be self-organized, organized with the involvement of state agencies, or representing new decentralized levels of formal authority, and are taking management action at the local level (Geoghegan, 2005). These institutions could be the channel through which watershed services are produced and environmental payments collected on behalf of local communities who manage the resources.

At regional levels, increasing populations present the most serious threat to water adequacy. A larger population leads to higher water demand for production of food and for domestic, municipal and industrial use. This situation results in the construction of dams or altering river courses hence redistributing water allotments among sharing countries (Gleick, 2000). The drive to secure shared water supplies has been noted as one of the urgent political issues on the UN's global agenda and the Nile River is an example of such a case in Africa (El-Fadel et al, 2003). The scarcity and security concerns over water will most likely result in the development of new markets for water and watershed services. Thus countries with a high proportion of river basin areas may want to argue for compensation for investment in management from downstream countries. An example of where this could be applicable is in the Nile River where Ethiopia has a large proportion of the Nile basin and any catchment management in the country will impact water availability in Egypt. Shared river basins, like those found in many parts of Africa, have the potential to generate payments for watershed services amongst nations, and portions of the payments could be directed towards the land managers, who are often local communities.

With increasing incidents of drought and levels of degradation of watershed areas, water scarcity is brought into sharper focus and interest to different stakeholders, countries and end-users. More dams will be constructed causing conflict of interest with downstream users (e.g. Nile river in Uganda, Ethiopia, Egypt; Zambezi river in Zambia, Zimbabwe and Mozambique). This water scarcity has heightened the value of watershed services as a provider of improved water supply and quality. This can explain the increasing development of integrated catchment management strategies between countries (e.g. Zambezi River Action Plan and Nile Basin Initiative) to improve water supply and minimize conflicts. The interest in water sharing (and conflict) and valuation of water provision has brought to the fore the need to value and market watershed services. It is therefore most likely that the market expansion will continue in the future.

Within a national developmental context, the critical issue to consider is how natural resource management can take into consideration the water services from forests and woodlands and how to compensate upstream resource managers or communities. Due to the lack of investments in sectors that might 'buy' watershed services in Africa, countries should consider allocating more fiscal resources to public programmes. The public programmes can then drive local activities aimed at curtailing deforestation and degradation, promoting conservation practices that build personal and community responsibility by giving economic value to watershed protection; and creating new livelihood options for those providing watershed protection services in the catchment. An example of this kind of incentive is the working for water programme in South Africa where communities are getting jobs from clearing alien invasive plants (van Wilgen et al, 2001).

Most of the river basins are trans-boundary and different countries pursue different management agendas, often in conflict with conservation priorities or water requirements of other countries. To address these potential issues of conflict, a number of countries have come together to form river basin management strategies. The initiatives are funded mostly by international organizations. Whilst the initiatives are mainly concerned with managing the water resources, they can also be good channels through which to promote the watershed services of forests and woodlands. Watershed services in the dry forest and woodlands areas are still developing and the challenge is for the countries to demonstrate the linkages between woodlands and the provision of water services.

Other Local Level Environmental Services from the Dry Forests and Woodlands of Africa

The dry forests and woodlands of Africa form a major frontier for agricultural expansion and play a crucial role in providing services to this sector. The majority of the rural people of Africa live in the dry forest and woodland areas and the major economic activities are crop farming and livestock husbandry. The dry forests and woodlands of Africa also vary considerably across a range of socio-economic settings, land use patterns and tenure systems and thus differentiating the types and levels of environmental services. The key support services include providing browse and fodder to livestock (see Chapter 8), soil amelioration, sand dune stabilization, shade and shelter, weed control, soil moisture and nutrient recycling. These support services are produced and consumed locally by communities and individual households. Because the demand is usually internal, no prices and markets have been developed, but their contribution to increasing and sustainable agricultural productivity has been widely researched and reported. Furthermore, environmental services such as water and nutrient cycling have the potential of attracting monetary markets and are very important in terms of livelihoods and rural economy.

Livestock production systems

The Sahelian, Sudanian and Somali-Masai vegetation types support livestock production systems that are the mainstay for the nomadic herdsmen in west, eastern and northern Africa. Pods and foliage from *Acacia* species are well acknowledged for the fodder value throughout the dry lands of Africa (Barnes, 2001). Chapter 8 in this book has described the livestock production systems in African dry forests and woodlands.

Soil moisture and nutrient recycling

Acacia species in the Sahel, the Horn of Africa, East, southern and north Africa are utilized for many products (fuel, timber, forage, gum, tannins, fibre, medicine, food, handicrafts, domestic utensils) and environmental services, such as sand dune stabilization, soil fertility improvement, shade and shelter, game refuge and amenity (Wickens et al, 1995). The agroforestry forest parks (forét-parcs), which are areas with moderate to densely wooded natural tree and shrub growth (Baumer, 1994), also play an important role in supporting agricultural production through nutrient and moisture recycling. Miombo woodlands of southern Africa support agricultural production through soil moisture and nutrient recycling through litter inputs (Stromgaard, 1988; Nyathi and Campbell, 1993).

Other services

The shrubs and scattered trees in the Sahel are well known for reducing wind speeds and sediment transport (Leenders, 2006; Leenders et al, 2007) while others control weeds. *Acacia albida* trees are known to completely control *Striga hermonthica* weeds in millet fields (Gworgwor, 2007). The pollination services of bees associated with honey production in the miombo woodlands of Zambia are well known (Mickels-Kokwe, 2006). These services from African dry forests and woodlands play an important role in supporting local economic activities. Individual households invest time, labour and technology in practices that produce these services. The main incentive is that the benefits accrue directly to the households. The markets for such services are non-existent, or if they are present, they could be operating at very small scales, for example, local level sand dune stabilization to protect an individual farm could benefit a wider area if such a practice were adopted by more members of the community. The challenge for countries in the dry forests and woodlands of Africa is how to up-scale these local level practices to yield landscape level environmental services that benefit the wider community and can be bundled together and sold to bigger markets.

Key Issues and Challenges

Whilst the dry forests and woodlands of Africa are well known for providing livelihood options to many millions of people in Africa, their role in providing environmental services has not been fully explored. A number of constraints appear to hinder the development of environmental services provided by African dry forests and woodlands. These include a lack of understanding of the value of services, insecure tenure, poorly defined markets, lack of institutional and technical capacity, and lack of a transparent benefit sharing mechanism. An elaboration of these issues follows.

Lack of understanding on the values of the resources

At national levels, statistics on the contribution of forests and woodlands to national economies are still very poorly understood (FAO, 2004). Such inadequacies in the valuation of ecosystem service are also apparent at the global level (Constanza et al, 1997). Apart from the very few carbon projects cited in this chapter, and the few examples of watershed services, it is still difficult to give estimates of the value of environmental services. This lack of monetary estimates creates the notion amongst policy-makers that dry forests are not very important. This is generally reflected by the declining budgetary allocations to the forestry sector in many countries. It is therefore recommended that countries in the dry forest areas should undertake studies that estimate the contribution of forest to livelihoods and national economies (Bond, 2008). There is a particular need to assess the contribution of environmental services to the total value of forests and woodlands.

Insecure tenure

Most of the dry forests and woodlands in Africa are open access areas and generally there is insecure access to forest resources and services. This tenure system, typical across most of Africa, is an important barrier to the production of environmental services and payment schemes (Muramira, 2005; Mwangi and Mutunga, 2005; Ochieng et al, 2007 cited in Dillaha et al, 2008). Consequently the major environmental services, such as CO_2 sequestration and watershed services, are perceived as public goods that come from common resources. As such, the investment in management practices by individuals does not guarantee maximum benefits to the individuals. Also at the very local level, it is not clear who should produce these services and who should benefit. A key question to the generation of ES is what is the role of the resource owners and users, communities and national governments? Most ES are viewed as free goods and these attract free riders, so some stakeholders could be benefiting from the services without investing capital, labour, finance, assets and human capital.

Achieving tenure security is an important first step in promoting production of environmental services. Because informal and formal laws on tenure systems overlap in many parts of Africa, enhancing tenure security should be regarded as a long-term process involving negotiations and conflict resolution. Some approaches that countries can adopt include supporting community based natural resource management programmes where a community is clearly defined, access rights are legally recognized and supported, forest management plans are put in place and equitable benefit sharing arrangements are agreed upon.

Lack of clearly defined markets

On the market side, the non-exclusive nature of environmental services has contributed to the lack of clearly defined markets for such services. A community or government might invest in good management practices that generate environmental services but because of the public nature of the good, there might be no clearly defined market. This is a major bottleneck in the development of markets for CO_2 sequestration since carbon cycling operates at a global scale due to its atmospheric form and is 'very poorly captured' in any set of prices or markets (Kinzig et al, unpublished data). Even with watershed services, the markets are not clearly defined and in most cases public funds (public payments schemes) are used to compensate upstream activities (e.g. Gouyon, 2002) or support communities to participate in catchment management projects (e.g. work for water programmes in South Africa).

The REDD is a new source of markets for countries in the dry forest regions experiencing high rates of deforestation and although the details of the programme remain to be finalized, it presents opportunities for countries with dry forests and woodlands to get involved in payment for carbon sequestration. Countries with dry forests and woodlands should therefore invest in creating an understanding of how these policies work at the global level so as to position themselves in negotiations.

Transaction costs

The transaction costs associated with negotiating, implementing, and monitoring environmental services can be high. Environmental services are being promoted as a mechanism for reducing poverty of rural communities. Because of the tenure systems in the dry forests and woodlands of Africa, there are multiple small-scale land-owners and users and potential beneficiaries. Transaction costs increase when multiple parties are involved (Kerr et al, 2006). As a result most investors usually avoid small-scale projects and dealing with many small landowners. This could be a significant barrier to the provision of environmental services, and countries need to develop strategies to reduce such costs, such as setting up intermediary organizations that can work directly with communities on environmental services (Jindal, 2006).

Governance and institutional barriers

A central tenet of environmental services is the need for good governance and transparency to ensure that the poor communities benefit. Historically forest and woodland resources have been associated with corruption and elite capture of benefits by a few powerful people. This negative perception that 'someone else will benefit' and the tragedy of the commons syndrome could be some of the barriers hindering the development of environmental services. Whilst it is important for governments to play an arbitration role in payment for environmental schemes, increasing government regulation could curb PES efficiency and implementation scale, and could eventually harm the poor (Wunder, 2007). A stable and well-defined regulatory environment is necessary to promote investment in environmental services and to ensure that the resource managers are fairly compensated for the management of resources. Enabling policies and laws that promote and protect community interests in natural resource management are therefore urgently needed to address these constraints.

References

Akinnifesi, F. K., Kwesiga, F., Mhango, J., Chilanga, T., Mkonda, A., Kadu, C. A. C., Kadzere, D., Daka, J. D., Ramadhani, T. and Dhliwayo, P. (2006) 'Towards the development of miombo fruit trees as commercial tree crops in southern Africa', *Forests, Trees and Livelihoods*, vol 16, pp103–121

Backéus, I., Patterson, B., Stromquist, L. and Ruffo, C. (2006) 'Tree communities and structural dynamics in miombo (*Brachystegia – Julbernadia*) woodland, Tanzania', *Forest Ecology and Management*, vol 230, pp171–177

Barbier, E. and Swanson, T. (eds) (1992) *Economics for the Wilds*, Earthscan, London

Barnes, R. D. (2001) 'The African acacias – a thorny subject', *Southern African Forestry Journal*, vol 190, pp9–18

Barrow, E. and Mlenge, W. (2003) 'Trees as key to pastoralist risk management in semiarid landscapes in Shinyanga, Tanzania and Turkana, Kenya', Paper prepared for the CIFOR International Conference on Rural Livelihoods, Forests and Biodiversity, Bonn, Germany, May

Batjes, N. H. (2001) 'Options for increasing carbon sequestration in West African soils: An exploratory study with special focus on Senegal', *Land Degradation and Development*, vol 12, pp131–142

Baumer, M. (1994) 'Forest parks or parklands?', *Bois et Forêts des Tropiques*, vol 240, pp53–68

Bond, I. (2008) *Payments for Ecosystem Services (PES) in East and Southern Africa: Assessing Prospects and Pathways Forward*, report prepared for The Katoomba Group and International Institute for Environment and Development (IIED), London

Campbell, B. M. (ed) (1996) *The Miombo Transition: Woodlands and Welfare in Africa*, CIFOR, Bogor

Campbell, B. M., Angelsen, A., Cunningham, A., Katerere, Y., Sitoe, A. and Wunder, S. (2007) 'Miombo woodlands – opportunities and barriers to sustainable forest management', A paper prepared for the CIFOR/World Bank Project: Miombo Woodlands: Policies and Incentive

Chidumayo, E. N. (1988) 'A re-assessment of effects of fire on miombo regeneration in the Zambian Copperbelt', *Journal of Tropical Ecology*, vol 4, pp361–372

Chidumayo, E. N. (1997) *Miombo Ecology and Management. An Introduction*, Intermediate Technology Publications, London

Chidumayo, E. N. (2002) 'Changes in the miombo woodland structure under different land tenure and use systems in central Zambia', *Journal of Biogeography*, vol 29, pp1619–1626

Chidumayo, E. N. (2004) 'Development of *Brachystegia-Julbernardia* woodland after clear-felling in central Zambia: Evidence of high resilience', *Applied Vegetation Science*, vol 7, pp 237–242

Costanza, R., d'Arge, R., de Groot, R., Farber, S., Grasso, M., Hannon, B., Limburg, K., Naeem, S., O'Neill, R. V., Paruelo, J., Raskin, R. G., Sutton, P. and van den Belt, M. (1997) 'The value of the world's ecosystem services and natural capital', *Nature*, vol 387, pp253–260

Daily, G. C. (1997) *Nature's Services: Societal Dependence on Natural Ecosystems*, Island Press,Washington, DC

Dillaha, T., Ferraro, P., Huang, M., Southgate, D., Upadhyaya, S. and Wunder, S. (2008) 'Payment for watershed services: Regional synthesis', USAID PES policy brief 7

Egadu, S. P., Mucunguzi, P. and Obua, J. (2007) 'The population of Acacia tree species producing gum arabic in the Karamoja region, Uganda', *African Journal of Ecology*, vol 35, pp236–241

El-Fadel, M., El-Sayegh, Y., El-Fadl, K. and Khorbotly, D. (2003) 'The Nile River Basin: A case study in surface water conflict resolution', *Journal of Natural Resources and Life Science Education*, vol 32, pp107–117

Engel, S., Pagiola, S. and Wunder, S. (2008) 'Designing payments for environmental services in theory and practice: An overview of the issues', *Ecological Economics*, vol 65, pp 663–674

Eswaran, H., Lal, R. and Reich, P. F. (2001) 'Land degradation: An overview', in E. M. Bridges, I. D. Hannam, L. R. Oldeman, F. W. T. Pening de Vries, S. J. Scherr and S. Sompatpanit (eds) *Responses to Land Degradation*, proceedings of the 2nd International Conference on Land Degradation and Desertification, Khon Kaen, Thailand, Oxford Press, New Delhi, India, pp1–20

FAO (2000) *Global Forest Resources Assessment, 2000*, FAO, Rome

FAO (2004) *Programme des products forestiers nonligneus: Contribution des insects de la forêt a la sécurité alimantaire – L`example des chenilles d`Afrique Centrale*, FAO, Rome

FAO (2005) *State of the World's Forests*, FAO, Rome

FAO (2006) *Global Forest Resources Assessment, 2005*, FAO, Rome

FAO (2007) 'Correspondence on carbon stocks in forests: Extract from database', *Forest Resources Assessments*, FAO, Rome

Frost, P. (1996) 'The ecology of the miombo woodlands', in B.M. Campbell (ed) *The Miombo Transition: Woodlands and Welfare in Africa*, CIFOR, Bogor

Gambiza, J., Campbell, B. M., Moe, S. R. and Mapaure, I. (2008) 'Season of grazing and stocking rate interactively affect fuel loads in *Baikiaea plurijuga* Harms woodland in northwestern Zimbabwe', *Journal of African Ecology* (online publication) doi:10.1111/j.136

Geoghegan, T. (2005) *Challenges to Establishing Markets for Watershed Services: Learning from Country Diagnostics*, IIED, London

Gleick, P. (2000) *Water Conflict Chronology*, Pacific Institute, Oakland, CA

Grieg-Gran, M., Porras, I. and Wunder, S. (2005) 'How can market mechanisms for forest environmental services help the poor? Preliminary lessons from Latin America', *World Development*, vol 33, pp1511–1527

Gouyon, A. (2002) 'Rewarding the upland poor for the environmental services: A review of initiatives from developed countries', Paper presented at RUPES Regional Inception/Planning Workshop, Puncak, Indonesia, 6–8 February

Gworgwor, N. A. (2007) 'Trees to control weeds in pearl millet', *Agronomy for Sustainable Development*, vol 7, pp89–94
Inamadar A., De Jode, H., Lindsay, K. and Cobb, S. (1999) 'Capitalizing on nature: Protected area management', *Science*, vol 283, pp1856–1857
IPCC (2001) *Climate Change 2001: Impacts, Adaptation and Vulnerability*, Contribution of working group II to the Third Assessment Report of the Intergovernmental Panel on Climate Change, Cambridge University Press, Cambridge, UK
Jindal, R. (2004) 'Measuring the socio-economic impact of carbon sequestration on local communities: An assessment study with specific reference to the Nhambita pilot project in Mozambique', MSc dissertation, University of Edinburgh, Edinburgh
Jindal, R. (2006) 'Carbon sequestration projects in Africa: Potential benefits and challenges to scaling up', Earthtrends, World Resources Institute, Washington, DC
Jindal, R., Swallow, B. and Kerr, J. (2008) 'Forestry-based carbon sequestration projects in Africa: Potential benefits and challenges', *Natural Resource Forum*, vol 32, pp116–130
Johnson, N., White, A. and Perrot-Maitre, D. (2001) *Developing Markets for Water Services from Forests: Issues and Lessons for Innovators*, Joint publication of Forest Trends, World Resources Institute and the Katoomba Group, Washington, DC
Kamwenda, G. J. (2002) 'Ngitili agrosilvipastoral systems in the United Republic of Tanzania',*Unasylva*, vol 53, pp46–50
Katila, M. and Puustjärvi, E. (2003) 'Impact of new markets for environmental services on forest products trade', FAO Impact Assessment of Forest Products Trade in the Promotion of Sustainable Forest Management (GCP/INT/775/JPN), INDUFOR in collaboration with Ecosecurities Ltd, Finland
Kerr, J., Foley, C., Jindal, R. and Chung, K. (2006) 'Reconciling environment and development in the Clean Development Mechanism', *Journal of Sustainable Forestry*, vol 23, no 1, pp1–18
Kokwe, M., Mickels-Kokwe, G., Moombe, K., Njovu, F. and Zulu, D. (2005) 'Tailoring sustainable management approaches for woodlands to social, economic and technological capacities. The case of Zambia', Report for the CIFOR: Stimulating policy dialogue on sustainable management of Africa's dry forests – a CIFOR/SIDA initiative
Ky-Dembele, C., Tigabu, M., Bayala, J., Ouédraogo, S. J. and Odén, P. C. (2007) 'The relative importance of different regeneration mechanisms in a selectively cut savanna-woodland in Burkina Faso, West Africa', *Forest Ecology and Management*, vol 243, pp28–38
Lal, R. (2004) 'Carbon sequestration in dryland ecosystems', *Environmental Management*, vol 33, pp528–544
Lal, R., Hassan, H. M. and Dumanski, J. (1999) 'Desertification control to sequester C and mitigate the greenhouse effect', in N. J. Rosenberg, R. C. Izaurralde and E. L. Malone (eds) *Carbon Sequestration in Soils: Science, Monitoring and Beyond*, Battelle Press, Columbus, Ohio
Landell-Mills, N. and Porras I. T. (2002) *Silver Bullet or Fool's Gold? A Global Review of Markets for Forest Environmental Services and their Impact on the Poor*, IIED, London
Leenders, J. K. (2006) 'Wind erosion control with scattered vegetation in the Sahelian zone of Burkina Faso', Doctoral thesis, Wageningen University, Wageningen, The Netherlands
Leenders, J. K., van Boxel, J. H. and Sterk, G. (2007) 'The effect of single vegetation elements on wind speed and sediment transport in the Sahelian zone of Burkina Faso', *Earth Surface Processes and Landforms*, vol 32, pp1454–1474

Le Maitre, D. C., van Wilgen, B. W., Gelderblom, C. M., Bailey, C., Chapman, R. A. and Nel, J. A. (2001) 'Invasive alien trees and water resources in South Africa: Case studies of costs and benefits of management', *Forest Ecology and Management*, vol 55, pp1–17

Lovett, P. N. and Haq, N. (2000) 'Diversity of the sheanut tree (*Vitellaria paradoxa* C.F. Gaertn. F.) in Ghana', *Genetics and Crop Evolution*, vol 47, 293–304

Malaise, F. (1997) *Se Nourrir en Foret Claire Africaine. Approche Ecologique et Nutrionnelle*, Presses Agron, Gemblous, Belgium

Matta, J. and Kerr, J. (2006) 'Can environmental service payments sustain collaborative forest management?', *Journal of Sustainable Forestry*, vol 23, pp63–79

Mbwambo, L. and Nshubemuki, L. (2007) 'Principal way to promote amelioration of degraded miombo ecosystems in semi-arid Tanzania', Working Papers of the Finnish Forest Research Institute, Helsinki, Finland

Mickels-Kokwe, G. (2006) 'Small-scale woodland-based enterprises with outstanding economic potential: The case of honey in Zambia', CIFOR, Bogor

Monela, G. C., Chamshama, S. A. O., Mwaipopo, R. and Gamassa, D. M. (2005) 'A Study on the Social, Economic, and Environmental Impacts of Forest Landscape Restoration in Shinyanga Region, Tanzania. Dar es Salaam, Tanzania', Forestry and Beekeeping Division of the Ministry of Natural Resources and Tourism of Tanzania and IUCN–The World Conservation Union, Eastern Africa Regional Office, Nairobi

Montagnini, F. and Nair, P. K. R. (2004) 'Carbon sequestration: An under-exploited environmental benefit of agroforestry systems', *Agroforestry Systems*, vol 61–62, pp281–298

Muramira, T. E. (2005) 'Overview of payment for ecosystem services in Uganda', presentation at the Building Foundations for Pro-Poor Ecosystem Service Payments in Africa workshop, 19–22 September, Queen Elizabeth National Park, Uganda

Murdiyarso, D. and Skutsch, M. (2006) 'Community forest management as a carbon mitigation option: Case studies', CIFOR, Bogor

Mwangi, S. and Mutunga, C. (2005) 'Overview of payment for ecosystem services in Kenya', presentation at the Building Foundations for Pro-Poor Ecosystem Service Payments in Africa workshop, 19–22 September, Queen Elizabeth National Park, Uganda

Nair, C. T. S and Tieguhong, J. (2004) 'African forests and forestry: An overview', Report prepared for the project: Lessons Learnt on Sustainable Forest Management in Africa, FAO and CIFOR, Bogor

Nhira, C. (1996) 'A preliminary prognosis of the incentive structure in a pilot resource sharing project in Zimbabwe', in B. M. Campbell (ed) *The Miombo Transition: Woodlands and Welfare in Africa*, CIFOR, Bogor

Nyathi, P. and Campbell, B. M. (1993) 'The acquisition and use of miombo litter by small-scale farmers in Masvingo, Zimbabwe', *Agroforestry Systems*, vol 22, pp43–48

Owino, F. (no date) 'Opportunities to improve dryland management in sub-Saharan Africa: Implications of international conventions and agreements', Forest Resources International, Nairobi, Kenya

Pagiola, S., Bishop, J. and Landell-Mills, N. (eds) (2002) *Selling Forest Environmental Services. Market-based Mechanisms for Conservation and Development*, Earthscan Publishing, London

Piearce, G. D. (1993) 'Natural regeneration of indigenous trees: The key to their successful management', in G. D. Piearce and D. J. Gumbo (eds) *The Ecology and Management of Indigenous Forests in Southern Africa*, Proceedings of an International Symposium, Victoria Falls, Zimbabwe, 27–29 July 1992, Forestry

Commission of Zimbabwe and SAREC, Harare
Sanchez P. A. (1999) 'Improved fallows come of age in the tropics', *Agroforestry Systems*, vol 47, pp3–12
San Jose, J. J., Montes, R. A. and Farinas, M. R. (1998) 'Carbon stokes and fluxes in a temporal scaling from a savanna to a semi-deciduous forest', *Forest Ecology and Management*, vol 105, pp 251–262
Shackleton, C. M. and Clarke, J. (2007) *Silvicultural Options for Strengthening the Use and Management of Miombo by the Poor*, GENESIS report for the World Bank, Washington, DC
Singh, A., Dieye, A. and Finco, M. (1999) 'Assessing the environmental conditions of major river basins of Africa as surrogates for watershed health', *Ecosystem Health*, vol 5, pp265–274
Skutsch, M. M. and Trines, E. (2008) 'Report from the UNFCC: Policy piece', *African Journal of Ecology*, vol 46, pp1–2
Sohngen, B. and Brown, S. (2006) 'The Cost and Distribution of Carbon Sequestration by Holding Trees Longer Than the Optimal Rotation Period in US Production Forestry Regions', invited presentation at the Finnish Forest Research Institute (METLA), Helsinki, Finland, 14 March
Stoffberg, H. and van Rooyen, G. (2006) 'A study on the carbon storage of *Jacaranda mimosifolia* street trees in the City of Tshwane's (Pretoria) urban forest, South Africa', www.greencarbon.co.za
Stromgaard, P. (1988) 'Soil and vegetation changes under shifting cultivation in the miombo of East Africa', *Human Geography*, vol 70, pp363–374
Tan, Z., and Lal, R. (2005) 'Carbon sequestration potential estimates with changes in land use and tillage practice in Ohio, USA', *Agriculture, Ecosystems and Environment*, vol 111, pp140–152
Touré, A. (2006) 'Carbon sequestration: Pulling off an ecological hat-trick?', *ICT Update – A Current Awareness Bulletin for ACP Agriculture*, no 31
Tschakert, P. and Tappan, G. (2004) 'The social context of carbon sequestration: Considerations from a multi-scale environmental history of the Old Peanut Basin of Senegal', *Journal of Arid Environments*, vol 59, pp535–564
Unruh, J. D., Houghton, R. A. and Lefebvre, P. A. (1993) 'Carbon storage in agroforestry: An estimate for sub-Saharan Africa', *Climate Research*, vol 3, pp39–52
UNEP (2006) *Africa's Lakes: Atlas of our Changing Environment*, Division of Early Warning and Assessment, UNEP, Nairobi
UNFCCC (2003) *Caring for Climate: A Guide to Climate Change Convention and the Kyoto Protocol*, United Nations Framework Convention on Climate Change (UNFCCC), Bonn, Germany
UNFCCC (2006) 'Reducing emissions from deforestation in developing countries', Background paper for the workshop on reducing emissions from deforestation in developing countries, 30 August–1 September, Rome, Italy
van Wilgen, B. W., Richardson, D. M., Le Maitre, D. C., Marais, C. and Magadlela, D. (2001) 'The economic consequences of alien plant invasions: Examples of impacts and approaches to sustainable management in South Africa', *Environment, Development and Sustainability*, vol 3, pp145–168
Wiegand, K., Ward, D. and Saltz, D. (2005) 'Multi-scale patterns and bush encroachment in an arid savanna with a shallow soil layer', *Journal of Vegetation Science*, vol 16, pp311–320
Wickens, G. E., Sief El Din, A. G., Sita, G. and Nahal, I. (1995) *Role of Acacia Species in the Rural Economy of Dry Africa and the Near East*, FAO, Rome

Williams, M., Ryan, C. M., Rees, R. M., Simbane, E., Fernado, J. and Grace, J. (2008) 'Carbon sequestration and biodiversity of re-growing miombo woodlands in Mozambique', *Forest Ecology and Management*, vol 254, pp145–155

Woomer, P. L., Touré, A. and Sall, M. (2004) 'Carbon stocks in Senegal's Sahel transition zone', *Journal of Arid Environments*, vol 69, pp499–510

Wunder, S. (2005) 'Payment for environmental services: some nuts and bolts', *Occasional Paper no 42*, CIFOR, Bogor

Wunder, S. (2007) 'The efficiency of payments for environmental services in tropical conservation', *Conservation Biology*, vol 21, pp48–58

Zahabu, E., Skutsch, M. M., Sosovele, H. and Malimbwi, R. (2007) 'Reduced emissions from deforestation and degradation. Policy Brief', *African Journal of Ecology*, vol 45, pp451–453

Zohlo, R. (2005) 'Effect of fire frequency on the regeneration of miombo woodland in Nhambita, Mozambique', MSc dissertation, University of Edinburgh, Edinburgh

11

Managing Dry Forests and Woodlands for Products and Services: A Prognostic Synthesis

Davison Gumbo and Emmanuel Chidumayo

MUSINGS OVER THE PAST AND THE FUTURE

Each of the preceding chapters has raised key issues and challenges concerning the importance and management of African dry forests and woodlands and the products and services that they provide. In planning for dry forest and woodland management for the future, it is crucial to learn from the past, including learning from anthropogenic effects (Bird and Cali, 1998). Hoffman (1997) claims that historic anthropogenic influences, through the demand for wood for iron smelting, have been a lot more extensive than previously accepted. In addition, Huffman (1982), Hall (1984) and Schmidt and Avery (1996), working on Iron Age archaeological sites, have confirmed that vast amounts of wood were used, most of which was extracted from surrounding forests and woodlands. Domestic livestock, in particular cattle and sheep, evidence of whose existence dates back to the 1st century AD, have also been mentioned as agents of dry forest and woodland change in Africa (Tlou and Campbell, 1997; Campbell and Ramsay, 1994; Denbow and Wilmsen, 1989; Garlake, 1978). Further, cycles of abandonment and settlement as well as slash and burn correspond to deforestation and reforestation in Africa (Schmidt and Avery, 1996; Goucher, 1981). These are a few but key examples that show how strongly African dry forests and woodlands were influenced by human activities over thousands of years ago; and the intensity of which has increased many fold in the last century as human and livestock populations grew, mines and road infrastructure were developed and cities expanded through rapid urbanization.

The use of dry forest and woodland resources should also be seen with a broader focus especially as these resources also provided goods for local and international markets in the same periods. Some scholars show how charcoal fueled local trade (for blacksmiths) and later entered into the vast trade routes of western Africa (Haaland, 1980; Goucher, 1981). In eastern and southern Africa a similar picture emerges (Schmidt and Avery, 1996; van der Merwe and Killick, 1979). The trees most sought after were *Burkea africana*, *Combretum imberbe*, *Prosopsis africana*, *Percopsis angolensis*, *Ziziphus mucronata* and a host of *Acacia* species, but the most preferred species across Africa was *Burkea africana* (van der Merwe and Killick, 1979; Goucher, 1981; Goucher and Herbert, 1996). In addition, and as noted in Chapter 7, trade in charcoal could also have been complemented by a thriving timber trade, and though not widely recorded, centred on mahogany and ebony (Taylor, 1960). The depletion of forests for charcoal to fuel the smelting and smithing processes, not only reduced the number of key species but also created a disturbance regime that allowed for the establishment of Africa's dry forests and woodlands. The disturbance regime was augmented by fire as well as low levels of slash-and-burn agriculture. Use of dry forests and woodlands has not changed that much but has gained in intensity as populations grow. As noted, vegetation formations have historically been transformed through charcoal making, slash and burn, livestock rearing and lately, mining and infrastructure development have been added to the fray and in combination present formidable challenges to the management of dry forests and woodlands in Africa. As the previous chapters have shown, these vegetation formations can significantly contribute not only to the maintenance and improvement of people's wellbeing and livelihood security, but also to the sustainable growth of national economies. Given the many products and services that African dry forests and woodlands provide, one major challenge that remains to be addressed is whether appropriate plans and strategies exist, or can be developed, for managing these vegetation types for multiple and sometimes competing products and services in a sustainable way. Yet it must be noted that sustainable use ideals are central to any planned management and conservation of African dry forests and woodlands.

Increasingly, there are calls for the application of sustainable-use management to bring socio-economic factors to the centre stage of issues pertaining to sustainable forests and woodland management (Hutton and Dickson, 2000; Child, 2004). Major threats to sustainable management of African dry forests and woodlands in the future exist at the intersection of local land use and global environmental changes. Solutions to the problems partially lie in the existence of strong local governance systems where such systems receive sufficient support from the state. This is critical as increasingly, the integration of social and ecological components are being used to analyse resource management, conservation and sustainability issues (Gunderson and Holling, 2002). The dependence of rural communities on the biodiversity, ecological processes and ecosystem services provided by dry forests and woodlands is enormous, yet it is precisely these forests and woodland types that are fast disappearing. For the hundreds of millions of poor and rural inhabitants of forest and woodland ecosystems, the losses in forest cover have had particularly severe consequences. Forests are embedded within larger-level socio-economic and political settings, which also have the capacity to significantly influence outcomes (Chapin et al, 2006).

The range of challenges presented in the preceding chapters requires working partnerships and these should be established on two fronts. Firstly, at the country level, the involvement of all the major stakeholders (e.g. public agencies, the private sector, NGOs, local communities and informal local institutions) is essential for efficient management of dry forest and woodland resources. It is important that governments:

- provide an enabling environment for economically efficient and environmentally sound forest sector activity;
- ensure better pricing to reflect the full scarcity of forest resources;
- tax resource users accordingly to internalize externalities.

The scaling down of government involvement in production will release scarce resources for those areas in which its intervention is more pertinent, e.g. sustainable dry forest and woodland management (where the market often fails), research, human resources development and donor coordination. Secondly, the global community should develop a common agenda for internalizing the priorities and needs of dry forest and woodland countries. The world community must also provide the strong financial and technical assistance needed by African countries for improvements in forest resource management. Other issues to be considered include:

- Strategizing for carbon sequestration in dry forest and woodland countries where basic baseline information and data are not available, in particular methods to be adopted for this purpose.
- How, from a research perspective, should the mismatch between tenure (state) and capacity to regulate access to forestry resources be tackled?
- It is critical that the profile of the forestry sector be raised at national government level so that sufficient funds are made available at that level. How can this profile be raised? Is it a question of showing contribution to GDP or the inclusion of the non-financial benefits?
- Do we have the right skills and numbers to address dry forest and woodland research?

Dry forests and woodlands need to be viewed as economically viable, multiple use systems that are inseparable from people's welfare, livelihoods, culture and development. The importance of dry forest and woodland products in providing housing, health care, employment, food, etc. and in contributing to all eight Millennium Development Goals (Petheram et al, 2006) needs to be accepted and mainstreamed by a much wider range of stakeholders and sectors than is presently the case. Important services and value are being lost on a daily basis through resource mismanagement and land transformation. This comes with considerable environmental and social costs and is jeopardizing the livelihoods of the poor as well as future economic streams from these systems. Forests and forest product use and management need to be seen as part of a holistic and integrated approach to rural development and poverty alleviation and be acknowledged in key development strategies (e.g. Poverty Reduction Strategies, Sustainable Development Strategies) and decentralized planning initiatives.

The use of products generated by forest resources needs to be integrated into national surveys for statistical documentation of volumes and values, including household income and expenditure surveys. This type of data could help to generate the policy responses required to address this sector.

Indigenous or traditional knowledge embedded in customary laws, norms, beliefs and practices relating to the use and management of specific dry forest species needs to be included in any efforts to improve forest and species management. Furthermore, the interaction, and potential synergies, between such informal controls and more formal mechanisms such as government imposed restrictions need to be understood. The links between culture and forest product conservation are particularly important in some contexts and should be more explicitly explored. There are indications that cultural value may be as, or more, important than economic value in providing an incentive for sustainable management (Sambou et al, 2002). Indeed, the latter is often controversial with a high market value sometimes resulting in precisely the opposite outcome to that desired, resulting in overexploitation rather than conservation.

The trade in lesser known forest products and services needs to be supported at all levels, from local to international, through the provision of appropriate business support services and a favourable policy environment. Local and regional trade should not be ignored in the pursuit of high value international markets, as all levels of trade have a vital role to play in the livelihoods of those involved, with support for the local level trade often being an important way to target some of the poorest people and women in particular.

Institutionalized support needs to be provided for the commercialization of forest products, including micro-credit provision, skills training, organizational capacity building and development of value added products to assist local producers to expand their trading activities. Some of this may require cross-departmental or cross-sectoral cooperation. Targeted subsidies and support programmes that can make the cash sale of forest products more viable need attention (Petheram et al, 2006). Work from Botswana has shown how investment in marketing and training substantially increased producers' incomes, providing a livelihood for people who would have otherwise been supported by more costly state welfare grants (Terry, 1999). Suitable models that involve partnerships between the private sector, NGOs and producer groups, with appropriate government oversight, need to be explored (Wynberg, 2006). Coordinated regional approaches and policies are required for important products traded across borders (e.g. mopane worms, woodcraft, medicines), or where products produced in adjacent countries are competing for the same markets (e.g. baskets).

The objective of this chapter is to bring together issues raised in the preceding chapters and propose plans and strategies for managing dry forests and woodlands for increasing the benefits of these vegetation types in a sustainable manner. To achieve this, the chapter presents three areas for discussion. Firstly, it examines the social-political context of sustainable forest management and the cross-cutting issues of policy and legislation that underpin the implementation of plans and strategies concerning natural resources management. Secondly, this chapter discusses the dynamics of dry forests and woodlands in the face of global change. Thirdly, the chapter briefly suggests ways of managing African dry forests and woodlands for multiple products and services before ending with a section on the way forward.

Policy and legislative reform

Key issues that have been repeatedly raised in the preceding chapters with implications for the management of African dry forests and woodlands relate to the need for policy and legislative reform. In this regard, issues pertaining to land and resource tenure, the associated property and access rights, and the need for integrated land-use planning must be addressed. It is our contention that institutional and political processes interact with economic and environmental or bio-physical factors depending on the management or conservation objectives. Further, we believe that macro level policies are very important in influencing forest management. In some cases, however, policy reform adds a layer of insecurity and uncertainty concerning the attainment of management objectives. A good example is land reform and 'degazettement' of forest reserves driven by the encroachment of local communities. Furthermore, it has been noted that uncertainties linked to long cutting cycles coupled with slow growth rates have made planning difficult, leading to overharvesting. This has been exacerbated by the lack of data and information.

Some of the key policy processes that are central to dry forest and woodland management relate to decentralization and devolution in governance. We note that decentralization and devolution of management for key resources may confer responsibility and ownership for stakeholders. While many dry forest and woodland countries recognize the need to devolve resource management responsibility to the lowest structures of local level institutions of the affected communities, in practice this is often not the case. Policy is often not matched with practice and for dry forests and woodlands where centralized decision-making and management is at play, communities are already alienated and this has promoted resource overuse that often lead to degradation. Centralization of management also inhibits resource access rights and benefits for local level user-groups from protected areas. Devolution increases resource access rights and benefits particularly to those most dependent on them (Ribot, 1999).

Policy and legislative reform is necessary to provide an enabling environment for managing dry forests and woodlands for multiple products and services. The burden of policy reform in the forestry sector in Africa is presaged on an understanding of the existing macro-policy context. Current knowledge shows that in Africa's dry forest and woodland countries the most dominant player is the public sector. Questions can be raised as to how the sector addressees and is addressed in the various national development plans and programmes if dry forests and woodlands are to be sustainably managed. In Table 11.1 we present a summary of the status of the forestry sector in selected dry forest and woodland countries in Africa. In this table we look at national forest policies, taking cognizance of the fact that FAO has been supporting the development and implementation of such initiatives in a number of countries. In the table we also show the principal supporting legislation for the sector as well as the institutional housing of the sector.

The table not only shows that there has been a widespread adoption of national forest policies across sub-Saharan Africa but also signals a shift from forests and woodlands as sources of timber to a wider range of forest goods and services and stakeholder needs. Clearly, these countries have embraced the concept of sustainable forest management and have acknowledged that good policies are needed, and it is hoped that this is not just a case of conformity but a realization of the importance of the sector. Besides working on NWFPs, most governments already had various pieces of regulation addressing the sector varying from acts of parliament to statutory instruments. In almost all the countries, the legislative framework for forests and woodlands is also supported by other sector-based pieces of legislation, especially environment and mining, which goes a long way to show that forests and woodlands are recognized in some countries. However, recognition alone is not enough as it does not translate into direct support for the forest sector.

While the above picture looks rosy in most countries, the institutional frameworks do not necessarily give priority and support to the sector. The general tendency is to place forests and woodlands within a ministry dealing with environmental issues as in Cameroon, Zimbabwe and Kenya. While the idea of linking forests and woodlands with environment sounds good in some cases, only a part of the environment is considered, e.g. water, as is the case in Burkina Faso. There are major variations in terms of where the forest sector is placed. In Ghana the sector is aligned with science and technology, while in Mozambique and Tanzania it is with agriculture and tourism, respectively. In other countries it is combined with land, natural resources or fisheries. Table 11.1 also shows the broad range of supportive legislation to the forest sector but this can also result in fragmentation and duplication of authority and responsibilities. For example, in Zimbabwe ten different ministries administer an estimated 20 environment-related laws. The situation is further complicated in countries with federal systems of government where responsibilities are shared by central and provincial authorities.

While Table 11.1 does not comprehensively show how countries are addressing emerging issues such as co-management of forests, it is clear that

Table 11.1 *Positioning and arrangements in the macro-policy context of the forest sector in selected African dry forest and woodland countries*

Country	*Status of NFP*	*Principal Act(s)*	*Support Legislation*	*Key Institutions*
Angola (2007)	National Forest, Wildlife and Conservation under discussion	The Forest, Wildlife and Conservation Areas Law	Environmental Basis Law Land law Biodiversity National Strategy and Plan of Action	Ministry of Agriculture and Rural Development (MINADER) • National Directorate of Agriculture, Livestock and Forest (DNAPF) • The Forest Development Institute (IDF)
Burkina Faso (2007)	The National Forestry Action Plan 1989 and 1991		Forest Code, Mining Code Environmental Code for sustainable natural resources management	Ministry of Environment and Standard of Living • National Council for the Environment and Sustainable Durable Development • Department of Water and Forests • Department of the Environment
Cameroon (2005)	The 1994 Forest Law	The 1994 Forest Law	The presidential decree 94/436 The frame law on Environmental Protection 1996	Ministry of Environment and Forestry (MINEF) • The Sub Directorate of forest inventories and management • The Wildlife Division • The Wood Industry Division
Ghana (2008)	Forest and Wildlife Policy 1994	The Forestry Commission Act 1999 Timber Resource Management Act 1997	Timber Resources Management (Amendment) Regulations 2003 Trees and Timber Amendment Act, 1994 Forest Protection Amendment Act 2002	Ministry of Lands, Forestry and Mines • The Forestry Commission • Forestry Research Institute of Ghana (FORIG)
Kenya (2009)	Draft Paper no 1 of 2007 on Forest Policy was published in 2007	Forests Act 2005	The Environmental Management and Co-ordination Act (EMCA) 1999 The Wildlife (Conservation and Management) Act	Ministry of Forestry and Wildlife • Kenya Forest Service • Kenya Forestry College • Kenya Forestry Research Institute • Forestry Extension Services
Malawi (2007)	National Forest Policy of 1996, Community Based Forest Management 2003	Forest Act 1997	Environmental Management Act 1996 Land use Management Act Parks and Wildlife Management Act	Ministry of Energy, Mines and Natural Resources • Department of Forestry • Forestry Research Institute of Malawi • Forestry Extension Services

Table 11.1 *continued*

Country	*Status of NFP*	*Principal Act(s)*	*Support Legislation*	*Key Institutions*
Mozambique (2007)	The Forestry and Wildlife Policy	Forestry and Wildlife Law 1999	The Land Law, 1997 The Environmental Law, 1997 Mining legislation	Ministry of Agriculture • National Directorate of Land and Forests • Forestry Research Centre (CEF)
Namibia (2007)	Approved 2001; Draft New Forest and Veld Fire Policy (2007)	Forest Act 2001 Forest Amendment Act 2005	Namibia Forestry Strategic Plan 1996 Water Act 1969 Nature Conservation of 1975 Environmental Protection Bill (in pipeline)	Ministry of Agriculture, Water and Forestry • Department of Water Affairs and Forestry • Directorate of Forestry – Research – Extension
Nigeria (2008)	Approved National Forest Policy 2006	A draft National Forestry Act is still in process by relevant government agencies	The Forestry Law	Federal Ministry of Environment, Housing and Urban Development • Federal Department of Forestry • Forestry Research Institute of Nigeria, Ibadan, Oyo State • Forestry Extension Services
South Africa (2007)	White Paper on Sustainable Forest Development 1998 (currently being reviewed as part of the SA National Forest Programme)	National Forest Act 1998	National Veld and Forest Fire Act 2001 Forestry Laws Amendment Act 2005 National Environmental Management Act 1998 National Environment Management: Biodiversity Act 2003	Ministry of Water Affairs and Forestry (MWAF) • Department of Water Affairs & Forestry (DWAF) • Council for Scientific & Industrial Research (CSIR) Natural Resources & Environment • Forestry Extension Service
Sudan (2007)	National Forestry policy statement 2006	The Forests and Renewable Natural Resources Act 2002	Environment Protection Act 2000 The Wildlife Conservation and National Parks Act 1986	Ministry of Agriculture and Forests • Forests National Corporation (FNC) • Forestry Research Centre • Forestry Extension Services
Tanzania (2007)	National Forest Policy 1998, (currently being reviewed)	Forest Act 2002	Wildlife Act Environment Act Fisheries Act Water Resources Act	Ministry of Natural Resources and Tourism (MNRT) • Forestry and Beekeeping Division (FBD) • Tanzania Forestry Research Institute (TAFORI) • Forestry Extension Services

Table 11.1 *continued*

Country	*Status of NFP*	*Principal Act(s)*	*Support Legislation*	*Key Institutions*
Uganda (2007)	National Forest Policy 2001	The National Forestry And Tree Planting Act 2003	National Environment Management Act 1995 Uganda Wildlife Act	Ministry of Water and Environment • Forest Inspection Division (FID) • National Forestry Resources Research Institute (Na FORRI) • District Forestry Services
Zambia (2007)	National Forest Policy 1998	Forests Act 1973	Environmental Protection and Pollution Control Act Zambia Wildlife Authority Act 1998 Forestry Act 1999 Land Act 1995	Ministry of Tourism, Environment and Natural Resources (MTENR) • Forest Department (FD) • Forestry Research Division • National Centre for Scientific and Industrial Research (NCSIR) • Forestry Extension Unit

Note: NFP = National Forest Policy, an agreed strategic framework of priorities and viable options for improving forestry management at national level.

Source: www.nfp-facility.org/45444/en/ago/

Table 11.2 *Policy and legislative reform objectives in improving management of African dry forests and woodlands for products and services*

Threat and resource use issues	*Policy and legislative reform objectives*	*Scale of institutional involvement*			
		1	*2*	*3*	*4*
Climate change and vulnerability	Enhancing adaptation to climate change	✔	✔		✔
Management of protected areas	Strengthening management of protected areas		✔		
Resource overharvesting	Reducing overharvesting and where possible promoting alternative products	✔			
Invasive species	Preventing and controlling invasive species		✔		
Conservation in modified landscapes	Encouraging tree conservation in cultural and modified landscapes		✔		
International agreements	Domesticating and enforcing international agreements and fostering cooperation with international NGOs		✔		
Community and private involvement	Promoting involvement of local communities and the private sector in the management and utilization of natural resources	✔			
Subsistence and cultural uses of products and services	Recognizing and promoting sustainable subsistence and cultural uses of natural resources	✔			
Product and service commercialization	Promoting commercialization of natural products, value addition and trade	✔			
International trade	Promoting international trade that is based on sustainable use of natural resources	✔			
Production and use technologies	Building capacity and promoting technologies that contribute to resources conservation and efficient use	✔			
Plantation and woodlots	Building capacity for and promoting investment in appropriate development of plantations and woodlots	✔			
Wildlife and livestock production	Promoting efficient use of rangelands for either wildlife or livestock or their integration, but having regard to animal health and climate change implications	✔			
Carbon sequestration and trade	Building capacity for monitoring carbon sequestration under UNFCCC mechanisms and ensuring fair trade in carbon				✔
Land-use integration	Promoting integrated land-use planning and conservation of forests and woodlands for multiple products and services	✔			
Local knowledge and practices	Safeguarding and promoting useful local knowledge and practices that contribute to sustainable natural resources management	✔			
Resource tenure	Empowering the poor and vulnerable to have secure tenure to natural capital and resources	✔			
Property and access rights	Securing property and access rights for the poor and vulnerable in society to natural capital and resources	✔			

Notes: 1 = Local and national; 2 = local, national, and regional; 3 = regional and international; and 4 = national, regional and international

Source: Authors

various forms of policy and legislative reforms are needed. Such reforms should focus on complementary laws and by-laws on land tenure, property and access rights that deliberately protect the rights of the poor and vulnerable in society, particularly in rural areas, where they are more dependent on dry forests and woodlands for livelihoods and income generation. Issues pertaining to legislative reform necessary for improving the management of African dry forests and woodlands are presented in Table 11.2.

Response to these critical aspects will take place at four levels: local, national, regional and international. Of the 18 threat and resource use issues identified in Table 11.2, 12 are focusing at the local level and these should be dealt with through participation, social differentiation, authority structures and legal statutes.

Stakeholder participation

One key element to resource management decentralization is stakeholder participation. Resource sharing arrangements and management are gaining currency. This started in the wildlife sector where the management of parks and wildlife resources has moved in this direction by the state entering into co-management and resource sharing arrangements with stakeholders around designated conservation areas. In this way, previously marginalized community groups, such as those found around wildlife estates in Botswana, Namibia, Zambia and Zimbabwe have developed these sharing arrangements to different degrees in the last two decades. Such 'partnership' models raised the conservation status of the protected areas and resources. In Zambia, attempts to try such partnership arrangements for forest resources carried out through joint forest management approaches have some promise but indications are that a lot could be achieved if the state were willing to allow issues of permits and licensing to be carried out by the communities concerned (Bwalya, 2007). What is central is to ensure that appropriate and pro-community policies are put in place so that the communities become part and parcel of the management and enjoy benefit streams that meaningfully contribute to their welfare.

To a great extent, stakeholders may be actively involved in policy interpretation and use, but less so in the formulation processes of these policies. Often key stakeholders in dry forest and woodland countries are not consulted in defining resource management strategies and this often alienates them from the resultant policies and legal instruments. We posit the view that local people tend to be the least involved in policy design but tend to be the most adversely affected by resource use policies. While consultations in the wildlife sector have been carried out in southern Africa, such consultative processes are as yet to take hold in the forestry sector. The level of participation in policy formulation by various stakeholders involved in dry forests and woodlands is a key process to ensure ownership. It should be noted that there are often conflicting objectives over land use between local communities who may view policy actions as disenfranchising them from protected areas, forest reserves included, and these need to be addressed to the satisfaction of all stakeholders.

This is not just an issue for dry forests and woodlands in Africa, the involvement of communities has been tried successfully in India under the joint forest management (JFM) framework. It was driven by the need to shift from a top-down approach, which was underpinned by the state's desire to maximize commercial revenue from forests, to bottom-up participatory approaches through policy change that fostered the involvement of villagers in the management of local forests (Poffenberger and McGean, 1998). According to Khare (2000), the objective of JFM is to achieve better forest resources conservation by creating partnerships between the forest department and forest protection committees. As of 2006, close to 17.3 million hectares (27 per cent) of Indian forest lands were under the management of 85,000 JFMs (Saito-Jensen, 2008). This is not to suggest that there have been no problems. As with trials with JFM in Zambia, India has experienced problems related to benefit sharing, capacity issues in forest protection committees, failure of the state(s) 'to let go' in terms of management (Bwalya, 2007). These are some of the lessons that African dry forest and woodland countries must learn and incorporate in their planning.

The ability of local level institutions to play a part in the management of dry forests and woodlands is also limited by social differentiation. This means that each person brings individual characteristics of gender, religion, age, status and ethnicity into the dry forest and woodland arena. In order to effectively use differentiation for the benefit of resource managers, governments must manage the individual differences of each person or group. The social categories that flow from social differences are rarely neutral. These categories mark differences in status and power among groups and determine specific groups' relative access to resources and power within communities and the broader social systems. Status and power differences get reproduced in resource arenas and are embedded in authority structures, norms and management systems at local levels. In this way, they subtly confer privilege to some groups and disadvantage to others. As a result, different identity groups have very different perceptions and values of resources that tend to widen over time. Such differences among people based on these categories are grounded within structures of power, inequalities and unequal access to resources, which often result in conflicts over resources at various levels.

Resource governance

Governance structures constitute the organizations responsible for enforcing forestry and woodland management. The processes by which policies and institutions, both informal at local level and formal at national levels, often play out depend on such systems. In most cases local level systems of governance do not have a stipulated mandate to cover forests but often do so from a pragmatic point of view. Dominated in part by traditional institutions such as chiefs, governance issues are subject to the contradictions that come with such institutions. For example, the allocation of land may not be forest and woodland sensitive or gender sensitive. In some countries, e.g. Malawi, explicit local level structures, village natural resources management committees (encompassing forests), have been set up but have not received sufficient support to carry out their work. Interestingly enough, a focus on forests and woodlands for such institutions is unavoidable as these are the most visible resources requiring attention. In other countries, the functions of forest management have been placed into the hands of local government structures e.g. village development committees in Zimbabwe and area development committees in Zambia. Linking issues of forests and woodland management to local government administrations, most of which are weak, means that they cannot be sufficiently addressed.

Whether these are the formal legal instruments or the informal local level institutions, such structures are central to enforcement. Enforcement by institutions is enhanced or undermined by the effectiveness of authority structures. At the formal level, one important part has to do with the level of funding that such structures are given by the state. Budgets have to be in place to allow such structures to respond and operate in an effective way. Without adequate funding, enforcement to ensure compliance by institutions becomes compromised by the structures in place and in some cases can become a recipe for corruption.

One of the challenges facing policy-makers, local authorities, practitioners and groups working in natural resource management in the dry forests and woodlands is the need to recognize and understand how institutions, and changes in institutions, affect people's interaction with their environment. Often the values of the resources that are violated are much higher than the penalties imposed on offenders that may act as a disincentive for resource conservation. 'Institutional factors are arguably the most important factors determining the success or failure' of natural resource management or conservation (Kayambazinthu et al, 2002). The long history of change in resource management arrangements that has resulted in disempowerment of local institutions means that those that have persisted should be supported and strengthened. It is interesting to note that successes have also been tainted by the problems of power where local politicians and elites have taken over control of such initiatives for their own good. This can be a danger where local communities are not sufficiently empowered and unsure of their rights. Thus, while examples of persistent or resilient institutions are to be found in sacred practices of various forms (Kajembe et al, 2002; Kayambazintu et al, 2002) care must be taken of the threat of elite capture. Such institutions lead to the conservation of key biodiversity resources, which may be woodlands, wetlands or species of flora or fauna.

Managing for Multiple Products and Services

Managing dry forests and woodlands for multiple products and services requires a critical assessment of the outcomes of interactions between products and services and the extent to which they are compatible (Table 11.3). Pair-wise combinations of the different products and services considered in the previous chapters indicates that nearly 71 per cent of the combinations are incompatible from the current management perspective, 20 per cent are compatible and 9 per cent are conditionally compatible. This scenario means that managing dry forests and woodlands for a combination of products and services may present challenges for managers. Maximizing the positive compatibilities requires astute management, especially at the local level. Thus, the role of management is very critical in influencing the outcomes of managing for multiple products and/or services where compatibility depends on conditionalities (Table 11.3); under such conditionalities, the technique for harvesting a combination of products and services may be key to influencing the outcome of particular management approaches.

For example, the harvesting of edible caterpillars does not require that the host tree be cut down but only occurs where there is weak tenure and enforcement of local rules and regulations for resource use. In a situation where regulations are observed, biodiversity conservation and caterpillar collection are compatible but when tree cutting is involved then compatibility is reduced or replaced by incompatibility. Management guidelines are therefore required for situations where compatibility is linked to a particular condition, especially with regards to the type of resource harvesting technique. In some cases where the use of one product is incompatible with the use for another product, management guidelines may focus more on utilizing the resource base for more than one product. For example, if a medicinal tree is cut for charcoal or timber or firewood, it can be recommended that bark be removed from a cut tree for medicinal purposes and the rest of the wood used for the other wood products, such as poles, timber or charcoal. For products and services that are completely incompatible, land use planning is important in ensuring that sufficient land is made available for the different products. For example, areas can be designated for charcoal production while other areas are set aside for watershed and biodiversity management as in the case of national parks.

Although degradation of African dry forests and woodlands is acknowledged, these vegetation formations still contain much of at least their plant diversity, perhaps because they have not been totally cleared for large-scale commercial agriculture and/or forestry as in Europe and northern Africa. Sub-Saharan Africa is relatively unique in this context and therefore opportunities exist for managing dry forests and woodlands for multiple land uses, products and services. The challenge therefore is for an integrated management approach at landscape level to deal with seemingly competitive land uses, such as agriculture and forestry, and conservation as traditionally practised. Such innovative

Table 11.3 *Compatibility in the management of dry forests and woodlands for different product and/or service combinations*

Product and service	*Biodiver-sity*	*Fruits*	*Veget-ables*	*Mush-rooms*	*Cater-pillars*	*Honey*	*Exud-ates*	*Bark & fibre*	*Roots*	*Poles*	*Timber*	*Uten-sils*	*Fire-wood*	*Char-coal*	*Herb-age*	*Browse*	*Carbon*	*Water*	*Spirit-ual*
Biodiversity	0	+	+	+	-/+	-/+	-	-	-	-	-	-	-	-	-	-	+	+	+
Fruits		0	+	+	+/-	+	-	-	-	-	-	-	-	-	-	-	+	+	+
Vegetables			0	+	+	+	+	-	-	-	-	-	-	-	-	-	+	+	-
Mushrooms				0	+	+/-	+/-	-	-	-	-	-	-	-	+	+/-	+	+	+
Caterpillars					0	+/-	-	-	-	-	-	-	-	-	-	-	+	+/-	-
Honey						0	-	-	-	-	-	-	-	-	+	-	+/-	+/-	-
Exudates							0	-	-	-	-	-	-	-	+	-	-	-	-
Bark & fibre								0	-	-	-	-	-	-	+	-	-	-	-
Roots									0	-	-	-	-	-	-	-	-	-	-
Poles										0	-	-	+/-	-	-	-	-	-	-
Timber											0	+/-	+/-	+/-	+	-	-	-	-
Utensils												0	+	+/-	+	-	-	-	-
Firewood													0	-	+	-	-	-	-
Charcoal														0	+	-	-	-	-
Herbage															0	-	-	-	-
Browse																0	-	-	-
Carbon																	0	+	+
Water																		0	+
Spiritual																			0

Note: + compatible; - incompatible; +/- conditional compatibility.
Source: After Chundama and Gumbo (2009)

approaches can achieve some level of commercial agriculture and forestry, and also conservation associated with the agriculture/forestry activities, as long as appropriate infrastructure exists. One key question is whether a combination of short-, medium- and long-term land management approaches can contribute to a phased production of multiple products and services from dry forests and woodlands. For example, the land could be cleared for crop cultivation in the short term. During clearing, the relevant timber component is harvested and the rest of the woody component is used for wood crafts, poles, firewood and/or charcoal. During cultivation the crop system is specifically managed to ensure recovery of the woody component with its associated NWFPs as often happens in the traditional shifting cultivation practices. The woody regeneration is then specifically managed towards a future crop of fruits, poles, timber and eventually firewood and/or charcoal over the longer term. The feasibility of such management systems needs to be investigated through modelling and long-term experimentation. Such management systems are a direct challenge to the status quo in which land management is dealt with either by the agriculture department or ministry which focuses on agricultural production, or by forestry that focuses on timber production, or by nature conservation/national parks that aim at preventing degradation with no integrated land management. However, with still enough land available in many African dry forest and woodland countries, it is possible to experiment with new innovative approaches that integrate different land use successions and objectives.

African Dry Forest and Woodland Dynamics under Global Change

There will undoubtedly be some marked changes in distribution and extent of African vegetation types during the 21st century due to climate and land-use changes. However, climate change and land-use change are likely to be intertwined, and it is difficult to predict what the consequences will be in detail. Table 11.4, although simplistic, shows probable responses of African dry forests and woodlands to climate and land-use changes. Generally, because the majority of tropical trees have a C_3 photosynthesis, they are more likely to be negatively affected by global warming; in contrast, the majority of tropical grasses have a C_4 photosynthesis and therefore are likely to perform better under a warmer climate. However, although Table 11.4 presents possible responses of dry forest and woodland vegetation and plants to future environmental scenarios, these outcomes are by no means certain. For example, these predicted responses might be modified or even reversed when interactions with CO_2 enrichment and nutrient availability are considered. In addition the question of the long-term impact of repeated fires and herbivory on dry forest and woodland productivity remains to be answered. Furthermore although we have many scattered data, we still do not have a comprehensive idea of dry forest and woodland productivity levels in relation to source of regeneration,

Table 11.4 *Probable responses of African dry forests and woodlands to climate and land-use changes*

Factor	*Regeneration*		*Productivity*		*Species richness*		*Carbon sequestration*		*Fire*	*Herbivory*
	Trees	*Grasses*	*Trees*	*Grasses*	*Trees*	*Grasses*	*Trees*	*Grasses*		
Decreasing rainfall	–	–	–	–	–	–	–	–	–	–
Global warming	–	–	–	+	–	+	–	+	+	+
Frequent droughts	–	–	–	–	–	–	–	–	–	–
Land clearing	–	+	–	–	–	–	–	–	–	–
Fire	–	–	–	+	–	–	–	+	NA	–
Herbivory	–	–	–	–	–	–	–	–	–	NA

Note: + positive response; – negative response; NA not applicable

Source: Authors

stage of growth, moisture availability, nutrient status and dry forest or woodland type or their species composition (Küper et al, 2006).

It follows therefore that we need to understand what the main determinants of productivity are for each vegetation type, so that management practices can be more strategically applied. Investment in capacity building and research will play a key role in addressing these issues. Forest and biomass inventories, as well as monitoring plant growth and phenology need to be undertaken at representative sites and the data analysed in relation to climate and land-use data to better understand their interactions. This will feed into the development of more appropriate management guidelines that will ensure that African dry forests and woodlands continue to provide the products and services that are crucial to the maintenance of livelihoods of millions of people that live in these vegetation formations, and to sustainable economic development of sub-Saharan African countries.

References

Bird, M. and Cali, J. (1998) 'A million-year record of fire in sub-Saharan Africa', *Nature*, vol 394, pp767–769

Bwalya, B. (2007) *Katanino Joint Forest Management Area, Masaiti District, Zambia: Challenges and Opportunities*, NORAGRIC, Department of International Environment and Development Studies

Campbell, A. C. and Ramsay, J. (1994) 'Rakops–Motopi Road: Archaeological implications', unpublished report, commissioned by Knight Piesold, Botswana

Chapin, F. S., Lovecraft, A. L., Zavaleta, E. S., Martin, J. N., Robards, D., Kofinas, G. P., Trainor, S. F., Peterson, G. D., Huntington, H. P. and Naylor, R. L. (2006) 'Policy strategies to address sustainability of Alaskan boreal forests in response to a directionally changing climate', *Proceedings of the National Academy of Sciences, USA, November 7 2006*, vol 103, pp16637–16643

Child, B. (2004) *Parks in Transition: Biodiversity, Rural Development and the Bottom Line*, Earthscan, London

Chundama, M. and Gumbo, D. (2009) *Preparing for REDD in dryland forests: Investigating the options and potential synergy for REDD payments in the miombo eco-region (Zambia country study)*, International Institute for Environment and Development (IIED), London

Denbow, J. R. and Wilmsen, E. N. (1989) 'Iron Age Pastoralist Settlement in Botswana', *South African Journal of Science*, vol 79, pp405–408

Garlake, P. S. (1978) 'Pastoralism and Zimbabwe', *Journal of African History*, vol 19, pp479–494

Goucher, C. L. (1981) 'Iron is iron 'til it is rust: Trade and ecology in the decline of West African iron-smelting', *Journal of African History*, vol 22, no 3, pp179–189

Goucher, C. L. and Herbert, E. W. (1996) 'The blooms of Banjeli: Technology and gender in West African iron smelting', in P. R. Schmidt (ed) *The Culture and Technology of African Iron Production*, University of Florida, Gainesville

Gunderson, L. H. and Holling, C. S. (2002) *Panarchy: Understanding Transformations in Human and Natural Systems*, Island Press, Washington, DC

Haaland, R. (1980) 'Man's role in the changing habitat of Mema during the Old Kingdom of Ghana', *Norwegian Archaeological Review*, vol 28, pp31–47

Hall, M. (1983) 'Early farming communities in southern Africa: A population discovered', *South African Historical Journal*, vol 15, pp1–9

Hall, M. (1984) 'Man's historical and traditional use of fire in southern Africa', in P. de V. Booysen and N. M. Tainton (eds) *Ecological Effects of Fire in South African Ecosystems*, Springer-Verlag, Berlin, pp39–52

Hoffman, M. T. (1997) 'Human impacts on vegetation', in R. M. Cowling, D. M. Richardson and S. M. Pierce (eds) *Vegetation of Southern Africa*, Cambridge University Press, Cambridge, pp507–534

Huffmann, T. N. (1982) 'Archaeology and Ethnohistory of the African Iron Age', *Annual Review of Anthropology*, vol 11, pp133–150

Hutton, J. and Dickson, B. (eds) (2000) *Endangered Species, Threatened Convention: The Past, Present and Future of CITES*, Earthscan, London

Kajembe, G. C., Mwaipopo, C. S., Mvena, Z. S. K. and Monela, G. C. (2002) 'The role of traditional institutions, leadership and ecological knowledge in sustainable management of miombo woodlands in Handeni District, Tanzania', Paper presented at the EU-CIFOR Miombo Project workshop on St Lucia Park, Marlborough, Harare, 2–3 May

Kayambazinthu, D., Matose, F., Kajembe, G. and Nemarundwe, N. (2002) 'Institutional arrangements governing natural resource management of the miombo woodland', Paper presented at the EU-CIFOR Miombo Project workshop on St Lucia Park, Marlborough, Harare, 2–3 May

Khare, A. (2000) *Joint Forest Management: Policy, Practice and Prospects*, WWF India and International Institute of Environment and Development, London

Küper, W., Sommer, J. H., Lovett, J. C. and Barthlott, W. (2006) 'Deficiency in African plant distribution data: Missing pieces of the puzzle', *Botanical Journal of the Linnean Society*, vol 150, pp355–368

Petheram, L., Campbell, B., Marunda, C., Tiveau, D. and Shackleton, S. (2006) 'The wealth of the dry forests. Can sound forest management contribute to the Millennium Development Goals in sub-Saharan Africa?', *Forest Livelihood Briefs*, no 4, October, CIFOR, Bogor

Poffenberger, M. and McGean, B. (1998) *Village Voices, Forest Voices: Joint Forest Management in India*, Oxford India

Ribot, J. C. (1999) Decentralization, participation and accountability in Sahelian forestry: Legal instruments of political-administrative control', *Africa*, vol 69, pp23–65

Saito-Jensen, M. (2008) *Do Local Villagers Gain from Joint Forest Management? Why and Why Not? Lessons from Two Case Study Areas from Andhra Pradesh, India*, JICA Visiting Fellow Research report, September, Japan International Cooperation Agency, www.jica.go.jp/english/publications/reports/study/topical/JFM/pdf/090113_0.pdf

Sambou, B., Goudiaby, A., Ervik, F., Diallo, D. and Camara, M. C. (2002) 'Palm wine harvesting by the Bassari threatens *Borassus aethiopum* populations in north-west Guinea', *Biodiversity and Conservation*, vol 11, pp1149–1161

Schmidt, P. R. and Avery, D. H. (1996) 'Complex iron smelting and prehistoric culture', in P. R. Schmidt (ed) *The Culture and Technology of African Iron Production*, University of Florida, Gainesville

Taylor, C. J. (1960) *Synecology and Silviculture in Ghana*, Thomas Nelson and Sons, Edinburgh

Terry, E. M. (1999) 'The economic and social significance of the handicraft industry in Botswana', PhD thesis, School for Oriental and African Studies, University of London, London

Tlou, T. and Campbell, A. (1997) *History of Botswana*, Macmillan, Gaborone

van der Merwe, N. J. and Killick, D. J. (1979) 'Square: An iron smelting site near Phalaborwa', *South African Archaeological Society, Goodwin Series*, vol 3, pp86–93

Wynberg, R. P. (2006) 'Identifying pro-poor best practice models for commercialisation of southern African non-timber forest products', PhD thesis, University of Strathclyde, Glasgow

Index